Power Systems

Electrical power has been the technological foundation of industrial societies for many years. Although the systems designed to provide and apply electrical energy have reached a high degree of maturity, unforeseen problems are constantly encountered, necessitating the design of more efficient and reliable systems based on novel technologies. The book series Power Systems is aimed at providing detailed, accurate and sound technical information about these new developments in electrical power engineering. It includes topics on power generation, storage and transmission as well as electrical machines. The monographs and advanced textbooks in this series address researchers, lecturers, industrial engineers and senior students in electrical engineering.

Power Systems is indexed in Scopus

Farhad Shahnia · Fushuan Wen
Editors

Power and Energy Resources

Modelling and Optimization

Springer

Editors
Farhad Shahnia
School of Engineering and Energy
Murdoch University
Perth, WA, Australia

Fushuan Wen
School of Electrical Engineering
Zhejiang University
Hangzhou, Zhejiang, China

ISSN 1612-1287 ISSN 1860-4676 (electronic)
Power Systems
ISBN 978-981-96-2611-3 ISBN 978-981-96-2612-0 (eBook)
https://doi.org/10.1007/978-981-96-2612-0

This Springer imprint is published by the registered company Springer Nature Singapore Pte Ltd.
The registered company address is: 152 Beach Road, #21-01/04 Gateway East, Singapore 189721, Singapore

Preface

Climate change is driving transition towards net-zero emissions in electrical power systems through a large-scale integration of various technologies of renewable energy resources. This integration introduces several challenges to electrical power systems. One of these challenges is the influence of weather conditions, which may lead to significant intermittency and variability in the generated power from these resources.

The inherent uncertainty in the generated power by renewable resources can lead to technical and operational challenges for power systems such as the grid instability. These uncertainties should be considered properly in modelling and analyzing the performance of power systems. Therefore, suitable mathematical tools should be employed to properly model and analyze these resources.

Renewable energy systems are usually collections of multiple parallelly connected generation units that operate under different operating conditions, making it extremely difficult to accurately model all units and aggregate them. This complexity increases with the rise of the share of renewables in power generation mix. Therefore, developing precise models for renewable energy resources is essential for promoting security, reliability and efficiency of power grid operation.

Power systems incorporating large bundles of thermal units such as coal-fired power plants with wind farms need advanced mathematical models for economic dispatch of those units in order to meet technical and operational requirements of the thermal units imposed by the intermittency and variability of the output power of the wind farms.

The volatility of renewable energy sources can be mitigated in power systems by retrofitting existing natural gas-fired power plants because of their fast response characteristics. This solution can be achieved by accurately modelling the dynamics of the natural gas systems in both short term and long term to compensate the fluctuations of the generated power by renewable resources.

Hydropower is another renewable energy resource that is recognized globally as a kind of clean, flexible and high-quality resource. Different technologies of hydropower have different characteristics that necessitate appropriate modelling, analysis and optimization.

Photovoltaic systems are a kind of dominant renewable energy resources. However, their intermittency and variability pose considerable impacts on the security, stability and economics of power grids. Therefore, developing suitable mathematical models that can accurately forecast their power output in short and long terms is crucial. In addition, proper placement of photovoltaic systems in electrical distribution networks can bring many benefits for system operation. The power electronic inverters of photovoltaic systems can provide reactive power support for effective voltage regulation at their connection point. This support can supplement the voltage regulation traditionally offered by other legacy voltage regulation devices. Proper mathematical modelling and optimization techniques are required to effectively increase the introduced conservation voltage reduction benefit by the photovoltaic systems in order to reduce load demands and power losses in the network.

The interplay between electrical distribution networks and their embedded distributed energy resources introduces complexity in their unified management. Machine learning and optimization techniques can be employed to ensure the secure and economic operation of the distribution system as well as the carbon emission management.

Modelling reactive power-voltage response characteristics of renewable energy sources and the electrical distribution network reconfiguration are other technical challenges which can be addressed by properly designed and solved mathematical techniques to improve their adaptivity to time-varying operational conditions while simplifying the process of solving a large set of nonlinear equations.

This book aims to provide a research-oriented and comprehensive platform for modelling and optimization of different kinds of power and energy systems. It covers their diverse aspects, features and applications, making it a valuable resource for postgraduate students and researchers. The book is divided into 9 chapters with their breakdown as follows:

- Chapter "Statistically Feasible Robust Power System Operation" introduces a statistically feasible robust power system operation strategy;
- Chapter "Modeling and Operational Optimization of Hydropower Generation" presents the modelling and operational optimization of hydropower generation systems;
- Chapter "Economic Dispatch of Wind and Thermal Bundled Power Systems" uses a look-ahead rolling economic dispatch approach for wind and thermal bundled power systems for operational flexibility enhancement;
- Chapter "Modeling of Natural Gas Dynamics Impact on Power Systems with High Renewables" investigates the impacts of natural gas system dynamics on power systems with high renewable generation penetration;
- Chapter "Short-Term Prediction of Photovoltaic Power Generation" addresses the optimal placement of photovoltaic systems to enhance conservation voltage reduction in multi-energy systems;
- Chapter "Optimal Placement of Photovoltaic Systems in Multi-Energy Systems" employs an effective regression model-based framework for short-term prediction of photovoltaic power generation;

- Chapter "Modeling of Reactive Power-Voltage Response Characteristics of Renewable Energy Resources" presents the modelling of reactive power-voltage response characteristics of renewable energy resources based on a data-knowledge-hybrid-driven method;
- Chapter "Carbon-Aware Distribution Network Operation and Optimization" focuses on carbon-aware distribution network operation via fusing learning and optimization;
- Chapter "Distribution Network Reconfiguration" highlights distribution network reconfiguration using a disjunctive convex hull approach.

As the book's editors, we thank all the contributors for their support and hard work. We also thank the reviewers who provided valuable comments for improving the quality of the book. Also, we are grateful to Springer Nature, the publisher, for agreeing to publish this book. Last but not the least, we would like to thank our families—Farhad thanks his parents (Ali and Nahideh) and spouse (Negar), and Fushuan thanks his wife (Xiaoli) and daughters (Blossom and Yuki) for their continuous encouragement and support.

Perth, Australia Farhad Shahnia
Hangzhou, China Fushuan Wen
April 2025

Contents

Editors and Contributors

About the Editors

Farhad Shahnia received his Ph.D. in Electrical Engineering from Queensland University of Technology, Brisbane, Australia, in 2011. Currently, he is a Professor of Electrical Engineering at Murdoch University, Perth, Australia. His professional experience includes three years at the Research Office-Eastern Azarbayjan Electric Power Distribution Company, Tabriz, Iran. Prior to joining Murdoch University, he was a Research Fellow at the Queensland University of Technology, Brisbane, Australia, and a Lecturer at Curtin University, Perth, Australia. He has published 8 books, 13 book chapters and over 200 research articles on power and energy systems.

Fushuan Wen joined Zhejiang University (China) in 1991 where he has been a Full Professor since 1997. He has also been a Full Professor with Hainan Institute, Zhejiang University since 2022. He is also a Visiting Principal Research Scientist of the Shenzhen Institute of Artificial Intelligence and Robotics for Society, Shenzhen, China. His research interests include power industry restructuring, power system alarm processing, fault diagnosis and restoration strategies, smart grids and electric vehicles, and artificial intelligence applications in power and energy systems. He has published 400+ SCI-indexed papers, 750+ EI-indexed papers and 870+ Scopus-indexed papers. His publications have been cited for 23,700+ times. He received several awards both at the national level and provincial level, including the most prestigious National Natural Science Award of China. He has been listed in "Most Cited Chinese Researchers" in 10 consecutive years since 2015 by Elsevier. He is the Editor-in-Chief of *Energy Conversion and Economics*, the Deputy Editor-in-Chief of *Journal of Automation of Electric Power Systems* and a Subject Editor in power system economics of *IET Generation, Transmission and Distribution*. He was elected IEEE Fellow for contributions to fault diagnosis in power grids.

Contributors

Reza Bayani Department of Electrical and Computer Engineering, San Diego State University, San Diego, USA;
Department of Electrical and Computer Engineering, University of California San Diego, San Diego, USA

Siqi Bu Department of Electrical and Electronic Engineering, The Hong Kong Polytechnic University, Hong Kong, China

Sherif O. Faried Department of Electrical and Computer Engineering, University of Saskatchewan, Saskatoon, Canada

Chenge Gao Tsinghua-Berkeley Shenzhen Institute, Tsinghua University, Shenzhen, China

Ye Guo Tsinghua-Berkeley Shenzhen Institute, Tsinghua University, Shenzhen, China

Wenqian Jiang School of Science and Engineering, The Chinese University of Hong Kong, Shenzhen, Shenzhen, China

Shahab Karamdel Department of Electrical and Computer Engineering, University of Saskatchewan, Saskatoon, Canada

Chao Lei Department of Electrical and Electronic Engineering, The Hong Kong Polytechnic University, Hong Kong, China

Xiaodong Liang Department of Electrical and Computer Engineering, University of Saskatchewan, Saskatoon, Canada

Yikui Liu School of Electrical Engineering, Sichuan University, Chengdu, China

Saeed Manshadi Department of Electrical and Computer Engineering, San Diego State University, San Diego, USA

Linwei Sang Tsinghua Shenzhen International Graduate School, Tsinghua University, Beijing, China

Jian Shi Department of Electrical and Computer Engineering and Department of Engineering Technology, The University of Houston, Houston, USA

Qianggang Wang School of Electrical Engineering, Chongqing University, Chongqing, China

Chenye Wu School of Science and Engineering, The Chinese University of Hong Kong, Shenzhen, Shenzhen, China

Lei Wu Department of Electrical and Computer Engineering, Stevens Institute of Technology, Hoboken, NJ, USA

Yinliang Xu Tsinghua Shenzhen International Graduate School, Tsinghua University, Beijing, China

Nan Yang College of Electrical Engineering and New Energy, China Three Gorges University, Yichang, China

Abbreviations

ADMM	Alternating direction method of multiplier
BFM	Branch flow model
BNN	Bayesian neural network
CB	Capacitor bank
CC-G	Chance-constrained with Gaussian assumption
CC-L	Chance-constrained method with Laplace assumption
CHS	Cascade hydropower system
CNN	Convolutional neural network
CVR	Conservation voltage reduction
DCHR	Disjunctive convex hull relaxation
DG	Distributed generation
DL	Deep learning
DNN	Deep neural network
DNR	Distribution network reconfiguration
DSC	Direct supply connection
FLO4DNO	Fusing learning and optimization for distribution network operation
GPR	Gaussian process regression
HVDN	High-voltage distribution network
IEGS	Interconnected electricity-gas system
ISO	Independent system operator
KG	Knowledge graph
LDF	Load distribution formula
LSTM	Long short-term memory
LTS	Least trimmed square
MAE	Mean absolute error
MGBD	Multi-cut generalized benders decomposition
MIP	Mixed-integer programming
MISOCP	Mixed-integer second-order cone programming
MSE	Mean squared error
nRMSE	Normalized root mean squared error
OGF	Optimal gas flow

OLTC	On-load tap changer
P2G	Power-to-gas
PCC	Point of common coupling
PGE	Portland general electric
PINN	Physics-informed neural network
PV	Photovoltaic
ReconRO	Reconstruction-based robust optimization
ReLU	Rectified linear activation function
RES	Renewable energy source
RMSE	Root mean squared error
RO	Robust optimization
ROR	Run-of-river
SARIMA	Seasonal autoregressive integrated moving average
SDP	Semi-definite programming
SG	Scenario generation
SOC	State-of-charge
SOCO	Second-order cone
SSC	Serial supply connection
SVM	Support vector machine
TCL	Thermostatically controlled load
VRC	Voltage-response characteristic
WPT	Wind power turbine
WTBPS	Wind-thermal-bundled power system

Statistically Feasible Robust Power System Operation

Wenqian Jiang, Jian Shi, and Chenye Wu

Abstract The escalating threat of climate change is driving the electric power system through a profound transition toward net-zero emissions. During this transition, the most noticeable characteristic is the large-scale integration of renewable generation sources, which has increased almost fourfold over the past ten years (from 6% to more than 30%). Although renewable energy brings clean power, its inherent uncertain generation leads to potential power grid instability. Hence, dealing with the newly integrated uncertainty in modern power system operations is vital. Classical approaches to handling uncertainties include chance-constrained programming and robust optimization. However, chance-constrained programming is effective often when the distribution knowledge is accessible, whereas solutions from robust optimization are often criticized for conservatism. To this end, distributionally robust optimization improves robust optimization and chance-constrained programming by considering distributional misspecification. Nevertheless, the misspecification may also arise during dataset collection. Intuitively, sample sets with different quality will lead to diverse estimations for uncertainties. This warrants considering the dataset collection misspecification issue in detail, which has only emerged recently. One promising solution is to adopt the notion of statistical feasibility to account for the misspecification in dataset collection. Specifically, on top of chance-constrained programming, guaranteeing statistical feasibility modifies the original chance constraints by incorporating a confidence level requirement concerning dataset collection uncertainty. The modified chance constraints are dealt with from a robust optimization view by using purely data-driven methods, which improves statistically feasible robust optimization practicability. Compared with robust optimization, considering the confidence level grants decision-makers the flexibility in balancing cost-effectiveness and conservativeness. This chapter introduces statistical feasibility and provides its implementation to three typical power system operation

W. Jiang · C. Wu (✉)
School of Science and Engineering, The Chinese University of Hong Kong, Shenzhen, Shenzhen, China
e-mail: chenyewu@yeah.net

J. Shi
Department of Electrical and Computer Engineering and Department of Engineering Technology, The University of Houston, Houston, USA

F. Shahnia and F. Wen (eds.), *Power and Energy Resources*, Power Systems,
https://doi.org/10.1007/978-981-96-2612-0_1

tasks with different scales, including thermostatically controlled load scheduling, economic dispatch, and unit commitment, showcasing the effectiveness and scalability of robust operation with a statistical feasibility guarantee. Looking into the future, it is evident that the statistically feasible robust methods will contribute to sample efficient data-driven modern power system operation with high penetration of renewable energy.

Keywords Statistical feasibility · Uncertainty characterization · Dataset collection misspecification · Stochastic optimization

1 Introduction

The global decarbonization transition necessitates a substantial reduction in greenhouse gas emissions over the coming decades [1]. Achieving success in reducing greenhouse gas emissions requires not only a shift toward clean energy sources but also consistent efforts to adapt to the low-carbon shifts that have been already occurring across society. With electrical power systems accounting for a quarter of global greenhouse gas emissions, their role in achieving ambitious decarbonization goals is crucial. To realize a low-carbon electricity sector, it is imperative to make electricity generation more sustainable and environmentally friendly. Consequently, the integration of renewable energy sources into power systems has steadily increased in the last decades. On the one hand, solar power and wind power, being virtually emission-free during operation, offer significant advantages in decarbonization. On the other hand, thanks to remarkable advancements in clean energy technologies, the costs of renewable generation have drastically decreased, while their efficiency continues to improve annually, making them increasingly competitive with conventional energy sources. Nonetheless, the integration of renewable energy into power systems is not without challenges. One significant challenge arises from the unpredictable nature of renewable energy output. Unlike conventional power generation, which can be dispatched and controlled as needed, renewable energy generation is influenced by weather conditions such as solar irradiation and wind speed. These fluctuations introduce uncertainty into the power supply, posing significant obstacles to ensuring reliable grid operations.

To address the increasing uncertainty in modern power system operation, stochastic optimization has emerged as a promising solution. Classical methods of stochastic optimization include chance-constrained programming, robust optimization, and distributionally robust optimization. Recent years have witnessed a rapidly growing body of research employing chance-constrained programming, robust optimization, and distributionally robust optimization to manage uncertainties in modern power system operations. Chance-constrained programming is an effective tool to handle uncertainty by specifying a confidence level at which it is desired that the chance constraint holds. Therefore, most chance-constrained programming formulations involve chance constraints to characterize uncertainties in random sources in power systems, e.g., electricity prices, renewable generation, and electricity demand. For example, Du et al. present a linear sequential optimization enhanced multi-loop algorithm to solve the home appliance commitment problem from a chance-constrained programming perspective, accounting for price uncertainties when scheduling electric water heaters [2]. Aghdam et al. and Sundar et al. characterize renewable generation fluctuations by chance constraints in an optimal energy storage management problem in [3] and a chance-constrained secure unit commitment problem in [4], respectively; Hou et al. develop chance-constrained programming frameworks to deal with electric vehicle charging uncertainties in [5], and Pozo et al. design a chance-constrained programming to address demand uncertainty in [6]. Chance-constrained programming often requires conversion of the chance constraints to their deterministic equivalents, which is only successful for special cases, e.g., when the random variables characterizing uncertainties follow the Gaussian distribution.

For the issue of reliance on the distribution information, the robust optimization approach resorts to constructing uncertainty sets, e.g., ellipsoids or polytypes, which deal with uncertainties by identifying their boundaries. Just to name a few, Hou et al. utilize data-driven robust optimization to address wind and photovoltaic power uncertainty in unit commitment in [7]. Chen et al. utilize uncertainty intervals to handle stochastic electricity prices when designing demand response schemes for residential appliances in [8]. Chen et al. propose a two-stage adaptive robust optimization formulation to aggregate the system-level multi-period power flexibility in [9]. Wang et al. develop a robust optimization strategy to handle the uncertainties in hot water demand and ambient temperature for load scheduling of domestic electric water heaters in [10]. Robust optimization methods are often criticized for their conservativeness because they require that the solutions are feasible even for the worst possible realization of uncertainties, even though such worst cases may happen very rarely.

To provide more robustness to chance-constrained programming and relieve the conservativeness of robust optimization, distributionally robust optimization is recommended. Distributionally robust optimization methods deal with uncertainties by using the concept of ambiguity set, which considers distributional misspecification in the given dataset. In power systems, many customized distributionally robust optimization methods are designed for the classical economic dispatch problem and unit commitment problem. For instance, Geng et al. propose a distributionally robust optimization approach to conduct the economic dispatch with fluctuating renewable energy generation in [11]. Duan et al. adopt Wasserstein ball-centered distributionally robust optimization to calculate the optimal power flow in [12]. Zhu et al. adopt the Wasserstein metric to address distributionally robust unit commitment in [13]. Zhou et al. design an ambiguity set using unimodality, mode skewness, and second-order moment information in [14]. Although distributionally robust optimization methods improve robust optimization and chance-constrained programming, they ignore potential misspecification during dataset collection, which has only emerged recently. In this chapter, dataset collection misspecification refers to the fact that the collected sample sets may be biased compared to the true, unknown distribution of random variables (i.e., uncertainties) across different levels based on multiple sample collections. Intuitively, due to the unavoidable randomness in dataset collection and the number of samples in each collection being finite, even if all samples are generated from the same distribution, the estimations for the underlying unknown distribution using various sample collections with varying quality frequently exhibit distinct trend characteristics. Due to the variability in uncertainty estimation caused by dataset collection misspecification, decisions made based on such estimations will also differ. Therefore, it is crucial to thoroughly address the issue of dataset collection misspecification. One promising solution is to involve the notion of statistical feasibility to help account for this misspecification in dataset collection.

The concept of statistical feasibility investigates the randomness from the sample set by putting an additional confidence level on top of the original chance constraint. More specifically, it modifies the original chance constraints by incorporating a confidence level requirement concerning dataset collection uncertainty. Statistical feasibility allows decision-makers to derive a robust solution from a purely data-driven perspective, thereby eliminating the need for assumptions about distribution estimation for given sample sets often found in the literature. To guarantee statistical feasibility, the uncertainty sets need to be properly calibrated from data so that the errors involved in the optimization process can be quantified in a probabilistic way.

2 Revisit Classical Approaches

The goal of this section is to warm up the readers already with the necessary knowledge related to stochastic optimization. For those who are not familiar with these topics, please refer to classical references, e.g., [15–17], for a more detailed introduction.

Classical convex optimization techniques predominately aim at addressing deterministic optimization problems, where decision variables are obtained without considering any uncertainty. Mathematically, the problem is often as follows:

$$(\mathbf{DOP})\min f(\boldsymbol{x}) \tag{1a}$$

$$s.t. h(\boldsymbol{x}) \leq 0, \tag{1b}$$

$$g(\boldsymbol{x}) \leq 0, \tag{1c}$$

where $\boldsymbol{x} \in \mathbb{R}^n$ denotes the decision variable vector with the dimention n; $f(\cdot) \in \mathbb{R}$ is the objective function which often minimizes the electricity cost or maximizes the profit; $h(\cdot) \in \mathbb{R}^{m_1}$ and $g(\cdot) \in \mathbb{R}^{m_2}$ with dimentions m_1 and m_2 are constraints. We seperate constraints h(·), and g(·) to explicitly distinguish between deterministic constraints (without uncertainties) and uncertain constraints in the subsequent problem formulations. Functions $f(\cdot)$, $h(\cdot)$, and $g(\cdot)$ are all characterized by deterministic parameters. As renewable energy becomes increasingly integrated into power systems, the generation output of the entire grid is substantially influenced by volatile weather conditions. In addition to the fluctuating factors of renewable generation, other uncertain factors also greatly impact modern power system operations, e.g., the uncertainties in various electricity demands, and the strategic operation of flexible resources. Strategies designed solely for deterministic cases are no longer applicable. Therefore, it is essential to account for potential uncertainties in advance. Denote the uncertainty as $\boldsymbol{\xi}$ (i.e., random variable). After considering potential uncertainties, the system operators are interested in solving the following stochastic optimization problem:

$$(\mathbf{SP}) \min f(\boldsymbol{x}) \tag{2a}$$

$$s.t. h(\boldsymbol{x}) \leq 0 \tag{2b}$$

$$g(\boldsymbol{x}; \boldsymbol{\xi}) \leq 0, \tag{2c}$$

where $g(\cdot)$ depends on $\boldsymbol{\xi} \in \mathbb{R}^m$ which is random variable vector representing uncertainties. Next, we will briefly introduce three classical solutions for Problem (SP): chance-constrained programming, robust optimization, and distributionally robust optimization.

To deal with the uncertainty in the Problem (**SP**), typical chance-constrained programming involves the chance constraint as follows:

$$(\mathbf{CCP}) \min f(\boldsymbol{x}) \tag{3a}$$

$$s.t. h(\boldsymbol{x}) \leq 0, \tag{3b}$$

$$\mathbb{P}_{\boldsymbol{\xi}}(g(\boldsymbol{x}; \boldsymbol{\xi}) \leq 0) \geq \rho. \tag{3c}$$

Solving chance-constrained programming often requires $\boldsymbol{\xi}$'s distribution information. In contrast, robust optimization requires less information about $\boldsymbol{\xi}$ (only boundaries of $\boldsymbol{\xi}$) and can produce an approximate solution of (**CCP**) while satisfying the chance constraints. Specifically, the corresponding robust optimization is as follows:

$$(\mathbf{RO}) \min f(\boldsymbol{x}) \tag{4a}$$

$$s.t. h(x) \leq 0, \tag{4b}$$

$$g(\boldsymbol{x}; \boldsymbol{\xi}) \leq 0, \forall \boldsymbol{\xi} \in \mathcal{U}, \tag{4c}$$

where $\mathcal{U}$ is the uncertainty set. The uncertainty set is a fundamental concept used to model the uncertainty in robust optimization. It is a set that contains all the possible values that $\boldsymbol{\xi}$ can take. The goal of robust optimization is to find a feasible solution for all possible (observed) values within the uncertainty set. The uncertainty set can be defined in various ways. Some common types of uncertainty sets include ellipsoidal sets, polyhedral sets, interval sets, convex hull sets, etc. The choice of the uncertainty set is critical in robust optimization, as it directly affects the robustness and performance of the solution. A larger uncertainty set allows for more uncertainty but may lead to more conservative solutions, while a smaller set may provide less robustness. Therefore, there is often a trade-off between robustness and solution performance, and the selection of the uncertainty set should be carefully considered based on the specific problem.

These solutions suffer from distinct challenges: The performance of chance-constrained programming heavily relies on the accuracy of distributional information about uncertainty, while robust optimization considers worst-case optimization, and its solution is often too conservative. To this end, distributionally robust optimization improves robust optimization and chance-constrained programming by considering distributional misspecification, i.e.,

$$(\mathbf{DRO}) \min f(\boldsymbol{x}) \tag{5a}$$

$$s.t. h(x) \leq 0, \tag{5b}$$

$$\sup_{\boldsymbol{\xi} \sim \mathbb{P} \in \mathrm{P}} \mathrm{E}_{\mathbb{P}}(g(\boldsymbol{x}; \boldsymbol{\xi})) \leq 0, \tag{5c}$$

where $\mathbb{P}$ is in a distribution cluster P (a.k.a. ambiguity set). The ambiguity set in distributionally robust optimization accounts for the lack of precise information about the underlying probability distribution by considering a family of distributions consistent with the available data or prior knowledge. The goal is to find an optimal decision across all distributions within this set, ensuring that the solution is robust to potential distributional shifts. The construction of the ambiguity set is important in distributionally robust optimization. Depending on the specific problem structure and the available information, different types of ambiguity sets can be used, often including a Wasserstein distance-based set, probability-based set, moment-based set, etc.

The success of chance-constrained programming, robust optimization, and distributionally robust optimization typically relies on uncertainty characterization (for random variables $\boldsymbol{\xi}$) through historical samples, i.e., chance-constrained programming estimates distributions ($\mathbb{P}$) of historical samples for uncertainty; robust optimization captures the boundaries of samples ($\mathcal{U}$) for uncertainty; and distributionally robust optimization characterizes the ambiguity set (P) through samples. All these methods focus on characterizing uncertainty given a historical sample set, which is visualized in Fig. 1.

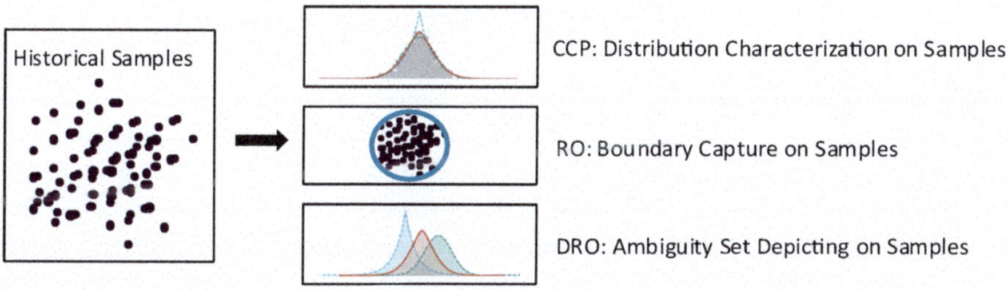

Fig. 1 Comparison of chance-constrained programming, robust optimization, and distributionally robust optimization

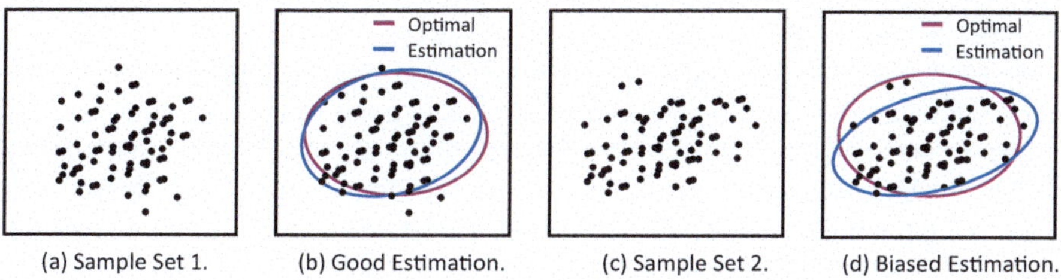

Fig. 2 Gaussian distribution estimation from sample sets of different qualities

However, the collected historical sample set may exhibit bias relative to the true unknown distribution of random variables at different levels for multiple data collections, which means that the dataset collection misspecification may appear, as indicated in the distribution estimation example in Fig. 2.

Sample sets with different qualities will lead to diverse estimations of random variables. This warrants considering the dataset collection misspecification issue in detail, which has only appeared very recently. To account for the misspecification in dataset collection, one promising solution is to adopt the notion of statistical feasibility.

3 Statistical Feasibility

Statistical feasibility was first proposed by Hong et al. in [18]. Its mathematical definition is as follows.

Definition 1 (Statistical Feasibility).

[18]: For any given dataset $\mathcal{D}_{\xi}$, an approach produces a solution $\widehat{g}$. This approach enjoys statistical feasibility guarantees if the solution $\widehat{g}$ is feasible for the constraints with confidence level η for any given dataset $\mathcal{D}_{\xi}$. Specifically, for a chance constraint $\mathbb{P}_{\xi}(g(\boldsymbol{x};\boldsymbol{\xi}) \leq 0) \geq \rho$, its statistical feasibility is defined as follows:

$$\mathbb{P}_{\mathcal{D}_\xi}(\mathbb{P}_{\xi}(g(\boldsymbol{x};\boldsymbol{\xi}) \leq 0) \geq \rho) \geq \eta. \tag{6}$$

Remark: Since the solution $\widehat{g}$ is based on dataset $\mathcal{D}_\xi$, symbol $\widehat{g}(\mathcal{D}_\xi)$ is used to demonstrate that $\widehat{g}$ is influenced by $\mathcal{D}_\xi$. Moreover, since the dataset samples may be randomly collected from the unknown distribution $\mathbb{P}_\xi$, the solution $\widehat{g}$ is also a random variable. The inside constraint in (6) is the classical chance constraint. The outside constraint in (6) further guarantees that for all resulted solutions $\widehat{g}$ (essentially, for different sample sets $\mathcal{D}_\xi$), the inside chance constraint can be satisfied with a probability no smaller than η. Intuitively, on top of chance-constrained programming, guaranteeing statistical feasibility modifies the original chance constraints by incorporating a confidence level requirement concerning dataset collection uncertainty. Compared with robust optimization, considering the confidence level grants flexibility in balancing cost-effectiveness and conservativeness.

Figure 3 illustrates the main idea of this concept. Specifically, the cube on the bottom characterizes all possible sample sets. Any point in the cube is a possible sample set with some limited number of samples, i.e., $\mathcal{D}_\xi = \{\boldsymbol{\xi}^{(i)}, \forall i\}$. . Statistically feasible robust optimization guarantees that except for a small part of biased sets ($\leq 1-\eta$, , potentially bringing huge estimation error), for any other sample sets, the solutions can ensure satisfying the classical chance constraints.

After involving the notion of statistical feasibility, the Problem (**CCP**) is transformed into the following problem:

$$(\mathbf{SFRO})\min f(\boldsymbol{x}) \tag{7a}$$

$$s.t. h(x) \leq 0, \tag{7b}$$

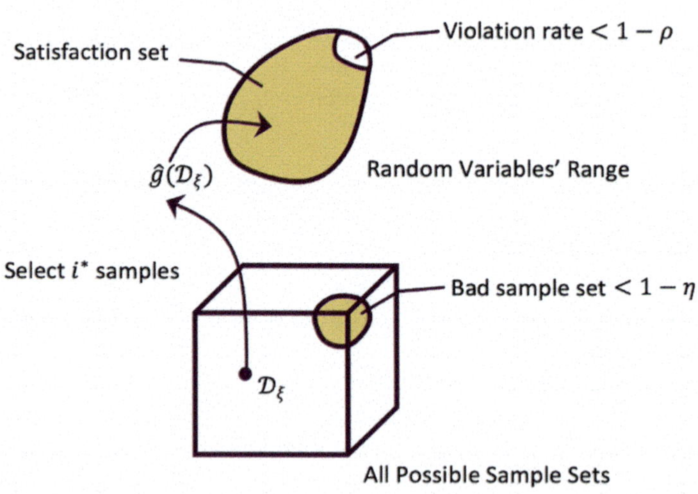

Fig. 3 Statistically feasible robust optimization

$$\mathbb{P}_{\mathcal{D}_\xi}(\mathbb{P}_{\xi}(g(\boldsymbol{x};\boldsymbol{\xi}) \leq 0) \geq \rho) \geq \eta. \tag{7c}$$

Now, the problem has turned to how to deal with Problem (**SFRO**) to obtain statistically feasible solutions. The main challenge in practice comes from the unknown distribution characterization of random variables $\boldsymbol{\xi}$. Generally, the decision-makers only have access to limited samples generated from unknown $\mathbb{P}_{\boldsymbol{\xi}}$. Even only considering the inner classical chance constraints, the analysis is already tricky given limited samples without more detailed distribution information of $\boldsymbol{\xi}$. Less to say, statistical feasibility adds another confidence guarantee for any given dataset $\mathcal{D}_{\boldsymbol{\xi}}$. Notice that Problem (**RO**) can connect with Problem (**CCP**) via the following fact: Any feasible solution to (**RO**) is also feasible to (**CCP**) if $\mathbb{P}_{\boldsymbol{\xi}}(\boldsymbol{\xi} \in \mathcal{U}) \geq \rho$.. Besides, the success of (**RO**) does not rely on the distribution characterization of ξ but only on its boundaries (which can be captured more effectively from a given dataset). Therefore, following this idea, an equivalent between (**RO**) and (**CCP**), transforming the modified chance constraints into an uncertainty set characterization problem, is established. This paves the way to dealing with the modified chance constraints using purely data-driven methods, improving statistically feasible robust optimization practicability.

4 Solving RO with Statistical Feasibility

Following Sect. 3, the problem has turned to how to design the uncertainty set construction procedure to guarantee statistically feasible solutions. This section first designs a two-stage uncertainty set construction strategy. This strategy chooses the high-probability density region of historical samples by dropping some extreme cases from a statistical perspective. To further relieve solutions' conservativeness, a self-improving uncertainty set reconstruction procedure is further designed, which integrates an initial approximate optimal solution into uncertainty set construction.

4.1 Robust Sample-Based Uncertainty Set Construction

To effectively solve the Problem (**SFRO**), a two-stage uncertainty set construction strategy can be designed. Specifically, for any given historical sample set $\mathcal{D}_{\boldsymbol{\xi}}$, it is divided into two parts $\mathcal{D}_1$, $\mathcal{D}_2$, with sizes of n_1 and n_2, respectively. $\mathcal{D}_1$ is utilized for the shape localization of the uncertainty set construction, while $\mathcal{D}_2$ is employed for shape calibration.

Shape Localization: Considering the tractability issue of the classical robust optimization, ellipsoids are adopted here as an example to approximate the uncertainty set and fix the position for the possibly smallest uncertainty set. Specifically, the ellipsoidal uncertainty set helps to characterize $\mathcal{D}_1$, which is defined as $\{\boldsymbol{\xi} | (\boldsymbol{\xi}-\boldsymbol{\mu})^T \boldsymbol{\sigma}^{-1}(\boldsymbol{\xi}-\boldsymbol{\mu}) \leq r\}$, for some $r > 0$. Parameters $\boldsymbol{\mu}$ and $\boldsymbol{\sigma}$ can be directly

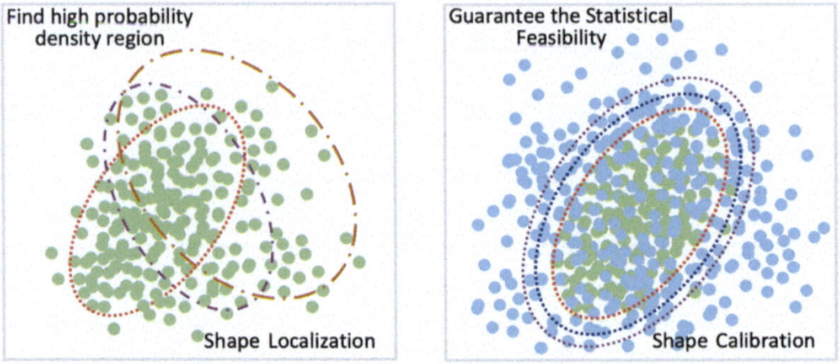

Fig. 4 Two-stage sample-based uncertainty set construction process [19]

chosen as the sample mean and variance of $\mathcal{D}_1$ and parameter r characterizes the size of the smallest ellipsoid covering all samples in $\mathcal{D}_1$.

Shape Calibration: After obtaining the approximated ellipsoidal set, $\mathcal{D}_2$ is involved to guarantee the statistical feasibility. Let $q(\boldsymbol{\xi}) = (\boldsymbol{\xi} - \boldsymbol{\mu})^T \boldsymbol{\sigma}^{-1}(\boldsymbol{\xi} - \boldsymbol{\mu})$, then the obtained uncertainty set from the Shape Localization stage can be denoted as $q(\boldsymbol{\xi}) \leq r$. Afterward, for $\boldsymbol{\xi} \in \mathcal{D}_2$, the values of $q(\boldsymbol{\xi})$ are sorted as $q(\boldsymbol{\xi})_{(1)} \leq q(\boldsymbol{\xi})_{(2)} \leq \cdots \leq q(\boldsymbol{\xi})_{(n_2)}$. Next, identifying the critical sample index i^*, which satisfies the condition:

$$i^* = \min\{n : \sum_{k=0}^{n-1} \binom{n_2}{k} \rho^k (1-\rho)^{n_2-k} \geq \eta\}, \tag{8}$$

for $1 \leq n \leq n_2$ if $1 - \rho^{n_2} \geq \eta$. Finally, let $\tilde{r}$ be equal to $q(\boldsymbol{\xi})_{(i^*)}$, which completes the ellipsoidal set calibration, i.e., $q(\boldsymbol{\xi}) \leq \tilde{r}$. Figure 4 illustrates the two-stage uncertainty set construction procedure.

4.2 *Self-Improving Uncertainty Set Reconstruction*

In Sect. 4.1, it is observed that the construction of the uncertainty set is distinctly decoupled from the resolution of the optimization problem (i.e., uncertainty set construction followed by solving the optimization problem). Intuitively, incorporating domain-specific insights pertinent to the optimization issue into the process of uncertainty set construction is likely to enhance the solutions' performance. For example, designing a customized uncertainty set of random variables in each constraint rather than a unified uncertainty set for all constraints may relieve the conservative issue of optimal solutions. This observation serves as an impetus for the development of an advanced self-improving construction method.

Suppose that the optimal solution $\boldsymbol{x}^*$ to Problem (CCP) was known. Clearly, $\boldsymbol{x}^*$ guarantees $1 - \rho$ tolerance to the Constraint (2c) due to the definition of optimality, which means

$$\mathbb{P}_{\xi}\left(g\left(x^{*};\xi\right)\leq 0\right)\geq\rho. \tag{9}$$

Then, one good uncertainty set candidate can be

$$\mathcal{U}^{*}=\left\{\xi \mid g\left(x^{*};\xi\right)\leq 0\right\}. \tag{10}$$

While it is impractical to ascertain the precise optimal solution for Problem (CCP) a priori, the derivation of an approximate solution may still yield valuable insights. Specifically, Problem (RO) is first solved based on an uncertainty set constructed by using the two-stage strategy in Sect. 4.1, therefore obtaining the approximated optimal solution $\hat{x}$. Then, the uncertainty set is constructed as follows:

$$\widehat{\mathcal{U}}=\left\{\xi \mid g\left(\hat{x};\xi\right)\leq 0\right\}. \tag{11}$$

Finally, the shape calibration can be conducted by tuning the size of $\widehat{\mathcal{U}}$ following the same procedure in Sect. 4.1.

5 Applications

Following the introduction of statistical feasibility, this section next introduces how to customize statistically feasible robust optimization techniques to handle distinct power system operation tasks, including thermostatically controlled load (TCL) scheduling (a prototype with single chance constraint) [19], economic dispatch (an extension to joint chance constraints) [20], and unit commitment (a further extension to joint chance constraints and mixed-integer programming) [21].

5.1 TCL Scheduling (a Prototype with Single Chance Constraint)

TCLs typically exhibit slow-varying thermal inertia, and their energy consumption can be modulated with minimal impact on users' comfort, rendering TCLs promising resources for delivering the critical flexibility required by modern power systems [22]. Consequently, the design of demand response programs targeting TCLs has garnered significant research interest [23–26]. Yet, much can be improved. The practical implementation of such programs is challenged by the inherent uncertainties associated with outdoor temperature fluctuations and electricity pricing dynamics.

In the classical deterministic TCL scheduling, control actions $\mathbf{u}=\{\mathrm{u_t};\forall \mathrm{t}\}$ from the time t_0 to t_0+N-1 are determined to maintain the room temperature in the comfort zone at the minimal cost in a rolling procedure. Mathematically, the deterministic TCL scheduling problem is as follows:

$$\min_{\mathbf{u}} \sum_{t=t_0}^{t_0+N-1} p_t \cdot |u_t| \cdot \frac{\Delta t}{\eta} \tag{12a}$$

$$s.t. T_{t+1} = \alpha T_t + (1-\alpha)(T_t^a - Ru_t), \forall t, \tag{12b}$$

$$T_{set} - \kappa \leq T_{t+1} \leq T_{set} + \kappa, \tag{12c}$$

$$-U_{tr} \leq u_t \leq U_{tr}, \tag{12d}$$

where p_t is the electricity price at time t; u_t is the control action of TCL, which, according to the convention, is positive for cooling and negative for heating; Δt is the simulation time step, which is usually set to 1 for simplicity; η is a TCL's coefficient of performance and typically larger η means less electricity loss and higher efficiency of the TCL. Constraints (12b) imply how the room temperature evolves where T_t denotes the room temperature at time t. The parameter $\alpha = e^{-\frac{\Delta t}{CR}}$, where C is the TCL's thermal capacitance and R is its thermal resistance. T_t^a represents the outdoor temperature at time t. Constraint (12d) limits the range of u_t, where U_{tr} is the TCL's energy transfer rate. Constraint (12c) denotes the goal of TCL scheduling to keep the room temperature within a certain range, where T_{set} is the consumer's desired room temperature and κ is the temperature deadband.

According to Eq. (12b), the outdoor temperature T_t^a, whose accurate predictions are often challenging, is a key factor in the evolution of the room temperature T_t. So is u_t. The uncertainty in the outdoor temperatures results in frequent violations of the room temperature requirement, i.e., constraint (12c). Although most consumers are willing to sacrifice a little comfort to reduce their electricity bills, such sacrifice has a bottom line. Such a bottom line can be formulated as the room temperature violation level. We are interested in designing more effective TCL scheduling to keep the room temperature in the comfort zone at the predefined violation level when considering the outdoor temperature uncertainty.

Jiang et al. employ the notion of statistical feasibility to design a robust sample-based scheduling framework for TCLs in [19]. The two-step uncertainty set construction theoretically ensures that the solutions can closely satisfy the room temperature requirements at a predefined confidence level. Then, the conservatism of feasible solutions is further relieved by combining approximate optimal solutions into uncertainty set construction. Numerical studies in [19] show that embedding such domain knowledge (i.e., approximate solutions and specific constraint information of the TCL scheduling problem) into the uncertainty set construction achieves 0.3% ~ 0.9% electricity cost reduction compared to other benchmark methods by better exploiting consumers' tolerance to the room temperature.

Table 1 compares the performance of the scenario generation method, chance-constrained method, robust sample-based method, and self-improving sample-based method when $\rho = 0.95$, $\eta = 0.95$, $n_1 = 200$, $n_2 = 1000$. The CPU time is the mean computation time for 1,000 experiments. The parameter $\hat{\rho}$ is defined as the estimated expected satisfaction rate of the obtained solutions, while the parameter $\hat{\eta}$ is the estimated value for the outer probability guarantee in Eq. (6). REP denotes the average savings in the electricity bills compared to the deterministic optimization with perfect predictions. Table 1 reveals that the robust sample-based method yields a conservative solution akin to those obtained by the scenario generation method and chance-constrained method, with parameters $\hat{\rho}$ and $\hat{\eta}$ to be one. In contrast, the self-improving sample-based method demonstrates superior performance, achieving a lower REP and effectively managing the room temperature satisfaction rate, with $\hat{\rho}$ and $\hat{\eta}$ close to the predefined threshold of 0.95. However, the self-improving sample-based method incurs a relatively greater computational time, attributable to its dependency on initial approximate solutions derived from the robust sample-based method.

Figure 5 shows the sensitivity analysis for sample size. The sample size plays a significant role in all data-driven methods because the only obtained information on random variables is from the finite samples. Intuitively, more samples yield a more accurate characterization of the uncertainty. Although the self-improving sample-based method already performs well on violation rate control given 1,200 samples, Jiang et al. further explore the effects of sample size on self-improving sample-based solutions' performance under the setting of $\rho = 0.95$, and $\eta = 0.95$ in [19]. As can be seen from Fig. 5, initially, with an increasing number of samples, the self-improving sample-based method enables the estimated satisfaction rates $\hat{\rho}$ and $\hat{\eta}$ converge to the target violation rate rapidly. However, as the sample size surpasses 5,000 (depicted in the lower left bubbles of Fig. 5), the convergence trend toward the target rate becomes less pronounced. This finding suggests that for the TCL scheduling case, a larger sample size does not inherently lead to superior solutions. There appears to be an optimal threshold of approximately 5,000 samples, beyond which the inclusion of additional samples does not yield further improvement. This observation underscores the imprudence of excessive sampling, indicating that there is no necessity to amass an overly large dataset of samples in the TCL scheduling scenario.

Table 1 Performance comparison of scenario generation method, chance-constrained method, robust sample-based method, and self-improving sample-based method with $\rho = 0.95$, $\eta = 0.95$ [19]

Method	REP (‰)	$\hat{\rho}$	$\hat{\eta}$
Scenario generation method	9.237	1	1
Chance-constrained method	7.825	1	1
Robust sample-based method	7.676	1	1
Robust self-improving sample-based method	4.073	0.932	0.929

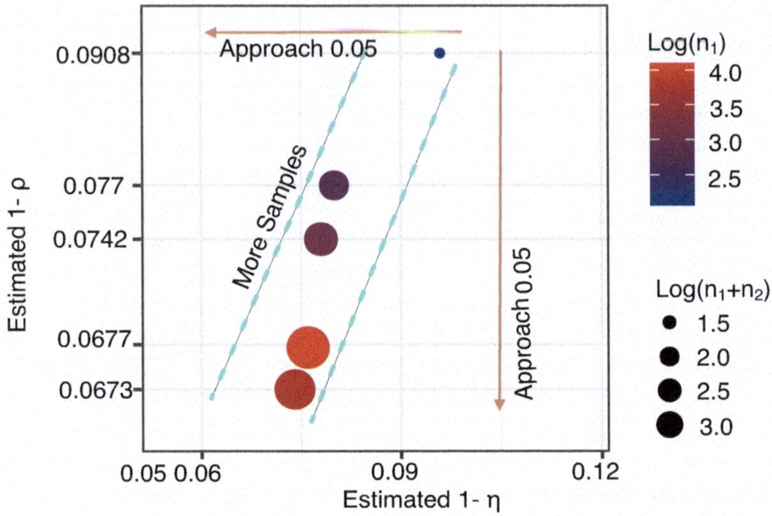

Fig. 5 The effects of sample sizes on violation rates of self-improving sample-based method [19]

5.2 *Economic Dispatch (an Extension to Joint Chance Constraints)*

Economic dispatch, a fundamental process in power system operation, has been extensively studied, particularly with the advent of improved forecasting techniques over the past few decades. Nonetheless, the growing integration of renewable energy sources presents formidable challenges to the efficient execution of economic dispatch. Specifically, the considerable prediction errors associated with renewable energy generation lead to increased system dispatch costs. Moreover, they introduce higher risks in the balancing of geographically dispersed loads and the enforcement of physical constraints, which are pivotal to the economic dispatch task.

While eliminating prediction errors entirely is not feasible, it is often practical to estimate the distributional information of renewable energy generation. This estimation enables the formulation of economic dispatch as a conventional chance-constrained programming. However, obtaining precise distributional information can be difficult in real-world applications, as we can only observe samples derived from the factual distribution. Implementing economic dispatch based solely on these observed samples is challenging, as economic dispatch frequently incorporates probabilistic constraints, such as chance constraints and, potentially, joint chance constraints, whose satisfaction is complex to validate from sample data.

Mathematically, the system model of economic dispatch considering renewable generation uncertainty is as follows:

$$\min \sum_{t=1}^{T} \left(\sum_{i \in \mathcal{N}} C_i(g_i^t) + R(G^t, D^t) \right) \tag{13a}$$

$$s.t. G^t = \sum_{i \in \mathcal{N}} g_i^t, D^t = \sum_{i \in \mathcal{M}} d_i^t, \forall t, \tag{13b}$$

$$S^t = D^t - G^t, \forall t, \tag{13c}$$

$$R(G^t, D^t) = \gamma_1 \operatorname{E}\left[(S^t)^+ \right] + \gamma_2 \operatorname{E}\left[(-S^t)^+ \right], \forall t, \tag{13d}$$

$$\boldsymbol{f}^t = \boldsymbol{H}_g \boldsymbol{g}^t - \boldsymbol{H}_d \boldsymbol{d}^t, \forall t, \tag{13e}$$

$$\mathbb{P}\left(\bigcap_{t=1}^{T} \{f_i^t \leq \overline{f}_i\} \right) \geq \rho_{i,1}, \forall i \in \mathcal{A}, \tag{13f}$$

$$\mathbb{P}\left(\bigcap_{t=1}^{T} \{f_i^t \geq -\overline{f}_i\} \right) \geq \rho_{i,2}, \forall i \in \mathcal{A}, \tag{13g}$$

$$g_i^{min} \leq g_i^t \leq g_i^{max}, \forall i \in \mathcal{N}, \forall t, \tag{13h}$$

$$-DR^i \leq g_i^{t+1} \leq UR^i, \forall i \in \mathcal{N}, \forall t, \tag{13i}$$

where $(\cdot)^+ = \max(\cdot, 0)$. The decision variables in (10) are g_i^t, the generation of generator i at time t, with the vector form $\boldsymbol{g}^t = [g_i^t, \forall i \in \mathcal{N}]$. The other parameters are as follows: C_i is the convex generation cost function for generator i; $R(G^t, D^t)$ is the risk cost with total generation G^t and total demand D^t; G^t is the total generation at time t; D^t is the total demand at time t; $\mathcal{N}$ denotes generator indexes and $\mathcal{M}$ is the demand indexes; S^t is the generation shortage at time t ($S^t < 0$ indicating generation excess); f_i^t is the power flow of transmission line i at time t, with the vector form $\boldsymbol{f}^t = [f_i^t, \forall i \in \mathcal{A}]$; d_i^t is the net demand for demand i at time t, with the vector form $\boldsymbol{d}^t = [d_i^t, \in \mathcal{M}]$; γ_1 and γ_2 are the unit generation shortage and excess risk cost; $\boldsymbol{H}_g$ and $\boldsymbol{H}_d$ are the shift factor matrix for generators and demands, respectively; $\overline{f}_i$ is the power flow capacity of transmission line i; $\rho_{i,1}$ and $\rho_{i,2}$ are the stability requirements of upper and lower limit capacities on transmission line i; g_i^{min} and g_i^{max} are the minimal and maximal generation capacities of generator i; DR^i and UR^i are the ramp up and down limits of generator i.

Constraint (13c) characterizes the generation shortage; constraint (13d) specifies the risk cost which includes the expected generation shortage and excess cost; constraint (13e) describes the power flow on the transmission lines; constraints (13f) and (13g) describe the joint chance constraints of upper and lower limit capacities on transmission lines during the whole time horizon; constraint (13h) characterizes the generation capacity of generators; and constraint (13i) characterizes the ramp up and down rate capacity of generators. This is a classical stochastic ED with joint chance constraints. The essential difficulty comes from the uncertainty of demands d_i^t, since both the risk cost $R(G^t, D^t)$ and the transmission line capacity constraints in (13f) and (13g) involve d_i^t. Without full knowledge of each d_i^t's distribution, we cannot accurately evaluate the objective function. Further, the feasibility of the chance constraints cannot be guaranteed without the distributional information. These two issues seriously deteriorate the economic performance and transmission line security. To tackle these challenges, Lu et al. introduce the notion of statistical feasibility to help establish a more practical formulation, which is only dependent on the samples in [20].

Specifically, Lu et al. first customize a sample-adaptive robust optimization to decouple the joint constraints in Problem (13). They then identify that the inaccurate uncertainty set leads to robust optimization's conservativeness and reconstruct the constraint-specific uncertainty sets. They design the corresponding sample-adaptive reconstruction-based robust optimization (ReconRO) based on the reconstructed uncertainty sets to further enhance the economic dispatch's effectiveness.

Table 2 illustrates the extra cost (refer to the gap to the optimal dispatch with accurate prediction) and empirical violation rate for different approaches. In Table 2, CC denotes the chance-constrained method where chance-constrained method with Laplace assumption (CC-L) uses Laplace distribution estimation and chance-constrained with Gaussian assumption (CC-G) adopts Gaussian distribution estimation; SG denotes the scenario generation method; RO is the classical robust optimization method. From Table 2, the ReconRO method achieves the lowest dispatch cost among all considered methods and reduces the extra cost by 1.39% compared with the optimal method.

Figure 6 presents the performance of different approaches when the stability requirement ρ decreases. The sample size $n = 1000$. It is evident from Fig. 6(b) that the solution obtained by the CC-G consistently results in infeasibility, as indicated by the curve positioned above the gray risk range curve. Furthermore, when the risk parameter ρ is less than 0.85, the solution derived from the CC-L is also infeasible. These observations suggest that the CC methods do not always ensure compliance with the chance constraints. Conversely, the scenario generation method yields feasible solutions across all scenarios and demonstrates improved economic performance as ρ decreases. Notably, the violation rate curve for the ReconRO method remains consistently below, yet closely aligned with the allowable risk range curve, indicating its ability to effectively utilize the flexibility reserve offered by the chance constraints. Additionally, the dispatch cost associated with ReconRO exhibits

Table 2 Performance comparison of different methods with $\rho = 0.95$, $\eta = 0.95$, and $n_1 + n_2 = 2000$ [20]

	CC-L	CC-G	SG	RO	ReconRO
Extra cost ($)	9315.2	9214.1	9253.3	17,396.7	9086.2
$\widehat{\rho}$	0.9999	0.9987	0.994	1	0.9639

a marked decline with a reduction in ρ. In summary, (1) the chance-constrained approach is heavily dependent on the underlying distributional assumptions. With a more precise distributional assumption, it can achieve favorable economic outcomes (as seen with CC-L). However, it does not consistently ensure adherence to the violation rate requirements. (2) RO emerges as the most conservative strategy, typically incurring the highest costs across all methods. Its forte lies in its guarantee of constraint satisfaction, ensuring no violations occur. (3) The ReconRO method reliably maintains the violation rate requirements.

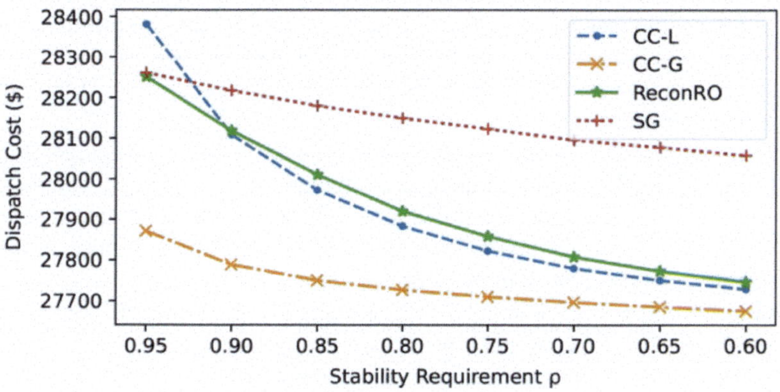

(a)The impact on dispatch cost.

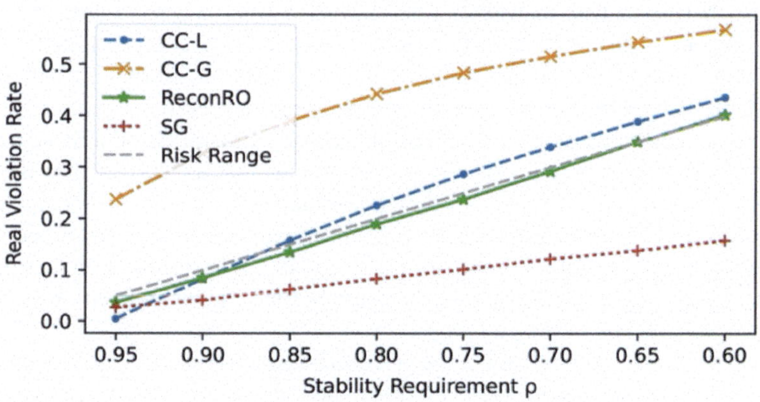

(b)The impact on violation rate.

Fig. 6 The impact of ρ on the performance [20]

5.3 Unit Commitment (an Extension to Joint Chance Constraints and Mixed-Integer Programming)

Unit commitment is a fundamental operational task conducted by Independent System Operators (ISOs) to devise a generator dispatch that minimizes the costs associated with satisfying predicted electricity demand. This optimization process takes into account various constraints related to generating resources and system reliability [27]. However, incorporating renewable energy sources into power generation has introduced novel challenges to the unit commitment problem. The unpredictable nature of renewable resources, characterized by variability and intermittency, gives rise to substantial forecasting inaccuracies [28]. These uncertainties necessitate a more complex unit commitment task, as they can significantly impact the optimal dispatch of generators.

Furthermore, the unit commitment problem is an NP-hard mixed-integer program (MIP) [29], which inherently entails high computational complexity. The necessity for multiple unit commitment optimizations within the stringent time constraints of daily electricity markets compounds this complexity, as ISOs must frequently reassess and adjust dispatch schedules to account for the dynamic nature of renewable generation and system demand [30]. Consequently, guaranteeing computational efficiency in the UC process is of paramount importance, posing a critical challenge for operators striving to maintain a balance between optimization accuracy and computational tractability.

To deal with the above-mentioned challenges, Liang et al. extend statistical feasibility into MIP and formulate the statistically feasible unit commitment in [21]. Unlike economic dispatch, binary variables lead to high computational costs in solving unit commitment. Hence, they consider the load characteristics to aid the unit status identification from a learning-to-optimization perspective. By identifying all unit statuses, they transform the MIP-based unit commitment into the LP-based unit commitment, effectively relieving the computational burden. Motivated by the properties of LP-based unit commitment, they exploit the optimization problem structure and customize an optimization kernel for unit commitment to boost its computational efficiency. By combining the notion of statistical feasibility with learning-to-optimization accelerating algorithms, the solution time can be reduced by more than half with tolerable performance loss.

Table 3 presents a comparison of the performance metrics across various methods. It is evident that the ReconRO method satisfies the stability requirements at the minimum cost, with a REP of 8.92%. The ReconRO outperforms its counterparts by achieving the lowest cost while adhering closely to the predetermined satisfaction rates (95%). Although the chance-constrained method incurs a lower cost than other methods, its violation rate exceeds the stability requirements. In contrast, the classical robust optimization method exhibits the highest REP of 42.95%, which is intrinsic to its conservative property. The scenario generation method and distributionally robust optimization method have notably reduced the conservativeness inherent in the classical robust optimization method, yet they do not fully capitalize on the potential

Table 3 Performance comparison of different methods [21]

	CC	RO	SG	DRO	ReconRO
REP (%)	5.83	42.95	13.45	11.43	8.92
$\hat{\rho}(\%)$	82.0	100	0	98.6	97.2
CPU time (s)	3.97	4.32	44.46	4.66	9.90

of the stability requirements. The scenario generation method is also characterized by its conservativeness, as it formulates constraints based on samples that may be skewed by numerous extreme scenarios. Furthermore, the scenario generation method is less time-efficient than its competitors, with a computational time of 44.46 s.

Figure 7 examines the sensitivity of various methodologies to the progressive integration of renewable energy penetration. More specifically, Fig. 7 showcases the performance of all methods within the context of an IEEE 39-bus system as the wind power penetration rate varies at 20%, 50%, and 80%. A notable trend is observed: the REP increases with the increasing penetration of renewable energy. This correlation is attributed to the heightened uncertainty associated with higher renewable energy integration, necessitating additional expenditures to mitigate this uncertainty. The ReconRO method consistently demonstrates superior performance in terms of satisfaction requirements and system costs. The distributionally robust optimization method emerges as the second-most effective method, providing an acceptable REP and satisfaction rate, albeit with a more conservative stance than ReconRO. While the classical robust optimization method and scenario generation method are characterized by their conservative nature, the conservativeness of the scenario generation method is relatively less pronounced than that of the classical robust optimization method with increasing renewable energy penetration. The CCP method yields the lowest costs; however, it does not consistently meet the chance constraints.

Table 4 provides a detailed comparison of the computational time for various power system cases by using the learning-to-optimization acceleration. Owing to the inherent structure of the tree-based model, the training of the unit status identifier is accomplished within a limited time. However, the inference phase exhibits a longer duration than the training phase. This is attributable to the necessity of generating predictions for each unit for every incoming sample, thereby creating a substantial computational loop.

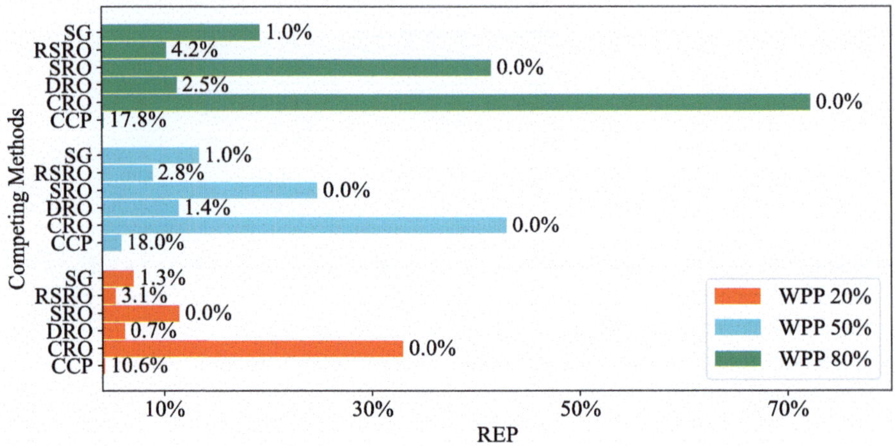

Fig. 7 The impact of renewable energy penetration [21]

Table 4 Running time for unit status identification [21]

	IEEE 39-bus system	IEEE 118-bus system	IEEE 300-bus system
Clustering time (s)	0.22	0.58	0.83
Training time (s)	6.43	23.34	27.79
Inference time (s)	136.17	783.31	937.88

References

1. C. Parmesan, M.D. Morecroft, Y. Trisurat, Climate change 2022: impacts, adaptation and vulnerability, PhD thesis (GIEC, 2022)
2. P. Du, N. Lu, Appliance commitment for household load scheduling. IEEE Trans. on Smart Grid **2**(2), 411–419 (2011)
3. F.H. Aghdam, N.T. Kalantari, B. Mohammadi-Ivatloo, A chance-constrained energy management in multi-microgrid systems considering degradation cost of energy storage elements. J. Energy Storage **29**, 101416 (2020)
4. K. Sundar, H. Nagarajan, L. Roald, S. Misra, R. Bent, D. Bienstock, Chance-constrained unit commitment with n-1 security and wind uncertainty. IEEE Trans. Control. Netw. Syst. **6**(3), 1062–1074 (2019)
5. W. Hou, Z. Liu, L. Ma, L. Wang, A real-time rolling horizon chance constrained optimization model for energy hub scheduling. Sustain. Cities Soc. **62**, 102417 (2020)
6. D. Pozo, J. Contreras, A chance-constrained unit commitment with an n-k security criterion and significant wind generation. IEEE Trans. Power Syst. **28**(3), 2842–2851 (2013)
7. W. Hou, H. Wei, Data-driven robust day-ahead unit commitment model for hydro/thermal/wind/photovoltaic/nuclear power systems. Int. J. Electr. Power Energy Syst. **125**, 106427 (2021)
8. Z. Chen, L. Wu, Y. Fu, Real-time price-based demand response management for residential appliances via stochastic optimization and robust optimization. IEEE Trans. Smart Grid **3**(4), 1822–1831 (2012)
9. X. Chen, N. Li, Leveraging two-stage adaptive robust optimization for power flexibility aggregation. IEEE Trans. Smart Grid **12**(5), 3954–3965 (2021)

10. J. Wang, Y. Shi, K. Fang, Y. Zhou, Y. Li, A robust optimization strategy for domestic electric water heater load scheduling under uncertainties. Appl. Sci. **7**(11), 1136 (2017)
11. X. Geng, L. Xie, Data-driven decision making in power systems with probabilistic guarantees: Theory and applications of chance-constrained optimization. Annu. Rev. Control. **47**, 341–363 (2019)
12. C. Duan, W. Fang, L. Jiang, L. Yao, J. Liu, Distributionally robust chance-constrained approximate AC-OPF with Wasserstein metric. IEEE Trans. Power Syst. **33**(5), 4924–4936 (2018)
13. R. Zhu, H. Wei, X. Bai, Wasserstein metric based distributionally robust approximate framework for unit commitment. IEEE Trans. Power Syst. **34**(4), 2991–3001 (2019)
14. A. Zhou, M. Yang, X. Zheng, S. Yin, Distributionally robust unit commitment considering unimodality-skewness information of wind power uncertainty. IEEE Trans. Power Syst. **38**(6), 5420–5431 (2023)
15. A. Shapiro, D. Dentcheva, A. Ruszczynski, Lectures on stochastic programming: modeling and theory. Soc. Ind. Appl. Math. (2021)
16. A. Ben-Tal, L. El Ghaoui, A. Nemirovski, Robust optimization (Princeton university press, 2009)
17. X.A. Sun, A.J. Conejo, Robust optimization in electric energy systems (Springer International Publishing, 2021)
18. L.J. Hong, Z. Huang, H. Lam, Learning-based robust optimization: procedures and statistical guarantees. Manage. Sci. **67**(6), 3447–3467 (2021)
19. W. Jiang, C. Lu, C. Wu, Robust scheduling of thermostatically controlled loads with statistically feasible guarantees. IEEE Trans. on Smart Grid **14**(5), 3561–3572 (2023)
20. C. Lu, N. Gu, W. Jiang, C. Wu, Sample-adaptive robust economic dispatch with statistically feasible guarantees. IEEE Trans. Power Syst. **39**(1), 779–793 (2024)
21. J. Liang, W. Jiang, C. Lu, C. Wu, Joint chance-constrained unit commitment: Statistically feasible robust optimization with learning-to-optimize acceleration. IEEE Trans. Power Syst. 1–13 (2024)
22. M.A. Zehir, A. Batman, M. Bagriyanik, Review and comparison of demand response options for more effective use of renewable energy at consumer level. Renew. Sustain. Energy Rev. **56**, 631–642 (2016)
23. A. Arteconi, D. Patteeuw, K. Bruninx, E. Delarue, W. D'haeseleer, L. Helsen, Active demand response with electric heating systems: impact of market penetration. Appl. Energy **177**, 636–648 (2016)
24. F. Hu, K. Sundar, S. Srinivasan, R. Bent, Demand response analogues for residential loads in natural gas networks. IEEE Access **9**, 161978–161990 (2021)
25. Y. Zhou, C. Wang, J. Wu, J. Wang, M. Cheng, G. Li, Optimal scheduling of aggregated thermostatically controlled loads with renewable generation in the intraday electricity market. Appl. Energy **188**, 456–465 (2017)
26. G. Chen, H. Zhang, H. Hui, N. Dai, Y. Song, Scheduling thermostatically controlled loads to provide regulation capacity based on a learning-based optimal power flow model. IEEE Trans. Sustain. Energy **12**(4), 2459–2470 (2021)
27. D. Bertsimas, E. Litvinov, X.A. Sun, J. Zhao, T. Zheng, Adaptive robust optimization for the security constrained unit commitment problem. IEEE Trans. Power Syst. **28**(1), 52–63 (2013)
28. C. Zhao, J. Wang, J.-P. Watson, Y. Guan, Multi-stage robust unit commitment considering wind and demand response uncertainties. IEEE Trans. Power Syst. **28**(3), 2708–2717 (2013)
29. Q.P. Zheng, J. Wang, A.L. Liu, Stochastic optimization for unit commitment—a review. IEEE Trans. Power Syst. **30**(4), 1913–1924 (2015)
30. A.V. Ramesh, X. Li, Feasibility layer aided machine learning approach for day-ahead operations. IEEE Trans. Power Syst. **39**(1), 1582–1593 (2024)

Modeling and Operational Optimization of Hydropower Generation

Yikui Liu, Lei Wu, and Nan Yang

Abstract Hydropower is recognized globally as a clean, high-quality, and flexible renewable energy source and is playing an increasingly crucial role in helping achieve carbon emission reduction and providing flexibility to power systems. Focusing on hydropower generation, this chapter discusses the development status and introduces the main types of hydraulic turbines and hydropower plants. The chapter focuses and discusses detailed modeling of the hydropower plants with reservoirs, run-of-the-river plants, cascaded hydropower systems, and pumped storage plants in the context of optimized operation.

Keywords Hydropower · Economic dispatch · Hydraulic turbine · Pumped storage hydropower · Cascade hydropower system

Hydropower is globally recognized as a clean, high-quality, and flexible renewable energy source and is playing an increasingly crucial role in helping achieve carbon emission reduction and providing flexibility to power systems. As reported by "2022 Hydropower Status Report—Sector Trends and Insights", worldwide hydropower generated is around 4,250TWh of clean energy in 2021. This amount is equivalent to around one and a half times the entire electricity consumption of the European Union. In 2021, 26 GW of new hydropower capacity was put into operation. Most of this growth came from China, which takes 21GW among the 26 GW of the newly online capacity.

Y. Liu (✉)
School of Electrical Engineering, Sichuan University, Chengdu, China
e-mail: yikuiliu@scu.edu.cn

L. Wu
Department of Electrical and Computer Engineering, Stevens Institute of Technology, Hoboken, NJ, USA

N. Yang
College of Electrical Engineering and New Energy, China Three Gorges University, Yichang, China

F. Shahnia and F. Wen (eds.), *Power and Energy Resources*, Power Systems,
https://doi.org/10.1007/978-981-96-2612-0_2

Hydropower is a mature renewable energy technology that can generally be divided into conventional hydropower plants and pumped storage power plants. The former includes hydropower plants with reservoirs and run-of-river hydropower plants, and their combinations along a basin can form cascade hydropower systems. In general, the uncontrollability and uncertainty of water inflow, the complicated power generation characteristics of hydraulic turbines, and the hydraulic and electrical coupling between upstream and downstream hydropower plants are among the many factors that challenge the hydropower optimal scheduling.

1 Hydropower Plants and Hydraulic Turbines

1.1 Structure of the Hydropower Plant

Hydropower plants essentially convert the potential and kinetic energy of water inflow into electrical energy. Figure 1 shows the typical structure and the main components of a hydropower plant.

Dams, built to block the river, raise the forebay level and form a reservoir. On the one hand, the reservoir endows the hydropower plant with a certain water storage capacity, so that it has the ability to regulate the power generation; and on the other hand, the augmented forebay level endows the same amount of water with higher potential energy which can produce more electricity.

Water with high potential energy flows through the water intake, which is a structure that directs the water into the penstock and most importantly is equipped with a trash rack trapping debris in the water, such as aquatic life, garbage, and gravel. If

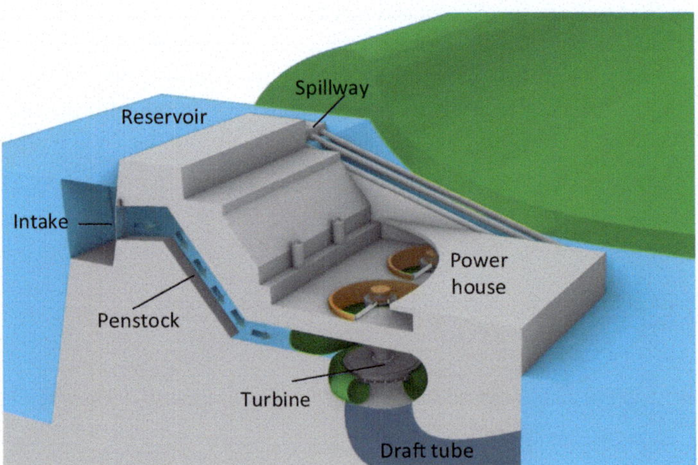

Fig. 1 The typical structure and the main components of a hydropower plant

debris passes through the trash rack and finally enters the hydraulic turbines along with the penstock, it could potentially damage the wicket gates, turbine nozzles, and turbine runners. In winter at high latitudes, the water may be mixed with ice. In this case, the trash rack is heated to melt the approaching ice. In addition, rakes and trolley arrangements are equipped to automatically clean the trash racks.

Guiding water from the water intake to hydraulic turbines, penstocks are large steel or concrete pipes that can withstand high-pressure water flow and the water hammer effect caused by opening and closing the penstock gates. A hydropower plant typically has multiple turbines. The penstocks can either be separate for each turbine or initially share a common segment before branching off to individual turbines at the end. Usually, the former is adopted when the intake is closer to the turbines, while the latter may be preferred when the intake is relatively farther away.

Hydraulic turbines are the core of energy conversion in hydropower plants. They convert hydraulic energy into mechanical energy to drive generators for ultimately producing electricity. There are two main types of hydraulic turbines: impulse turbines and reaction turbines. Impulse turbines, also known as velocity turbines, utilize the kinetic energy of water. An example of an impulse turbine is the Pelton turbine. Reaction turbines harness both the kinetic energy and the pressure of the water. Examples of reaction turbines include the Francis turbine and the Kaplan turbine. Impulse turbines are particularly suitable for applications with high water heads and relatively low flow rates, while reaction turbines are suitable for moderate to low water heads and relatively high flow rates. Practically, the Francis turbine and the Kaplan turbine, which are both in the category of reaction turbines, are most commonly adopted in hydropower plants. In fact, these two types of hydraulic turbines cover a broad spectrum of hydraulic conditions encountered in hydropower projects, making them the most frequently used types globally. The choice between Francis and Kaplan turbines depends on factors such as the specific water head and flow characteristics of the site, as well as economic considerations and environmental factors.

If reaction turbines are employed, a draft tube becomes essential as it connects the turbine outlet to the tailrace. The draft tube is designed with a gradually increasing diameter to ensure the safe discharge of water into the tailrace at a controlled velocity. At the end of the draft tube, outlet gates are installed, which can be closed when maintenance or repair work is required. The tailrace is a conduit through which water flows downstream after passing through the turbines.

The powerhouse is the central facility of a hydropower plant and is typically located at the base or near the base of a dam, where water is diverted from the reservoir through penstocks to drive the hydraulic turbines. It houses hydraulic turbines, generators, and associated equipment (e.g., control equipment, transformers, and switchgear), and can vary in size and complexity depending on the scale of the hydropower plant. Large hydropower plants, such as Three Gorges Dam and Itaipu Dam, have huge powerhouses housing dozens of generating units, complemented by complex control and measurement equipment, while smaller plants may have simpler configurations.

Spillway is a structure designed to safely release excessive water for preventing the water head in the reservoir from rising too high exceeding the safety operating threshold, which could potentially cause risks of damage to the dam and flooding. There are different types of spillways:

- Surface spillways bury tunnels in the dam, often at the crest or near the top, through which excessive water can flow;
- Gated spillways have switchable gates installed to regulate the flow of water downstream;
- Siphon spillways use the siphon effect to draw water from the reservoir through conduits to downstream.

In general, operators of hydropower plants usually strive to avoid water spillage as it wastes hydraulic resources, especially when downstream requirements such as ecology, irrigation, and city water usage are already met.

1.2 Hydraulic Turbine

Impulse turbines and reaction turbines are two main types of hydraulic turbines. Impulse turbines are also known as velocity turbines. An example of an impulse turbine is the Pelton turbine. Examples of reaction turbines are the Kaplan turbine and the Francis turbine.

1.2.1 Impulse Turbine

As shown in Fig. 2, an impulse turbine drives the rotor to rotate by directing high-pressure and high-velocity water jets that contain significant kinetic energy through the nozzles onto the blades of the turbine, thereby powering the generator to produce electricity. The high-velocity water is the primary source of energy that drives the turbine blades. The water is then exhausted from the turbine through the exhaust system. The primary function of the nozzle is to accelerate the flow of water to a high velocity by narrowing the passage. The blades of an impulse turbine are particularly arranged with suitable angles to ensure that the steam jet strikes them passing the maximum amount of force.

One of the advantages of impulse turbines is their simplicity in design and operation, making them suitable for use in smaller power generation. Impulse turbines, such as Pelton turbines, have quite high efficiency under appropriate operating conditions. The efficiencies of Pelton turbines can reach up to 80% or more under optimal conditions, making them one of the most efficient types. Impulse turbines are commonly used in mountainous regions and areas with high water heads, where they can efficiently harness the potential energy of fast-flowing water to generate electricity, but they generally have a lower power output capacity, making them typically used for smaller scale hydropower plants.

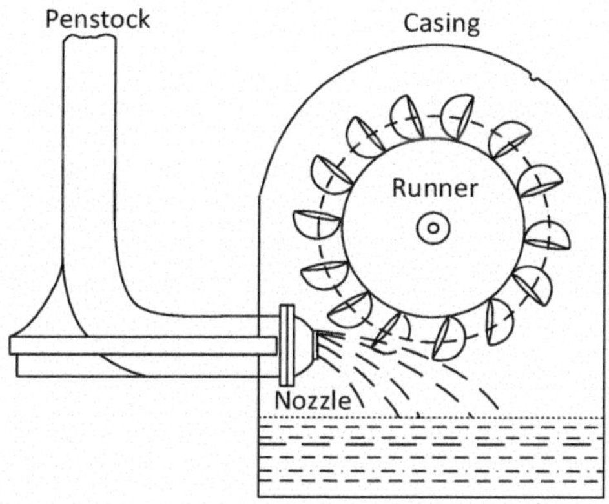

Fig. 2 Impulse turbine

The disadvantages of impulse turbines are generally caused by the high-velocity water that they simultaneously rely on. For example, due to the impact of high-velocity water flow, more frequent maintenance and inspection are required to ensure that components including blades, rotor, and bearings can operate normally and the entire turbine can operate safely and efficiently. Moreover, there are more strict requirements on the quality of the water stream. Impurities contained in high-velocity water, when impacting the blades, can cause damage to the turbine blades and other components, significantly reducing the turbine's efficiency and lifespan.

1.2.2 Reaction Turbine

While the impulse turbine harnesses the kinetic energy converted from the potential energy of flowing water, a reaction turbine utilizes both the kinetic energy and pressure of water. Taking Francis Turbines shown in Fig. 3 as an example, its components, including a runner, wicket gates, turbine blades, a shaft, and so on, are housed by a casing, which guides the water from the penstock to arrive at the wicket gates in a spiral pattern with a controlled velocity and pressure.

Before driving the turbine runner, the water passes through the wicket gates that can adjust the positions of installed blades to control the inflow of water. Adjusting the wicket gates is the primary manner to optimize the performance of the reaction turbine in response to varying water flow conditions. Properly positioned wicket gates can ensure that the water enters the turbine runner at a suitable angle and velocity, thereby enhancing the efficiency of the turbine by maximizing the energy transfer from the water to the turbine blades. In addition, during start-up and shutdown procedures, wicket gates can gradually increase or decrease the water inflow, helping manage pressure changes and smooth transitions in turbine operation.

As water pressure on the blades drives the runner to rotate, both its pressure and kinetic energy are converted into mechanical energy which is transferred by a shaft that connects the turbine and the generator. The shaft spins to drive the generator, which ultimately converts the mechanical energy into electricity. After

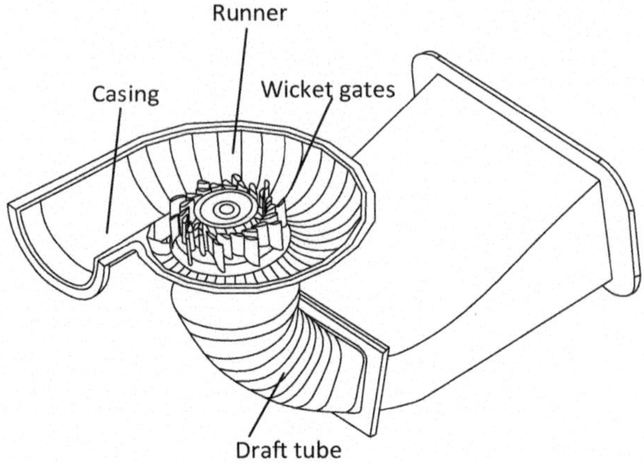

Fig. 3 Reaction turbine

passing through the runner, the water is then carried away through a draft tube. The draft tube helps maintain a steady flow and pressure at the outlet of the turbine.

Much higher complexity and costs of manufacture and installation compared to impulse turbines are the disadvantages of reaction turbines. However, reaction turbines can operate efficiently in a variety of conditions, including varying water flows and water heads, making them suitable for medium to low water head applications where a steady flow of water is available. The Francis turbine and the Kaplan turbine are the two commonly used types of reaction turbines, with the Francis turbine being the most common, occupying more than three-quarters of the proportion in practical applications. However, for the sites of relatively low water heads, Kaplan turbines are often preferred.

The main difference between the Francis turbine and the Kaplan turbine lies in their different water entry manners. For Francis turbines, water enters through a scroll case and is guided tangentially toward the runner through a controlled pathway, which allows for efficient energy conversion and adaptability to varying flow conditions. For Kaplan turbines, also known as Propeller Turbines, water enters into the turbine through a draft tube axially or nearly axially. The angles of turbine blades can be adjusted based on the flow conditions, allowing them to efficiently capture energy from water. The axial or near-axial entry of water is a distinguishing characteristic of Kaplan turbines compared to Francis turbines that typically receive water tangentially.

1.3 Modeling of Water-To-Power Conversion

Compared to thermal power units, modeling the operation of hydropower units is more complex. The power output of a hydropower unit is jointly determined by its water discharge and the net water head of the reservoir, which forms strong nonlinearity and non-convexity. The net water head refers to the difference between the water head and the water tail, in which the former is nonlinearly dependent on the current water level of the reservoir, while the latter is influenced by the water discharge and the water spillage. Usually, the greater the water discharge plus the water spillage, the stronger the elevation effect on the water tail, which causes a reduction in the net water head. It is evident that the power output explicitly exhibits a strong nonlinear coupling with the net water head and water discharge, as well as an implicit nonlinear coupling with the reservoir water volume, the water spillage, and the water tail.

Figure 4 illustrates the concepts of water head, water tail, water discharge, and other related terms. The water head is the vertical distance from the water level at the upstream reservoir to the water level at the turbine's intake. The higher the water head, the more potential energy the water contains, which, for the same amount of discharged water, can be converted into more mechanical energy and consequently more electrical energy. The water tail, also known as tailwater, refers to the water level of the downstream river where the water exits the turbine after passing through it to generate power. As mentioned, the net water head is equal to the difference between the water head and the water tail. The water discharge refers to the rate at which water flows through the turbines and exits the plant. The water discharge can refer to an individual turbine or the entire hydropower plant, where the latter is equal to the sum of water discharges of all turbines.

As in (1), the water-to-power conversion function relates the net water head h, water discharge q_i, and water-to-power efficiency η_i, where i is the index of hydropower units and $g = 9.81 \times 10^{-3}$ is the gravitational acceleration. Referring to reference [1], the relationship between the water head h^H and water volume v in the reservoir can be fitted by a fourth-degree polynomial curve as in (2), and the water tail h^T can be represented by another fourth-degree polynomial curve with respect to the water discharge of dam q and the water spillage s as in (3). The net water head can be calculated as in (4):

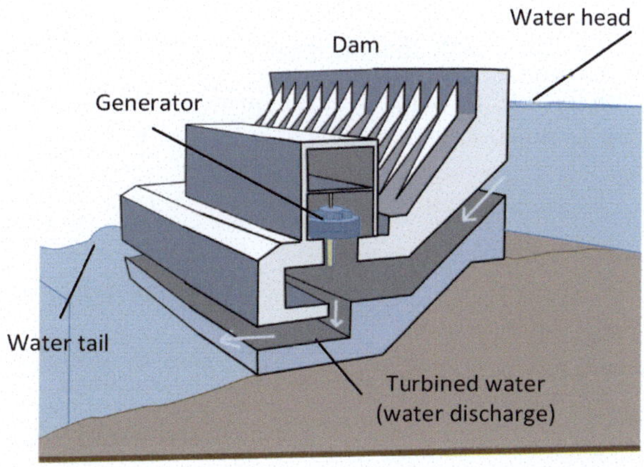

Fig. 4 The illustration of some concepts

$$p = g \cdot \eta_i \cdot h \cdot q_i; \tag{1}$$

$$h^H = a_0 + a_1 \cdot v + a_2 \cdot v^2 + a_3 \cdot v^3 + a_3 \cdot v^4; \tag{2}$$

$$h^T = b_0 + b_1 \cdot (q + s) + b_2 \cdot (q + s)^2 + b_3 \cdot (q + s)^3 + b_3 \cdot (q + s)^4; \tag{3}$$

$$h = h^H - h^T. \tag{4}$$

The hydraulic efficiency, which depends on the water discharge of hydropower unit i and the net water head h, , can be modeled as in (5):

$$\eta_i = d_{0,i} + d_{1,i} \cdot q_i + d_{2,i} \cdot h + d_{3,i} \cdot q_i \cdot h + d_{4,i} \cdot {q_i}^2 + d_{5,i} \cdot h^2. \tag{5}$$

In (5), $d_{0,i}$, $d_{1,i}$, $d_{2,i}$, $d_{3,i}$, $d_{4,i}$, and $d_{5,i}$ are coefficients determined by the inherent characteristics of the hydropower unit and can be obtained through experiments [1].

It is not difficult to see that by substituting (2)–(5) into (1), a four-dimensional high-order polynomial function (6) can be obtained. Moreover, as $q = \sum_i q_i$, including q indeed couples water discharges of all hydropower units:

$$p = f(q_i, q, s, v). \tag{6}$$

The roughest simplification is to assume that the water head h^H remains constant and the water tail effect can be ignored, namely $h^T = 0$, and thereby the net water head degenerates to a constant, i.e., $h = \tilde{h}^H$, denoted as $\tilde{h}$. On this basis, by further assuming a constant conversion efficiency $\tilde{\eta}_i$, Eq. (1) can be simplified as in (7), where $\tilde{h}$ and $\tilde{\eta}_i$ are constant parameters:

$$p_i = g \cdot \tilde{\eta}_i \cdot \tilde{h} \cdot q_i. \tag{7}$$

It is foreseeable that the accuracy of (7) could be far from satisfactory.

With fixed h, η_i could become a quadratic function of q_i as in (8), and (7) turns to be a cubic function with respect to q_i as in (9):

$$\eta_i = D_{0,i} + D_{2,i} \cdot \tilde{h} + D_{5,i} \cdot \tilde{h}^2 + D_{1,i} \cdot q_i + D_{3,i} \cdot \tilde{h} \cdot q_i + D_{4,i} \cdot {q_i}^2;$$

$$= E_{0,i} + E_{1,i} \cdot q_i + E_{2,i} \cdot {q_i}^2; \tag{8}$$

where $E_{0,i} = D_{0,i} + D_{2,i} \cdot \tilde{h} + D_{5,i} \cdot \tilde{h}^2$;

$$E_{1,i} = D_{1,i} + D_{3,i} \cdot \tilde{h};$$

$$E_{2,i} = D_{4,i};$$

$$\begin{aligned} p_i &= g \cdot \tilde{h} \cdot \left(E_{0,i} + E_{1,i} \cdot q_i + E_{2,i} \cdot {q_i}^2\right) \cdot q_i \\ &= \alpha_{0,i} + \alpha_{1,i} \cdot q_i + \alpha_{2,i} \cdot {q_i}^2 + \alpha_{3,i} \cdot {q_i}^3. \end{aligned}; \tag{9}$$

Equation (9) can be piecewise linearized with respect to q_i. Because (9) is a cubic function and cannot be guaranteed to be convex within the possible range of q_i, , from an optimization perspective, binary variables become necessary to indicate the selection of linearized pieces. There are many forms of linearization expression. The following (10)–(13) is the most intuitive one:

$$p_i = \sum_k B_{i,k} \cdot z_{i,k} + \sum_k C_{i,k} \cdot q_{i,k}; \tag{10}$$

$$q_i = \sum_k Q_{i,k} \cdot z_{i,k} + \sum_k q_{i,k}; \tag{11}$$

$$\sum_k z_{i,k} = 1; \tag{12}$$

$$0 \le q_{i,k} \le Q^{SEG} \cdot z_{i,k}. \tag{13}$$

In (10), k represents the index of the segments; $z_{i,k}$ is an ancillary variable that indicates the selection of the segments; $B_{i,k}$ and $C_{i,k}$ are the segment point and the slope of segment k. . Constraint (11) relates $q_{i,k}$ and q_i. . Constraint (12) requires that there is only one segment being selected. Constraint (13) limits the range of $q_{i,k}$ in the selected segment. If $z_{i,k} = 0$, $q_{i,k}$ is forced to 0. When unit commitment is included, (12) is modified as in (14), where I_i is the ON/OFF indicator of hydropower unit i. When $I_i = 0$, p_i and q_i are both enforced to be zeros:

$$\sum_k z_{i,k} = I_i. \tag{14}$$

In practice, Eq. (9) is not necessarily derived directly from Eqs. (1)–(5). Typically, a series of operational points are obtained through experiments by adjusting the water discharge while maintaining the water head to be constant. Afterward, a cubic function is often used to fit these operational points. In some cases, if needed, operators may use quadratic or quartic polynomial functions for fitting to improve the accuracy.

When the net head varies greatly during the actual operation of a hydropower plant, considering a fixed head, typically either a typical operating head or an average head, may lead to unacceptable errors. One solution involves discretizing the net water head and constructing (9) for different ranges of water heads, and thereby (15) can be obtained:

$$p_i = \begin{cases} \alpha_{0,i,1} + \alpha_{1,i,1} \cdot q_i + \alpha_{2,i,1} \cdot {q_i}^2 + \alpha_{3,i,1} \cdot {q_i}^3;\ h \in [\underline{H}_1, \overline{H}_1] \\ \alpha_{0,i,2} + \alpha_{1,i,2} \cdot q_i + \alpha_{2,i,2} \cdot {q_i}^2 + \alpha_{3,i,2} \cdot {q_i}^3;\ h \in [\underline{H}_2, \overline{H}_2] \\ \cdots \\ \alpha_{0,i,j} + \alpha_{1,i,j} \cdot q_i + \alpha_{2,i,j} \cdot {q_i}^2 + \alpha_{3,i,j} \cdot {q_i}^3;\ h \in [\underline{H}_j, \overline{H}_j] \end{cases}; \tag{15}$$

Similar to (10)–(13), within a certain net water head range, piecewise linearization can be processed with respect to q_i. It is worth noting that from an optimization model perspective, ancillary binary variables need to be introduced to indicate which segment h belongs to, while corresponding constraints are activated and deactivated.

On the basis of (10)–(13), the constraints considering the net water head range can be formulated as in (16)–(20):

$$p_i = \sum_j \left(\sum_k B_{i,k,j} \cdot z_{i,k,j} + \sum_k C_{k,j} \cdot q_{i,k,j} \right); \tag{16}$$

$$q_i = \sum_j \sum_k Q_{i,k} \cdot z_{i,k,j} + \sum_j \sum_k q_{i,k,j}; \tag{17}$$

$$\sum_k z_{i,k,j} = y_j; \tag{18}$$

$$0 \leq q_{i,k,j} \leq Q^{SEG} \cdot z_{i,k,j}; \tag{19}$$

$$\sum_j y_j = 1. \tag{20}$$

In (16)–(19), y_j indicates the selection of the net water head range and can be determined by other constraints that are related to the reservoir water volume evolution over time. When $y_j = 0$, namely the current net water head is not within the range $[\underline{H}_j, \overline{H}_j]$, the corresponding $z_{i,k,j}$ will all be enforced as zero and $\sum_k B_{k,j} \cdot z_{i,k,j} + \sum_k C_{k,j} \cdot q_{i,k,j}$ becomes zero as well.

The effect of tailwater is not accounted for by the simplification above, namely $\tilde{h} = \tilde{h}^H$ is assumed. In the next approach, the water tail is partially considered which leads to a more complex representation.

In η_i, the effect of water tail is still neglected, which means in η_i, h is set as $\tilde{h}^H$, , but (1) is rewritten as in (21). The net water head is represented as $h^H - h^T$. Equation (21) can be further written as in (22), by submitting (8) into (21):

$$p_i = g \cdot \eta_i \cdot \left(h^H - h^T\right) \cdot q_i; \tag{21}$$

$$p_i = g \cdot \left(E_{0,i} + E_{1,i} \cdot q_i + E_{2,i} \cdot {q_i}^2\right) \cdot \left(h^H - h^T\right) \cdot q_i. \tag{22}$$

Further considering (3), namely equation of h^T, p_i can be written as a high-order polynomial in terms of q_i, q, s, and h^H. Similarly, h^H can be discretized, and on this basis, constraints for linearizing multivariate functions can be constructed.

A specific case is introduced where only one hydropower unit or an equivalent unit is considered. Equation (22) can be simplified into a high-degree polynomial in terms of q_i, s, and h^H, as $p_i = f(q_i, s, h^H)$. If water spillage is disregarded, i.e., $s = 0$, then $f(q_i, s, h^H)$ becomes $f(q_i, h^H)$ and forms a surface that can be visualized as in Fig. 5. On this basis, two-dimensional linearization techniques can be applied to linearize it, which ultimately can be represented by a set of linear constraints. It is worth mentioning that although s is referred to as water spillage, it incorporates downstream water demands, such as ecological and irrigation demands. When the water discharge q is insufficient to meet downstream ecological, production, and domestic needs, s will be non-zero.

In summary, the specific level of detail in hydropower unit modeling is determined by combining hydroelectric and operational characteristics. For example, for some hydropower plants where water head variations are tiny, water head variations may be disregarded. Similarly, for hydropower plants unaffected by impacts of water tail, their elevation may be considered negligible. Multidimensional linearization techniques can always be applied to linearize the water-to-power conversion function, avoiding the introduction of non-convexity and nonlinearity into the optimal scheduling problem of hydropower plants, which would otherwise lead to difficulties in problem-solving [2].

However, there is always a trade-off between modeling accuracy and solving efficiency. Linearizing high-dimensional and high-order polynomial functions requires fine-grained segmentation accompanied by a large number of auxiliary variables and constraints, which can also result in an unacceptable computational burden.

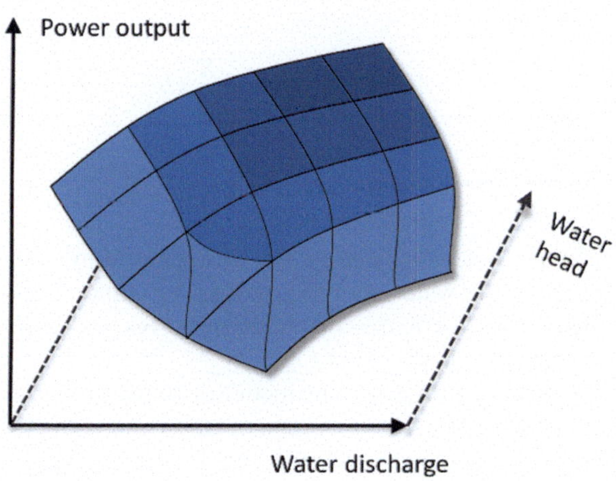

Fig. 5 Visualized surface

1.4 Modeling of Forbidden Zones

Similar to thermal power units, hydraulic turbines also have forbidden zones that refer to ranges of operating conditions within which an extended period of operation could cause excessive wear and potential damage to the shaft and turbine blades. The specific ranges of operating conditions can vary depending on the design and type of hydraulic turbines, but they typically include certain ranges of power outputs, flow rates, and water heads.

The forbidden zones of thermal power units can be typically expressed in terms of power ranges, and so is the hydraulic turbine, as in (23)–(25). In (23)–(25), $[P_{i,k}^{LB}, P_{i,k}^{UB}]$ defines feasible zones rather than forbidden zones and intuitively, and the forbidden zones are regions outside of them. Constraint (23) describes where the hydraulic turbine can operate. Constraint (24) enforces that if the turbine is ON, it should be in one of the feasible zones. Constraint (25) connects $p_{i,k}$ to p_i. Except for the activated feasible zone, $p_{i,k}$ in the other zones are forced to be zeros. $I_{i,k}$ and I_i are binary variables. Similarly, constraints (26)–(28) can be formulated with respect to the feasible zones defined by water discharge, $[Q_{i,k}^{LB}, Q_{i,k}^{UB}]$:

$$P_{i,k}^{LB} \cdot I_{i,k} \leq p_{i,k} \leq P_{i,k}^{UB} \cdot I_{i,k}; \tag{23}$$

$$\sum_k I_{i,k} = I_i; \tag{24}$$

$$\sum_k p_{i,k} = p_i; \tag{25}$$

$$Q_{i,k}^{LB} \cdot I_{i,k} \leq q_{i,k} \leq Q_{i,k}^{UB} \cdot I_{i,k}; \tag{26}$$

$$\sum_k I_{i,k} = I_i; \tag{27}$$

$$\sum_k q_{i,k} = q_i. \tag{28}$$

More meticulously, the forbidden zones of hydraulic turbines can be represented jointly by the water net head and power output. Moreover, the water net head and power output form irregular areas on the two-dimensional plane constituted by these two variables as shown in Fig. 6.

Reference [3] studied the modeling of two-dimensional irregular forbidden zones and used methods such as segmentation and convexification to divide the irregular forbidden zones into multiple convex sub-regions. Subsequently, multiple sets of linear constraints are employed to exclude the vibration zone sub-regions from the feasible domain of net water head and power output.

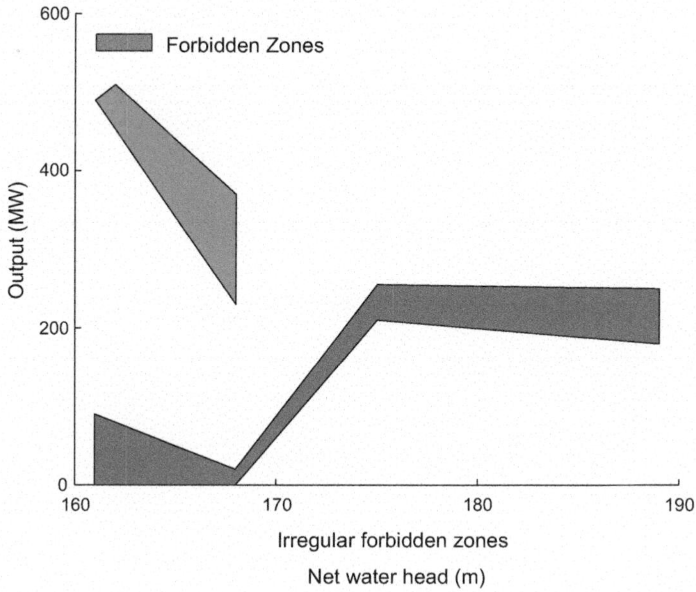

Fig. 6 Irregular forbidden zones

In fact, the power output in (23)–(25) can also form a two-dimensional plane jointly with the water head, similar to the one described in reference [3], as shown in Fig. 7. However, the corresponding forbidden zones in the two-dimensional plane are regular rectangles, indicating that they are not related to the water head.

1.5 Modeling of Penstock

For a hydropower plant with multiple turbines, the penstocks may either be dedicated to each turbine separately, or initially shared through a common penstock that later splits into individual segments leading to each turbine. The kinetic energy loss of water flowing through the penstock increases nonlinearly as the flow rate rises.

The structural differences in penstocks lead to different considerations in modeling the penstocks. In the case of separate penstocks, kinetic energy loss can be simply counted into the efficiency of individual hydraulic turbine. However, for a common penstock setup, this method easily introduces significant errors, because penstock sharing creates accumulation on water intake to individual hydraulic turbines and the cumulative water intake nonlinearly increases the overall kinetic energy loss compared to the case that individual turbines operate separately. For instance, with two turbines operating simultaneously, the kinetic energy loss in the common penstock is greater than the sum of kinetic energy losses when the two turbines operate separately with the same water intake.

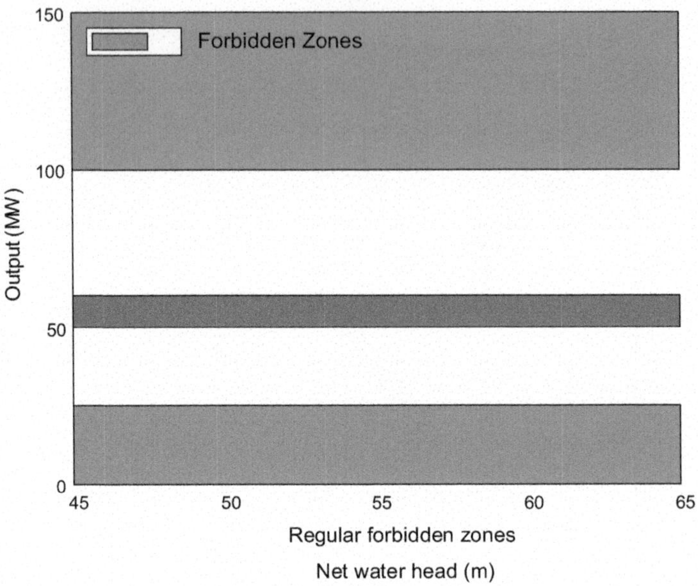

Fig. 7 Regular forbidden zones

Taking an actual hydropower plant, Round Butte, located in Portland, Oregon, U.S. as an example, as shown in Fig. 8, the shared penstock limits the ability of its three units to reach their full nameplate generation capacities at the same time. Additionally, the water flow congestion within the common penstock amplifies the energy loss.

The Round Butte Dam's water intake system features a 23-foot diameter steel-lined tunnel that splits into three 13½-foot penstocks, each feeding one of the three hydraulic turbines (as illustrated in Fig. 8). Friction along the pipe walls results in

Fig. 8 The common penstock to three units

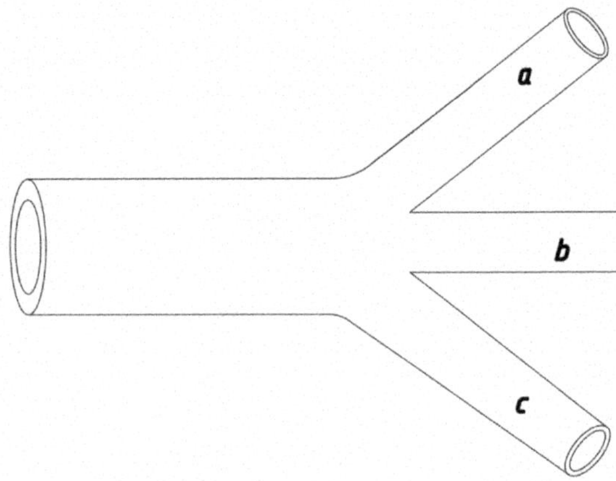

kinetic energy losses, which increase nonlinearly with higher water flow rates. The shared penstock creates an interdependence between the power generation efficiencies of the three units. For instance, activating Unit 3 additional to running Units 1 and 2 raises the total water flow through the common penstock, which amplifies overall energy losses, reduces the kinetic energy available per unit of water, and consequently decreases the power generation efficiency of Units 1 and 2. Due to the coupling effect, the maximum achievable power output is approximately 5.9% below the combined installed capacity of the three turbines. Notably, energy loss correlates positively with water flow—greater flow rates lead to higher losses.

The water-to-power function $g_i(\cdot)$ of unit i in Round Butte can be written as in (29). The integrated conversion coefficient η_i is fundamentally a function of $p_{i,t}$ and H, namely the water head. Since the water head H is consistent across all three units and the forebay level of the Round Butte Dam typically experiences minimal variation within a day, it can be treated as constant:

$$p_i = g_i(q_i) = g \cdot \eta_i \cdot H \cdot q_i. \tag{29}$$

For taking the common penstock caused energy loss into account, a term h_i^{LS} representing the additional water head loss is introduced, which modifies the net water head for unit i to $H - h_i^{LS}$. The revised power generation $p_i\prime$ is expressed as in (30), where the second term quantifies the reduction in power generation compared to the original value $p_i = g_i(q_i)$ for the same amount of water q_i. The reduction is defined as the power generation reduction r_i. Based on reference [4], the total water head loss, $\sum_i h_i^{LS}$, is calculated as $\beta \cdot q^2$, where $q = \sum_i q_i$. . This loss is proportionally distributed among the individual hydropower units, as in (31), based on q_i. Consequently, the power generation reduction r_i is reformulated as in (32), where $\delta = \beta/H$ is an integrated parameter and $l_i = q_i \cdot g_i(q_i)$ is an auxiliary variable. The observed energy loss data in the hydropower generation can be used to calibrate parameter δ:

$$p_i' = g \cdot \eta_i \cdot \left(H - h_i^{LS}\right) \cdot q_i = g_i(q_i) - \frac{h_i^{LS}}{H} \cdot g_i(q_i) = p_i - r_i; \tag{30}$$

$$h_i^{LS} = \frac{q_i}{q} \cdot \beta \cdot q^2 = \beta \cdot q_i \cdot q; \tag{31}$$

$$r_i = \frac{h_i^{LS}}{H} \cdot g_i(q_i) = \frac{\beta}{H} \cdot q_i \cdot q \cdot g_i(q_i) = \delta \cdot q \cdot l_i. \tag{32}$$

The nonlinear term $q \cdot l_i$ in (32) can be linearized using McCormick Relaxation as in (33a–33d). With this approach, $Q^{LB/UB}$ and $L_i^{LB/UB}$ respectively represent the lower and upper limits of q and l_i. Furthermore, given original cubic function $g_i(\cdot)$, the term $l_i = q_i \cdot g_i(q_i)$ becomes a quartic function of q_i. To further simplify the complexity, it can be piecewise linearized for practical implementation:

$$r_i \geq \delta \cdot \left(Q_t^{LB} \cdot l_{i,t} + q \cdot L_i^{LB} - Q_t^{LB} \cdot L_i^{LB}\right); \tag{33.a}$$

$$r_i \geq \delta \cdot \left(Q^{UB} \cdot l_i + q \cdot L_i^{UB} - Q^{UB} \cdot L_i^{UB}\right); \tag{33.b}$$

$$r_i \leq \delta \cdot \left(Q^{LB} \cdot l_i + q \cdot L_i^{UB} - Q^{LB} \cdot L_i^{UB}\right); \tag{33.c}$$

$$r_i \leq \delta \cdot \left(Q^{UB} \cdot l_i + q \cdot L_i^{LB} - Q^{UB} \cdot L_i^{LB}\right). \tag{33.d}$$

The tightness of the McCormick Relaxation is heavily influenced by the bounds of the two variables involved. Specifically, if $[Q^{LB}, Q^{UB}]$ and $[L_i^{LB}, L_i^{UB}]$ are set improperly loose, constraint (33a–33d) may become ineffective. Intuitively, to ensure the relaxation remains active and meaningful, the upper bound Q^{UB} should be carefully defined based on the operational status of the units at Round Butte Dam. When all the three units are ON, Q^{UB} should be set as $\sum_{i \in \mathcal{I}} Q_i^{UB}$, where Q_i^{UB} can be determined uniquely by inverting the monotonically increasing function $g_i(\cdot)$. If only two units, such as Units 1 and 2, are running, Q^{UB} should be set as Q_1^{UB}+Q_2^{UB}. Similarly, Q^{UB} shall be set as 0 for the situation when all units are OFF.

To address these varying scenarios, specific lower and upper bounds for q and l_i are established for each of the eight possible ON/OFF combinations of the three units. These eight cases, denoted by the set $\mathcal{J} \triangleq \{1, \ldots, 8\}$ and indexed by j each corresponds to a distinct counterpart of constraint (33a–33d). Table 1 lists the eight combinations along with their associated Q_j^{LB} and Q_j^{UB}. The lower and upper bounds for l_i can be respectively calculated as $L_i^{LB} = Q_i^{LB} \cdot P_i^{LB} \cdot I_i$ and $L_i^{UB} = Q_i^{UB} \cdot P_i^{UB} \cdot I_i$, where P_i^{LB} and P_i^{UB} represent the bounds of the power generation P_i and I_i is a binary variable indicating the operational status of unit i. This formulation ensures that each scenario is accurately modeled, maintaining the effectiveness of the relaxation.

The constraints (33a–33d) can be extended to (34a–34d) by incorporating a slack term M_j, which ensures that the constraints are active for only one specific scenario at a time. For example, the slack terms of M_3 and M_5 are defined in (35a, 35b), with similar formulations for the other cases. Here, M represents a sufficiently large

Table 1 The summary of all situations

j	Units' ON/OFF			Q_j^{LB}	Q_j^{UB}
	I_1	I_2	I_3		
1	0	0	0	0	0
2	1	0	0	$\sum_{i \in \{1\}} Q_i^{LB}$	$\sum_{i \in \{1\}} Q_i^{UB}$
3	0	1	0	$\sum_{i \in \{2\}} Q_i^{LB}$	$\sum_{i \in \{2\}} Q_i^{UB}$
4	0	0	1	$\sum_{i \in \{3\}} Q_i^{LB}$	$\sum_{i \in \{3\}} Q_i^{UB}$
5	1	1	0	$\sum_{i \in \{1,2\}} Q_i^{LB}$	$\sum_{i \in \{1,2\}} Q_i^{UB}$
6	1	0	1	$\sum_{i \in \{1,3\}} Q_i^{LB}$	$\sum_{i \in \{1,3\}} Q_i^{UB}$
7	0	1	1	$\sum_{i \in \{2,3\}} Q_i^{LB}$	$\sum_{i \in \{2,3\}} Q_i^{UB}$
8	1	1	1	$\sum_{i \in \{1,2,3\}} Q_i^{LB}$	$\sum_{i \in \{1,2,3\}} Q_i^{UB}$

positive constant. To further refine the constraints, (36) is introduced to tighten (34a–34d), hat the non-negative reduction term r_i equals zero, when the corresponding unit is OFF. This additional condition enforces that r_i is only active when the unit is operational, enhancing the model's accuracy. Notably, for scenarios j where unit i is OFF, constraints (34a–34d) become redundant and can be omitted to simplify the problem. By applying these linearization methods, the model is reformulated as a mixed-integer linear programming (MILP) problem, making it computationally tractable for optimization while maintaining the necessary level of detail and precision:

$$r_i \geq \delta \cdot \left(Q_j^{LB} \cdot l_i + q \cdot L_i^{LB} - Q_j^{LB} \cdot L_i^{LB}\right) - M_j; \forall j \in \mathcal{J}, \forall i \in \{1, 2, 3\} \tag{34.a}$$

$$r_i \geq \delta \cdot \left(Q_j^{UB} \cdot l_i + q \cdot L_i^{UB} - Q_j^{UB} \cdot L_i^{UB}\right) - M_j; \forall j \in \mathcal{J}, \forall i \in \{1, 2, 3\} \tag{34.b}$$

$$r_i \leq \delta \cdot \left(Q_j^{LB} \cdot l_i + q \cdot L_i^{UB} - Q_j^{LB} \cdot L_i^{UB}\right) + M_j; \forall j \in \mathcal{J}, \forall i \in \{1, 2, 3\} \tag{34.c}$$

$$r_i \leq \delta \cdot \left(Q_j^{UB} \cdot l_i + q \cdot L_i^{LB} - Q_j^{UB} \cdot L_i^{LB}\right) + M_j; \forall j \in \mathcal{J}, \forall i \in \{1, 2, 3\} \tag{34.d}$$

$$M_3 = M \cdot (I_1 + 1 - I_2 + I_3); \tag{35.a}$$

$$M_5 = M \cdot (1 - I_1 + 1 - I_2 + I_3); \tag{35.b}$$

$$M \cdot I_i \geq r_i \geq 0; \forall i \in \{1, 2, 3\}. \tag{36}$$

1.6 Numerical Simulation of Modeling the Penstock

The actual hydropower plant, Round Butte, located in Portland, Oregon, U.S., is taken as an example to show the effect of the common penstock.

The dispatchable power ranges for the units at Round Butte Dam are summarized in Table 2. Under ideal conditions, when the common penstock losses are neglected, the combined power output of the three units should reach their total installed capacity of 372 MW (112 MW + 130 MW + 130 MW) during peak demand periods. However, historical data reveals that the maximum output of simple summation has never been achieved in practice due to the losses incurred within the shared penstock. In reality, the observed maximum actual power output for the Round Butte Dam is approximately 350 MW. Based on the analysis and calibration of the historical data, the parameter δ has been determined to be 8.339×10^{-10}. This value reflects the

Table 2 Parameters of hydropower units

Dispatchable ranges of units	Round butte dam		
	Unit #1	Unit #2	Unit #3
Lower output bound	1 MW	1 MW	1 MW
Upper output bound	112 MW	130 MW	130 MW

impact of the common penstock losses on the power generation efficiency of the dam.

When accounting for the common penstock loss using constraint (34a–34d), the actual maximum power output decreases to 354.33 MW, reflecting a loss of 17.66 MW. This result aligns closely with the observed practical maximum of 350 MW. Furthermore, the common penstock loss calculated using (32) is 17.69 MW, which closely matches the 17.66 MW derived from constraint (34a–34d). This near agreement highlights the tightness and accuracy of the formulation in (34a–34d). The slight discrepancy between the two values arises from two primary sources: (1) Linearization error: The process of linearizing $l_i = q_i \cdot g_i(q_i)$ introduces minor inaccuracies; and (2) Calibration error: The calibration of δ can slightly influence the precision of the loss estimation. These factors account for the small gap, demonstrating the robustness of the model while acknowledging the inherent limitations of the approximations used.

2 Hydropower Plants with Reservoirs and Run-Of-River Hydropower Plants

Hydropower plants with reservoirs, also known as storage hydropower plants, and run-of-river (ROR) hydropower plants are two types of hydropower plants that use the energy of flowing water to produce electricity.

The former uses a dam to store water in a reservoir, and releases water to generate electricity typically during periods of high demand. The latter, without a large reservoir, utilizes the natural flow and elevation drop of a river and is less controllable. In general, storage hydropower plants are often with higher capacities, while ROR hydropower plants are smaller in capacity and less flexible in power generation, but more environmentally friendly.

The efficient operation of a hydropower plant hinges on two essential components: a reliable water inflow forecasting model and a well-structured optimal operation model. The water inflow forecasting model is tasked with producing accurate forecasts that account for inflow uncertainties, thereby providing critical insights into water availability. Meanwhile, the optimal operation model should effectively represent the physical and operational constraints of the hydropower plant, ensuring the efficient allocation and utilization of water resources across cascaded hydropower plants.

2.1 Water Inflow Forecasting

Due to the complex factors such as precipitation, temperature, environment, and climate, the water inflow in the basin exhibits high nonlinearity and volatility, making its prediction challenging. Water inflow prediction models can generally be classified into two categories: conceptual models and data-driven models. Conceptual models aim to simulate in detail the physical processes involved, such as evaporation, infiltration, transpiration, and precipitation, by using a large set of complex equations that represent underlying physical laws and interactions. In contrast, data-driven models bypass physical mechanisms and instead mine the potential relationships in measured data. Representative models include artificial neural networks [5], recurrent neural networks [6], long short-term memory networks, and various derived variants that integrate different mechanisms and enhanced structures, all of which have been applied in water inflow prediction. As an example, reference [5] established a prediction model based on artificial neural networks and used data from several meteorological stations in the basin to predict the flow in the Han River basin in Korea. Reference [6] combined an encoder-decoder-based long short-term memory network with an empirical mode decomposition algorithm to predict the flow at the Hankou hydrological station of the Yangtze River. On the basis of time-series point prediction, to describe the uncertainty of water inflow, reference [7] introduced a prediction model based on stochastic recurrent neural networks, achieving interval predictions of water inflow.

Most current water inflow forecasting models provide single-point predictions, offering limited insight into water inflow uncertainties. While this approach aligns well with deterministic optimization models, it is insufficient for stochastic optimization. To address the limitation, probabilistic forecasting models, including parametric and nonparametric methods, have been developed. Parametric models rely on predefined assumptions about water inflow distributions, which may not always accurately represent the actual uncertainty. Nonparametric models, such as those utilizing quantile regression, estimate cumulative distribution functions with a finite number of discrete points [8, 9] are widely adopted. These probabilistic models generate distributions of water inflow forecasts, which can then be used to create scenarios through sampling techniques, enhancing the ability to incorporate uncertainty into optimization processes effectively [10].

In the context of stochastic optimization, relying on sampling density functions or cumulative distribution functions to connect forecasting and scenario generation can introduce implicit accuracy losses. Therefore, ensuring a seamless integration between forecasting outputs and scenario generation is critical to fully enhance the effectiveness of both components.

A Bayesian Neural Network (BNN) shares a similar architecture with deterministic artificial neural networks but incorporates Bayesian principles, leveraging prior knowledge to enhance predictions. Unlike deterministic networks, where fixed weights connect the layers, BNNs use weight distributions, allowing the model to capture epistemic uncertainty arising from limited training data. Weight distributions enable BNNs to produce multiple outputs for the same input by sampling these weight distributions multiple times. In contrast to other deterministic models, such as convolutional neural networks, recurrent neural networks, or long short-term memory networks, the feature allows BNNs to generate diverse outputs. Although primarily a forecasting model, the BNN's ability to account for uncertainty makes it well-fitted for generating water inflow scenarios. These scenarios, after scenario reduction, can be directly utilized in stochastic optimization-based day-ahead operation models, ensuring an efficient transition from forecasting to operational decision-making.

The BNN model is designed to function as a scenario generator for the stochastic optimization-based model. Its architecture, illustrated in Fig. 9, features a multi-output structure with multiple hidden layers, each employing a rectified linear unit (ReLU) activation function. When the BNN model is tailored to support the day-ahead optimal operation model, its outputs correspond to hourly water inflow predictions for the next day, providing the necessary input for determining hourly operation schedules. The model's hyperparameters must be fine-tuned based on the characteristics of the dataset to ensure optimal performance.

Some related features include (1) streamflow of the river; (2) historical daily water inflow to the hydropower plant; (3) water temperature; (4) month of the year; and (5) day of the week. The finally adopted ones can be selected with the feature selection process. Before being input into the BNN, numerical features will undergo Min–Max normalization to scale them within a specific range. Categorical features, such as the day of the week and the month of the year, which capture seasonal trends in water inflow variations, will be converted into numerical representations to ensure compatibility with the model. For instance, the days of the week from Monday to Sunday can be encoded as integers {1, 2, ..., 7}, while the months from January to December can be represented as {1, 2, ..., 12}. This preprocessing ensures all inputs are appropriately formatted for the BNN.

Once the BNN is trained, it can generate multiple predictions for a given input feature by executing the model multiple times. During each execution, the weights are sampled from the distributions learned during training, resulting in distinct

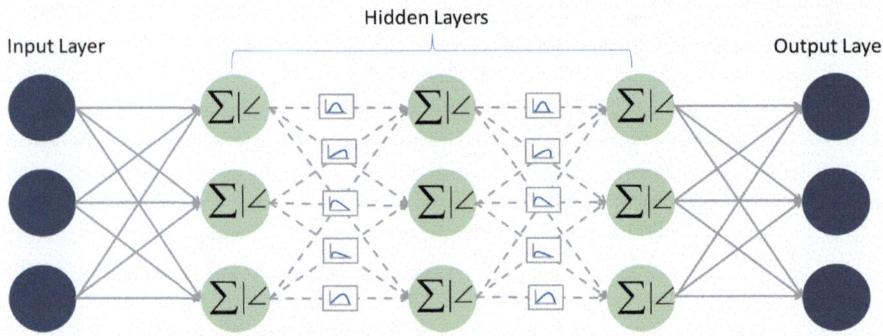

Fig. 9 The structure of a BNN forecast model

predictions for the same input. This variability arises because the sampled weights differ across executions. From a stochastic programming perspective, each prediction corresponds to a unique scenario. By running the model repeatedly, a set of scenarios can be generated. These scenarios can then be refined through scenario reduction techniques to identify the most representative ones, ensuring that the generated set effectively captures the variability and uncertainty in the data.

2.2 Hydropower Plants with Reservoirs

Hydropower plants with reservoirs, also known as storage hydropower plants, harness the potential energy stored in water of a reservoir. The reservoirs endow hydropower plants with the ability to adjust their power outputs against water inflow uncertainty and meet system demand. An example of a hydropower plant with a reservoir is shown in Fig. 10.

The day-ahead optimal scheduling model of a storage hydropower plant can be formulated as in (37)–(51) with the granularity being hourly to maximize energy generation. In the objective function (37), the first term represents the total electricity generation, and the second term represents the equivalent energy cost for the startup of hydropower units in the plant.

The water volume evaluation constraint of the reservoir can be formulated as in (38), where inflow W_t represents the forecasted water inflow and t is the index of time intervals. The total outflow, namely $\sum_i q_{i,t} + s_t$, is the sum of the water discharged through the turbines and the water spillage. To ensure physical feasibility, constraint (39) forces that the water spillage must be non-negative. Additionally, constraint (40) forces the terminal water volume v_T to restore to the initial value V, maintaining the reservoir's water balance over the planning horizon. SC_i represents the equivalent energy cost of starting up unit i.

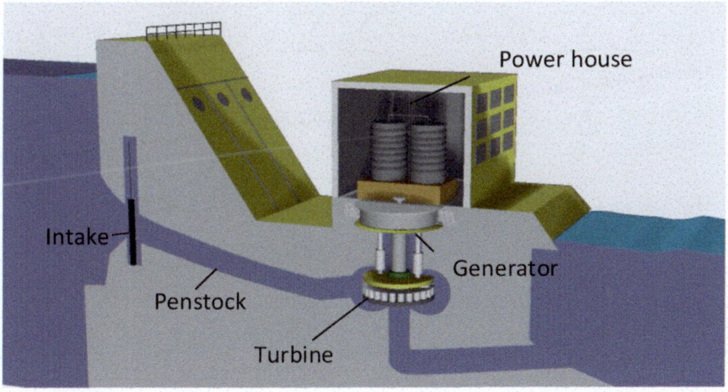

Fig. 10 A hydropower plant with a reservoir

Constraint (41) restricts the forebay level, namely the water head, to be within the allowed range $[H^{LB}, H^{UB}]$. Constraint (42) limits the power dispatchable ranges of hydropower units. Constraints (43)–(46) together determine the power output with the corresponding water discharge under the water head h_t referring to Sect. 1. Constraint (47) activates $y_{j,t}$ with respect to water head h_t. When the water head is within the range $[\underline{H}_j, \overline{H}_j]$, the corresponding $y_{j,t}$ further indicates the active water-to-power curve via (45). Constraint (46) implies that water head h_t should be within one and only one of the preset ranges. Constraints (49) and (50) are the classic "3-bin" form startup/shutdown constraints, where $I_{i,t}$ indicate ON/OFF status of unit i at hour t; $Y_{i,t}$ and $Z_{i,t}$ are respectively startup and shutdown indicators of unit i at hour t. Constraint (51) calculates the total water discharge:

$$max \sum_t \left(\sum_i p_{i,t} - \sum_i SC_i \cdot Y_{i,t} \right); \tag{37}$$

$$v_t = v_{t-1} + (W_t - q_t - s_t); \forall t \in \mathcal{T} \tag{38}$$

$$s_t \geq 0; \forall t \in \mathcal{T} \tag{39}$$

$$v_T = V; \tag{40}$$

$$H^{LB} \leq h_t \leq H^{UB}; \forall t \in \mathcal{T} \tag{41}$$

$$P_i^{LB} \cdot I_{i,t} \leq p_{i,t} \leq P_i^{UB} \cdot I_{i,t}; \forall i \in \mathcal{I}, \forall t \in \mathcal{T} \tag{42}$$

$$p_{i,t} = \sum_j \left(\sum_k B_{i,k,j} \cdot z_{i,k,j,t} + \sum_k C_{i,k,j} \cdot q_{i,k,j,t} \right); \forall i \in \mathcal{I}, \forall t \in \mathcal{T} \tag{43}$$

$$q_{i,t} = \sum_j \sum_k Q_{i,k,j} \cdot z_{i,k,j,t} + \sum_j \sum_k q_{i,k,j,t}; \forall i \in \mathcal{I}, \forall t \in \mathcal{T} \tag{44}$$

$$\sum_k z_{i,k,j,t} = y_{j,t}; \forall i \in \mathcal{I}, \forall j \in \mathcal{J}, \forall t \in \mathcal{T} \tag{45}$$

$$0 \leq q_{i,k,j,t} \leq Q^{SEG} \cdot z_{i,k,j,t}; \forall i \in \mathcal{I}, \forall j \in \mathcal{J}, \forall k \in \mathcal{K}, \forall t \in \mathcal{T} \tag{46}$$

$$h_t = f(v_t); \forall t \in \mathcal{T} \tag{47}$$

$$\underline{H}_j \cdot y_{j,t} - M \cdot (1 - y_{j,t}) \leq h_t \leq \overline{H}_j \cdot y_{j,t} + M \cdot (1 - y_{j,t}); \forall j \in \mathcal{J}, \forall t \in \mathcal{T} \tag{48}$$

$$\sum_j y_{j,t} = 1; \forall t \in \mathcal{T} \tag{49}$$

$$I_{i,t} - I_{i,t-1} = Y_{i,t} - Z_{i,t}; \forall i \in \mathcal{I}, \forall t \in \mathcal{T} \tag{50}$$

$$Y_{i,t} + Z_{i,t} \leq 1; \forall i \in \mathcal{I}, \forall t \in \mathcal{T} \tag{51}$$

$$q_t = \sum\nolimits_i q_{i,t}; \forall t \in \mathcal{T}. \tag{52}$$

With the generated water inflow scenarios, the stochastic day-ahead optimal scheduling model can be extended from (37)–(67). Most of the constraints are similar, differing only by the introduction of the subscript s to indicate the scenarios. Importantly, $W_{t,s}$ represents the inflow forecast for different scenarios at hour t. ρ_s is the occurrence probability or weight of scenario s and thereby (53) calculates the mean (37) over all generated scenarios.

It is emphasized that in (53)–(67), the forebay level is assumed to be scenario-independent, while water discharge, power generation, and water spillage variables are treated as scenario-dependent variables, which could ensure that the dam adjusts merely the scenario-dependent ones—namely, water discharges, power generation, and water spillage—to account for uncertain water inflows. By doing so, the stochastic programming model addresses the limitations of strict anticipativity, enabling a consistent water volume schedule that can effectively mitigate inflow uncertainties while satisfying water regulation requirements. Additionally, the unit commitment of the hydropower units is scenario-dependent, which allows for greater operational flexibility, leveraging the hydropower units' ability to quickly startup and shutdown as needed to respond to varying scenarios:

$$max \sum_s \rho_s \cdot \sum_{t \in \mathcal{T}} \left(\sum_i p_{i,t,s} - \sum_i SC_i \cdot Y_{i,t,s} \right); \tag{53}$$

$$v_t = v_{t-1} + \left(W_{t,s} - q_{t,s} - s_{t,s} \right); \forall t \in \mathcal{T}, \forall s \in \mathcal{S} \tag{54}$$

$$s_{t,s} \geq 0; \forall t \in \mathcal{T}, \forall s \in \mathcal{S} \tag{55}$$

$$v_T = V; \tag{56}$$

$$H^{LB} \leq h_t \leq H^{UB}; \forall t \in \mathcal{T} \tag{57}$$

$$P_i^{LB} \cdot I_{i,t} \leq p_{i,t,s} \leq P_i^{UB} \cdot I_{i,t}; \forall i \in \mathcal{I}, \forall t \in \mathcal{T}, \forall s \in \mathcal{S} \tag{58}$$

$$p_{i,t,s} = \sum_j \left(\sum_k B_{i,k,j} \cdot z_{i,k,j,t,s} + \sum_k C_{i,k,j} \cdot q_{i,k,j,t,s} \right); \forall i \in \mathcal{I}, \forall t \in \mathcal{T}, \forall s \in \mathcal{S} \tag{59}$$

$$q_{i,t,s} = \sum_j \sum_k Q_{i,k} \cdot z_{i,k,j,t,s} + \sum_j \sum_k q_{i,k,j,t,s}; \forall i \in \mathcal{I}, \forall t \in \mathcal{T}, \forall s \in \mathcal{S} \tag{60}$$

$$\sum_k z_{i,k,j,t,s} = y_{j,t}; \forall i \in \mathcal{I}, \forall j \in \mathcal{J}, \forall t \in \mathcal{T}, \forall s \in \mathcal{S} \tag{61}$$

$$0 \leq q_{i,k,j,t,s} \leq Q^{SEG} \cdot z_{i,k,j,t,s}; \forall i \in \mathcal{I}, \forall j \in \mathcal{J}, \forall k \in \mathcal{K}, \forall t \in \mathcal{T}, \forall s \in \mathcal{S} \tag{62}$$

$$\underline{H}_j \cdot y_{j,t} - M \cdot (1 - y_{j,t}) \leq h_t \leq \overline{H}_j \cdot y_{j,t} + M \cdot (1 - y_{j,t}); \forall t \in \mathcal{T} \tag{63}$$

$$\sum_j y_{j,t} = 1; \forall t \in \mathcal{T} \tag{64}$$

$$I_{i,t,s} - I_{i,t-1,s} = Y_{i,t,s} - Z_{i,t,s}; \forall i \in \mathcal{I}, \forall t \in \mathcal{T}, \forall s \in \mathcal{S} \tag{65}$$

$$Y_{i,t,s} + Z_{i,t,s} \leq 1; \forall i \in \mathcal{I}, \forall t \in \mathcal{T}, \forall s \in \mathcal{S} \tag{66}$$

$$q_{t,s} = \sum_i q_{i,t,s}; \forall t \in \mathcal{T}, \forall s \in \mathcal{S}. \tag{67}$$

2.3 Run-Of-River Hydropower Plants

ROR hydropower plants harness the natural flow of rivers to produce electricity. Unlike hydropower plants with reservoirs, they have little to no water storage ability, which leads to less environmental impact compared to large reservoirs, as there is less need for inundating large areas of land. Because ROR hydropower plants do not require the construction of dams, block rivers to store water, or alter river flows, they are more environmentally friendly to aquatic life. For the same reason, the construction cost of the ROR hydropower plant is much lower compared to dam-based hydropower plants. However, ROR hydropower plants typically have limited generation capacities and mostly serve as distributed energy resources. An example of a ROR hydropower plant is shown in Fig. 11.

Due to the absence of water storage capacity, the water intake of the ROR hydropower plant diverts a portion of the inflow into the plant while allowing the rest to continue its natural course. This also makes ROR hydropower plants highly dependent on the river inflow, resulting in intermittent power generation.

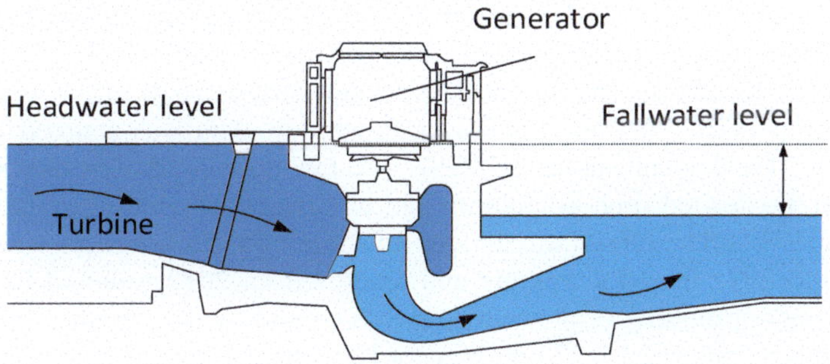

Fig. 11 A ROR hydropower plant

In terms of modeling, ROR hydropower plants with no water storage capacity are often considered as uncontrollable renewable energy sources, similar to solar and wind energy, from a system perspective. Therefore, system operators generally predict the outputs of ROR hydropower plants before the scheduling phase, and dispatch controllable resources to mitigate the fluctuations caused by varying water inflow in the real-time operation phase.

For ROR hydropower plants with limited water storage capacities rather than having none at all, they can be traded the same as hydropower plants with reservoirs but with close H^{LB} and H^{UB} in constraint (41). If the objective is to maximize energy generation, the model (37)–(52) can be used. From a power grid's perspective, constraints (38)–(52) remain valid, while the objective function will be adjusted to reflect the operation goals of the power grid.

2.4 Numerical Simulation of Scenario Generation

The scenario generation model built on the basis of BNN, abbreviated as BNN-SG, illustrated in Fig. 9 is validated. In the subsequent section on Cascaded Hydropower Systems, a stochastic optimization model will be introduced and validated using the generated water inflow scenarios. The BNN-based scenario generation model is implemented in Python using PyCharm with the TensorFlow 2 library. The model is constructed with Keras, now integrated into TensorFlow, utilizing its features for efficient development. The hidden layers depicted in Fig. 9 are implemented using the DenseVariational layer from Keras (Keras.layers.DenseVariational), which leverages predefined prior and posterior functions to model uncertainty. For training, the RMSprop optimizer, available in Keras, is employed to apply the RMSprop algorithm. The loss function for each output label (hourly water inflow) is defined with MeanSquaredError (Keras.losses.MeanSquaredError). The model is trained and validated using actual operational data from Round Butte Dam provided by Portland General Electric (PGE) and river streamflow data obtained from the United States

Geological Survey (USGS). This data ensures the BNN model's ability to generate realistic water inflow scenarios and assess its performance effectively.

12 years of historical operational data spanning from 2010 to August 2022 is available from PGE. Time-matched river streamflow data was sourced from the United States Geological Survey (USGS). The raw data underwent preprocessing to adjust to the desired time granularity and was then converted into time-aligned features and labels. The dataset was divided into training and testing sets, with records from 2010 to 2021 used for training and records from 2022 reserved for testing. After filtering out missing or abnormal records caused by measurement errors, the training dataset comprised 3,274 daily records, while the test dataset included 215 daily records. Each record corresponds to a single day in the historical dataset. Due to PGE's data confidentiality policies, the detailed datasets used are not publicly accessible.

The BNN scenario generation model consists of three hidden layers, each comprising 10 neurons. The hyperparameters are configured as follows: a learning rate of 0.0005, a batch size of 10, and 500 epochs. To evaluate the model, two weeks from the test dataset, each containing 7 consecutive days without missing or abnormal records, are selected:

- Winter Week: January 24, 2022, to January 30, 2022 (starting on a Monday).
- Summer Week: June 20, 2022, to June 26, 2022 (starting on a Monday).

Generally, the water inflow during the summer week is slightly higher than that in the winter week, reflecting seasonal variations. For each day in these two weeks, the model generates 100 scenarios. The lower and upper envelopes of these scenarios are then concatenated to form comprehensive visual representations, as shown in Figs. 12 and 13. These envelopes provide insights into the variability and uncertainty captured by the generated scenarios.

The two figures illustrate that the hourly water inflows exhibit slight fluctuations throughout the days of the two selected weeks. For the majority of hours in these weeks, the actual hourly water inflows fall within the range of the generated

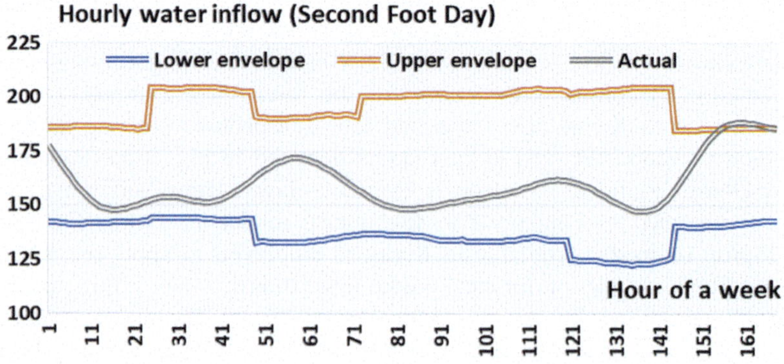

Fig. 12 The generated scenario envelopes for the winter week

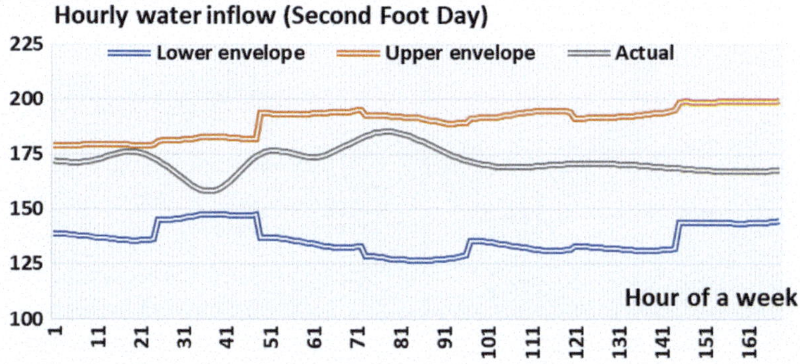

Fig. 13 The generated scenario envelopes for the summer week

scenarios. An exception occurs during the last few hours of the winter week, where the water inflow surges abruptly, slightly exceeding the scenario envelopes. The upper and lower envelopes display a zigzag pattern. This is due to the daily generation of 24-h water inflow scenarios, which introduces day-to-day variability. Two factors contribute to this zigzag shape: (1) Categorical Features—The inclusion of month-of-the-year and day-of-the-week features influences the scenario generation, as these features reflect seasonal and weekly trends in water inflow, and (2) Randomness in BNN-Based Generation—The stochastic nature of the BNN-based scenario generation occasionally produces relatively extreme cases, contributing to variations in the envelopes. These characteristics highlight the model's ability to capture both the underlying patterns and the randomness in water inflow behavior.

3 Cascaded Hydropower Systems

3.1 Modeling of Cascaded Hydropower Systems

Cascade hydropower systems (CHSs), as shown in Fig. 14, are systems where multiple hydropower plants, which could be hydropower plants with reservoirs or ROR hydropower plants, are strategically located along a river at different elevations to maximize the use of the gravitational flow of water.

Coordinating the water usage of multiple cascaded dams is a complex task due to several factors: the variability and unpredictability of upstream river streamflow, the intricate and tightly coupled water interactions among the dams, and the stringent regulations coupled with the dynamic water usage requirements of individual dams. Specifically, the coupling of cascaded reservoirs exits—the water released from upstream dams can be reused by downstream dams, exploiting the value of reuse, while high output and rapid water discharge from the upstream dams may lead to

Fig. 14 Cascade hydropower systems

forced water spillage of the downstream. These challenges necessitate sophisticated models and strategies to ensure efficient and compliant operation of the entire system.

To address these challenges, the daily optimal operation of a CHS relies heavily on an optimal operation model that accurately reflects the system's physical and operational characteristics. Such a model is essential for effectively coordinating water usage across the cascaded hydropower plants. In addition to maximizing the utilization of water resources for hydropower generation, CHS systems must also fulfill critical social responsibilities, including agricultural irrigation and flood control. Consequently, hydro facility operators depend on advanced optimization tools to enhance water resource efficiency, maximize hydropower benefits, and ensure compliance with regulatory and societal obligations.

In practice, CHS operators often rely on empirical policies to balance the trade-off between modeling complexity and solution quality. For example, they may maintain a fixed ratio between the water discharges of upstream and downstream dams to keep the downstream forebay level relatively constant. While this approach is simple and practical, it imposes significant limitations. Such strategies inherently restrict the operational flexibility of the CHS, reduce scheduling accuracy, and lead to inefficiencies by overlooking critical factors, including (1) The dynamic operational efficiency of hydropower units, (2) Time-varying water inflows, and (3) Fluctuations in system load demand. As a result, these simplifications can lead to suboptimal operations and, over time, cause considerable economic losses for CHS owners. Adopting more sophisticated optimization models can help address these shortcomings and improve the overall efficiency and profitability of CHS operations.

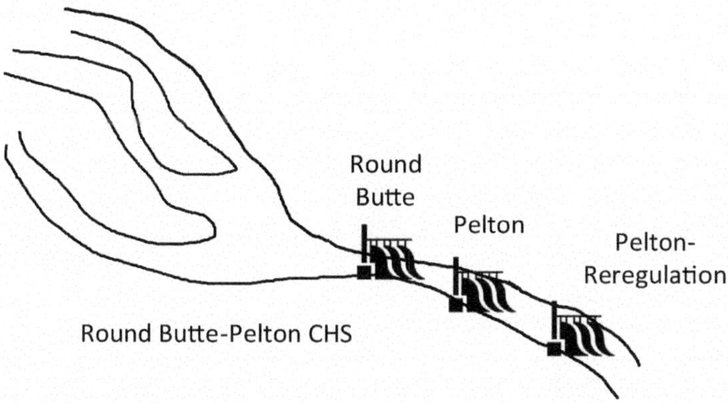

Fig. 15 Round Butte-Pelton CHS

The Round Butte-Pelton CHS project, located on the Deschutes River in Oregon, U.S., serves as a real-world example to develop the optimal scheduling model. The CHS comprises three dams: Round Butte, Pelton, and Pelton-Reregulation, as illustrated in Fig. 15. The upstream Round Butte Dam features three hydropower units and the midstream Pelton Dam also has three hydropower units, but smaller installed capacity. The downstream Pelton-Reregulation Dam, owned by a tribal entity, includes a single hydropower unit and a three-gate spillway. Considering it is primarily designed to maintain the natural flow of the Deschutes River rather than for power generation, consequently, its operation is excluded from the scheduling considerations. This setup highlights the operational and structural complexity of the CHS, making it an ideal case study for optimizing water and energy usage within cascaded hydropower systems.

The day-ahead optimal scheduling model for the Round Butte-Pelton CHS is formulated with an hourly time granularity, as shown in (68)–(70). The objective function, (68), aims to maximize the total electricity generation over a single day. In (68), the first term represents the total electricity generated by all the hydropower units; and the second term accounts for the energy consumed during the startup of hydropower units at these dams, reflecting the operational cost of bringing units online.

The constraints of the Round Butte Dam are formulated as in (69a–g). The water volume evaluation constraint (69a) calculates the dam's water volume balance, where the inflow W_t^{RB} is derived from forecasts, and the outflow is the sum of the total water discharge from all hydropower units, $\sum_{i\in\mathcal{I}^{RB}} q_{i,t}^{RB}$ plus the water spillage s_t^{RB}. Constraint (69b) ensures that the water spillage s_t^{RB} remains non-negative, while constraint (69c) enforces that the terminal water volume v_T^{RB} returns to the initial value V^{RB} at the end of the operating period. Additionally, SC_i^{RB} and SC_i^{PL} represent the energy consumption required to start up unit i in the Round Butte and Pelton Dams, respectively.

Constraint (69d) imposes limits on the water volume of the Round Butte Dam. The forebay level-to-water volume conversion functions $f^{RB}(\cdot)$ and $f^{PL}(\cdot)$ are utilized to determine the dam's water volume upper and lower bounds, which are derived from the forebay level limits, namely $F^{RB,LB}$ and $F^{RB,UB}$. Constraint (69e) defines the dispatchable power ranges of the hydropower units, where $g_i^{RB}(\cdot)$ calculates the power generation with respect to water discharge. It's important to note that the water discharge-to-power generation conversion functions $g_i^{RB}(\cdot)$ and $g_i^{PL}(\cdot)$ can vary significantly depending on the net water head. However, for the Round Butte and Pelton Dams, the forebay levels are regulated within a narrow range throughout each season due to strict operational requirements. The stability in forebay levels results in relatively consistent net water heads, rendering the dynamics of $g_i^{RB}(\cdot)$ and $g_i^{PL}(\cdot)$ negligible for this particular CHS. Thus, the water head, i.e., the forebay level in this case, is not necessarily to be dynamically calculated, which means $f^{RB}(\cdot)$ and $f^{PL}(\cdot)$ will only be applied with known $F^{RB,LB}$ and $F^{RB,UB}$, and thus their linearization becomes necessary as well. Constraints (69f) and (69g) are a classic form of startup/shutdown constraints with three binary variables, where, for Round Butte Dam, $I_{i,t}^{RB}$ indicate ON/OFF status of unit i at hour t; and $Y_{i,t}^{RB}$ and $Z_{i,t}^{RB}$ are respectively startup and shutdown indicators of unit i for hour t.

The constraints for the Pelton Dam are similarly structured as in (70a–h), with a key distinction in the water volume evaluation constraint (70a). The inflow into the Pelton Dam, w_t^{PL}, is explicitly equal to the earlier outflow from the Round Butte Dam. The interdependence reflects the cascading nature of the hydropower system, where the operations of the upstream dam directly influence the downstream dam's water availability. In the current version of (70a), the time delay is one time interval, namely one hour. However, in fact, the actual time delay between the two dams is around 20 min according to historical data. As the optimal operation model is built based on discrete time intervals, the water time delay represented by constraint (70a) can only be an integer multiple of the time interval, i.e., an integer multiple of one hour, which severely limits the accuracy of representing water time delays:

$$max\sum_{t\in\mathcal{T}}\left(\sum_{i\in\mathcal{I}^{RB}} p_{i,t}^{RB} + \sum_{i\in\mathcal{I}^{PL}} p_{i,t}^{PL}\right) - \left(\sum_{i\in\mathcal{I}^{RB}} SC_i^{RB}\cdot Y_{i,t}^{RB} + \sum_{i\in\mathcal{I}^{PL}} SC_i^{PL}\cdot Y_{i,t}^{PL}\right). \quad (68)$$

Constraints of Round Butte Dam:

$$v_t^{RB} = v_{t-1}^{RB} + \left(W_t^{RB} - \sum_{i\in\mathcal{I}^{RB}} q_{i,t}^{RB} - s_t^{RB}\right); \forall t \in \mathcal{T} \quad (69.a)$$

$$s_t^{RB} \geq 0; \forall t \in \mathcal{T} \quad (69.b)$$

$$v_T^{RB} = v_0^{RB} = V^{RB}; \quad (69.c)$$

$$f^{RB}\left(F^{RB,LB}\right) \leq v_t^{RB} \leq f^{RB}\left(F^{RB,UB}\right); \forall t \in \mathcal{T} \quad (69.d)$$

$$P_i^{RB,LB} \cdot I_{i,t}^{RB} \le g_i^{RB}(q_{i,t}^{RB}) \le P_i^{RB,UB} \cdot I_{i,t}^{RB}; \forall i \in \mathcal{I}^{RB}, \forall t \in \mathcal{T} \tag{69.e}$$

$$I_{i,t}^{RB} - I_{i,t-1}^{RB} = Y_{i,t}^{RB} - Z_{i,t}^{RB}; \forall i \in \mathcal{I}^{RB}, \forall t \in \mathcal{T} \tag{69.f}$$

$$Y_{i,t}^{RB} + Z_{i,t}^{RB} \le 1; \forall i \in \mathcal{I}^{RB}, \forall t \in \mathcal{T}. \tag{69.g}$$

Constraints of Pelton dam:

$$v_t^{PL} = v_{t-1}^{PL} + \left(w_t^{PL} - \sum_{i \in \mathcal{I}^{PL}} q_{i,t}^{PL} - s_t^{PL} \right); \forall t \in \mathcal{T} \tag{70.a}$$

$$w_t^{PL} = \sum_{i \in \mathcal{I}^{RB}} q_{i,t-1}^{RB} + s_{t-1}^{RB} - W^{LS}; \forall t \in \mathcal{T} \tag{70.b}$$

$$s_t^{PL} \ge 0; \forall t \in \mathcal{T} \tag{70.c}$$

$$v_T^{PL} = v_0^{PL} = V^{PL}; \tag{70.d}$$

$$f^{PL}\left(F^{PL,LB}\right) \le v_t^{PL} \le f^{PL}\left(F^{PL,UB}\right); \forall t \in \mathcal{T} \tag{70.e}$$

$$P_i^{PL,LB} \cdot I_{i,t}^{PL} \le g_i^{PL}(q_{i,t}^{PL}) \le P_i^{PL,UB} \cdot I_{i,t}^{PL}; \forall i \in \mathcal{I}^{PL}, \forall t \in \mathcal{T} \tag{70.f}$$

$$I_{i,t}^{PL} - I_{i,t-1}^{PL} = Y_{i,t}^{PL} - Z_{i,t}^{PL}; \forall i \in \mathcal{I}^{PL}, \forall t \in \mathcal{T} \tag{70.g}$$

$$Y_{i,t}^{PL} + Z_{i,t}^{PL} \le 1; \forall i \in \mathcal{I}^{PL}, \forall t \in \mathcal{T}. \tag{70.h}$$

It is not difficult to see that the two dams, Round Butte and Pelton, are coupled through constraints (70a)–(70b), which is in a specific form. Indeed, (70a) can be generalized to represent different connections of reservoirs as in (71). Vector $\boldsymbol{v}_t$ collects the water volumes of the reservoirs along the river of the CHS; vectors $\boldsymbol{q}_t$ and $\boldsymbol{s}_t$ respectively collect the total water discharge and water spillage; $\boldsymbol{w}_t$ collects the forecasted water inflow to the reservoirs. For those without water inflow, the corresponding elements in $\boldsymbol{w}_t$ are set as zeros:

$$\boldsymbol{v}_t = \boldsymbol{v}_{t-1} + \boldsymbol{w}_t + \boldsymbol{A} \cdot \left(\boldsymbol{q}_t + \boldsymbol{s}_t\right). \tag{71}$$

The adjacency matrix $\boldsymbol{A}$ of reservoirs is a matrix that represents the connection between different reservoirs. For a CHS with n reservoirs, $\boldsymbol{A}$ is an $n \times n$ matrix, where $\boldsymbol{A}_{i,j} = 1$ indicates that the outflow of reservoir j directly becomes the inflow to reservoir i and $\boldsymbol{A}_{i,j} = 0$ indicates t that the outflow of reservoir j does not directly become the inflow to reservoir i. $\boldsymbol{A}_{i,i}$ are always -1 representing the water discharge and water spillage of the reservoir itself.

As the example shown in Fig. 16, the $\boldsymbol{A}$ matrix is

$$\begin{bmatrix} -1 & 0 & 0 \\ 1 & -1 & 0 \\ 0 & 1 & -1 \end{bmatrix}.$$

As another example shown in Fig. 17, the $\boldsymbol{A}$ matrix is

$$\begin{bmatrix} -1 & 0 & 0 \\ 0 & -1 & 0 \\ 1 & 1 & -1 \end{bmatrix}.$$

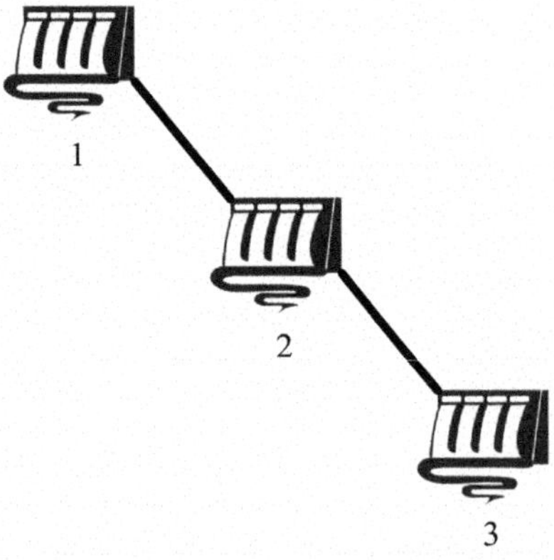

Fig. 16 Connection of the first example

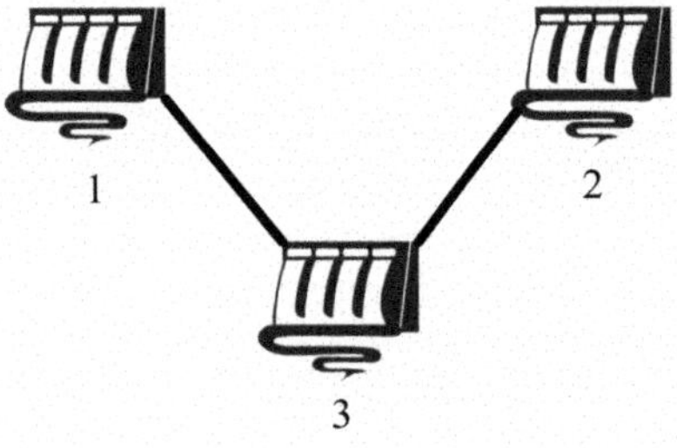

Fig. 17 Connection of the second example

3.2 Modeling of Water Delay

As pointed out in the above section, the actual time delay between the two dams is around 20 min, but since the discrete time intervals are adopted, the water time delay represented by constraint (70.a) has to be an integer multiple of the time interval. For the case of 20 min, setting the delayed time interval neither as 0 nor 1 is suitable.

To account for the time delay in water flow between adjacent dam, a factor $\lambda \in [0,1]$ is introduced, which captures the proportionality of water flow that has reached the downstream Pelton Dam at a given time, reflecting the delay in water transfer from the upstream Round Butte Dam. Adopting a constant 20-min water flow time delay from Round Butte Dam to Pelton dam leads to λ in (72) being 1/3, i.e., 20 min/60 min. In fact, λ represents the proportion of water discharged and spilled by the upstream hydropower station in the previous time interval $t-1$ that will arrive at the dam in time interval t. When the time delay is 20 min, the proportion is 1/3, and thereby $\lambda = 1/3$. It is easy to see that in two special cases, namely when $\lambda = 0$ and $\lambda = 1$, the delayed time intervals are respectively 0 and 1:

$$w_t^{PL} = \lambda \cdot \left(\sum_{i \in \mathcal{I}^{RB}} q_{i,t-1}^{RB} + s_{t-1}^{RB} \right) + (1-\lambda) \cdot \left(\sum_{i \in \mathcal{I}^{RB}} q_{i,t}^{RB} + s_t^{RB} \right) - W^{LS}; \forall t \in \mathcal{T}. \tag{72}$$

For cases where the water time delay is greater than one time interval, it can be expressed as a combination of an integer number of time intervals and the lambda. For example, assuming the time delay is 150 min, it can be written as $t-2$ plus $\lambda = 1/2$. Then (72) can be written as in (73):

$$w_t^{PL} = \lambda \cdot \left(\sum_{i \in \mathcal{I}^{RB}} q_{i,t-3}^{RB} + s_{t-3}^{RB} \right) + (1-\lambda) \cdot \left(\sum_{i \in \mathcal{I}^{RB}} q_{i,t-2}^{RB} + s_{t-2}^{RB} \right) - W^{LS}; \forall t \in \mathcal{T}. \tag{73}$$

In practice, the water time delay is more complicated and dynamic instead of constant. It has a highly nonlinear relationship with the water outflow from the upstream power station, i.e., water discharge plus water spillage. Simulating dynamic water delay inevitably introduces nonlinearity, which requires sophisticated treatment for the introduced nonlinearity under a MILP framework. For interested readers, reference [11] suggested a model of dynamic water delay. Within the framework of (72), the impact of the upstream Round Butte's water outflow can be accounted for via λ, which makes it a function of $\sum_{i \in \mathcal{I}^{RB}} q_{i,t-1}^{RB} + s_{t-1}^{RB}$, and then (72) can be rewritten as in (74). However, how to elegantly handle the introduced nonlinearity of λ_{t-1} becomes a new challenge:

$$w_t^{PL} = \lambda_{t-1} \cdot \left(\sum_{i \in \mathcal{I}^{RB}} q_{i,t-1}^{RB} + s_{t-1}^{RB} \right) + (1 - \lambda_{t-1}) \cdot \left(\sum_{i \in \mathcal{I}^{RB}} q_{i,t}^{RB} + s_t^{RB} \right) - W^{LS}; \forall t \in \mathcal{T} \tag{74}$$

where $\lambda_{t-1} = f(\sum_{i \in \mathcal{I}^{RB}} q_{i,t-1}^{RB} + s_{t-1}^{RB})$.

3.3 Ecology and Water Usage Demand

Hydropower plants, especially CHSs, have various impacts on ecology. Dams and reservoirs will change the natural streamflow of rivers, which can disrupt the ecosystems that depend on it. As an example, dams create barriers to fish and other aquatic organisms, preventing migration and breeding; reservoirs may flood large areas, destroying terrestrial habitats and displacing wildlife; the operation of CHSs may alter natural runoff to meet power generation needs, causing the disappearance of original rising and falling of river levels, affecting fish spawning and reproduction. In addition, CHSs may change water temperature and water oxygen levels, which could considerably affect aquatic life.

Typically, downstream ecological, agricultural, and urban water demands are usually simply represented by constraint (75), which requires that the sum of turbine water discharge and the water spillage from a CHS remains larger than a certain value, ensuring that these demands are met. However, such a simple constraint cannot adequately satisfy ecological needs, such as the fish spawning requirement. Some references point out that many fish species require higher water flows and rising water levels during spawning seasons to trigger spawning behavior and create suitable habitats for eggs and larvae:

$$q_t + s_t \geq Q_t^{DE}; \forall t \in \mathcal{T}. \tag{75}$$

As an example, in the early stages of the operation of the Three Gorges Dam, the hydrological stimuli required for the spawning and reproduction of the four major fish species weakened, leading to a continuous decline in their population.

The reproduction of the four major Chinese fish species in the Yangtze River mainly occurs from late April to early July each year. When the water temperature rises above 18 degrees Celsius, the spawning behavior of fish can be triggered by the rising and falling river levels. Fish spawning is closely related to the flow rate and water level of the river. Most spawning behavior occurs during the rising water period, and the scale of spawning depends on the duration of the rising water and the relative increase in water level, rather than the starting water level. The longer the duration of rising water and the greater the relative increase in water level, the larger the spawning scale.

Based on the analysis of ecological hydrological indicators during the spawning period of the four major Chinese fish species before the impoundment of the Three

Gorges Reservoir, the hydrological requirements for spawning are proposed as follows: 2 to 3 water level rising events in the spawning period, with each lasting 3 to 8 days.

To this end, constraints (76)–(78) could be added to ensure that within the scheduling time horizon, there will be a flooding process lasting for N time intervals, with a rise in water levels of no less than r between two consecutive time intervals. It is worth mentioning that, from the Three Gorges Reservoir example, the hydrological requirements include a water level rise lasting 3 to 8 days during the spawning period, which means the time interval is better to be daily. The discussed day-ahead scheduling model in the previous section can hardly incorporate constraints (76)–(78), as its time granularity is hourly and only focuses on a single day. Therefore, a larger time granularity and a longer scheduling time horizon should be chosen, switching the day-ahead scheduling problem to a medium-term or long-term scheduling problem of CHSs.

In constraints (76)–(78), binary variable u_t indicates that the activation of a water level rise process can start at time interval t within the scheduling time horizon but only once. Constraint (77) activates and deactivates $u_{t,n}$ with u_t. $u_{t,n}$ are sub-indicators of the water level rise process that starts at time interval t, where $n \in [0, \cdots, N-1]$. $u_{t,0}$, $u_{t,1}$, …, $u_{t,N-1}$ respectively indicate the first, second, …, and last time interval of the water level rise process that starts at time interval t. Once $\sum_{n=0}^{N-1} u_{t-n,n} = 1$, (76) becomes active and $q_t + s_t \geq r \cdot (q_{t-1} + s_{t-1})$ is enforced; otherwise (76) will be relaxed. Constraints (77) and (78) together can ensure $\sum_{n=0}^{N-1} u_{t-n,n}$ to be equal to 0 or 1:

$$\left(1 - \sum_{n=0}^{N-1} u_{t-n,n}\right) \cdot M + q_t + s_t \geq r \cdot (q_{t-1} + s_{t-1}); \tag{76}$$

$$\sum_{n=0}^{N-1} u_{t,n} = N \cdot u_t; \tag{77}$$

$$\sum_t u_t = 1. \tag{78}$$

Spawning period may require multiple water level rise events to repeatedly stimulate fish to spawn. In this case, constraints (79)–(82) can be used, where L represents the number of water level rise required within the scheduling time horizon. Compared to constraints (76)–(78), only constraint (82) is newly added to ensure that there is no overlap between two water level rise events:

$$\left(1 - \sum_{n=0}^{N-1} u_{t-n,n}\right) \cdot M + q_t + s_t \geq r \cdot (q_{t-1} + s_{t-1}); \tag{79}$$

$$\sum_{n=0}^{N-1} u_{t,n} = N \cdot u_t; \tag{80}$$

$$\sum_{t}^{T} u_t = L; \tag{81}$$

$$\sum_{\tau=t}^{t+N-1} u_\tau \leq 1. \tag{82}$$

3.4 Modeling of Future Value of Stored Water

The energy generation potential of reservoir stored water can be represented by the reservoir water volume future value. The reservoir water volume future value refers to the maximum equivalent electricity that can be converted by depleting the stored water of all cascaded reservoirs, and it is influenced by multiple factors.

Firstly, the cascade coupling is an influential factor: the water released from upstream power dams can be reused by downstream dams, generating additional energy.

Secondly, the assessment period is an influential factor: the length of the assessment period determines the rate of reservoir stored water consumption, and high output and rapid water discharge from upstream may force downstream dams to spill water.

Lastly, the future natural inflow is an influential factor: on the one hand, future natural inflow impacts the rate of reservoir stored water consumption; on the other hand, when future natural inflows are scarce, the value of reservoir stored water increases, and conversely, it decreases.

The evaluation of the reservoir water volume future value is crucial for guiding CHS to balance the current and future benefits of energy generation. Without a proper assessment of the future energy generation potential of reservoir stored water, it may lead to an irrational setting of reservoir capacity at the end of the hydropower scheduling period. When maximizing energy generation is set as the scheduling objective, not limiting the reservoir capacity at the end of the scheduling period could cause the cascade hydropower system shortsightedly exhausting the reservoir stored water to increase the current energy generation. To this end, the reservoir capacity at the end of the scheduling period returning to its initial value, i.e., adding the constraint $v_T = V_0$ is always explicitly or implicitly enforced, where v_T represents the reservoir capacity at the end of the period and V_0 represents the known initial reservoir capacity.

However, the scheduling period does not necessarily coincide with the reservoir water volume scheduling cycle, making the constraint $v_T = V_0$ unreasonable and economically detrimental, due to its restriction on reservoir water volume flexibility. Some works suggest that the reservoir water volume at the end of the scheduling period can be determined in the problem with a longer scheduling period. For example, the reservoir water volume at the end of a day-ahead scheduling period can be determined on a weekly scale, and the latter can be further determined on a quarterly scale. However, this essentially creates a nested problem, as the issue of determining the reservoir water volume at the end of the period still exists with a larger time scale.

Research on the future value of reservoir water volume is still relatively scarce. The concept was mentioned in an early reference [12], which simply converted the stored water of cascade reservoirs into energy using a fixed conversion factor, without accounting for the impact of cascade coupling or future natural inflow. The method of reference [13] generated a series of future water inflow scenarios, then used production simulation to calculate the energy generation corresponding to the depletion of reservoir stored water based on these scenarios, obtaining a series of sample points of reservoir water volume-to-energy generation. These sample points are then linearly fitted to form a reservoir water volume-to-energy generation conversion curve. However, this method focuses on a single-stage hydropower plant, which means the production simulation considers no impacts from the cascade coupling. On the basis of reference [13], references [14] and [15] extended a similar approach to cascade hydropower systems, using a generation coefficient as a weight to combine the reservoir water volumes of individual cascaded reservoirs, and thereby obtaining the overall reservoir water volume of the cascade hydropower system. Production simulation is then used to obtain a series of sample points of overall reservoir water volume-to-power generation, and a quadratic function is fitted to form the conversion curve. It is evident that these methods reflect the impact of future natural inflow through scenarios and consider the cascade coupling effect through production simulation.

However, some challenges still exist. First, the time horizon of production simulation needs to be preset, so it cannot participate in the joint optimization as a variable, causing difficulties in reflecting the impact of the assessment period; second, linear and quadratic functions have poor fitting accuracy for the sample points, resulting in significant bias in the assessment values; last, production simulation merely outputs the energy generation, which is unable to reflect the value of reservoir stored water in the context of future natural inflow. For example, if future natural inflow is expected to be abundant, the value of reservoir stored water may decrease, and vice versa.

3.5 Numerical Simulation of the Cascaded Hydropower System Model

In this section, a stochastic optimization model is validated with the generated water inflow scenarios.

June 24, 2022, is selected for a detailed analysis. For this day, 100 generated scenarios are processed using a scenario reduction method, resulting in two sets of representative scenarios: one with 20 scenarios, referred to as Case-20, and another with 30 scenarios, referred to as Case-30. The number of representative scenarios is intentionally limited, as the intraday variation in water inflow is relatively minor. Additionally, using an excessive number of scenarios would increase the computational burden of solving the stochastic optimization model without significant added value.

For Round Butte and Pelton Dams, their units' dispatchable ranges are listed in Table 3. In Table 3, SFD is short for Second Foot Day, meaning. Table 4 shows their forebay level limits and corresponding water volumes. Notably, PGE tightly controls the forebay levels, keeping them within narrow bounds, which may result in relatively stable net water heads at both dams during daily operations, allowing the water discharge-to-power generation functions, namely $g_i^{RB}(\cdot)$ and $g_i^{PL}(\cdot)$ to be treated as univariate functions of water discharge. For Round Butte Dam, the cubic approximation of $g_i^{RB}(\cdot)$ for each of the three units is linearized, yielding 12, 13, and 12 segments of the piecewise linearized functions, respectively. Pelton Dam's $g_i^{PL}(\cdot)$ functions, which are identical for the three units, are originally quadratic approximations, as per PGE's design, and are linearized into 15 segments. The scheduling time horizon is set as 24 h, meaning T=24 in $\mathcal{T}$.

It is worthwhile to mention that the objective (68) is revised to include the hourly electricity prices, which changes the energy generation maximization to benefit maximization.

The MILP gap is set at 0.5%, as further reductions would yield minimal objective value improvements relative to the computational cost. For Case-20, the optimal objective value is $99,163.07 with a final MILP gap of 0.43%, equating to an absolute gap of $428.78. At hour 14, the water volume in the Round Butte Dam reaches its upper limit, storing water for increased power generation during the high electricity price period starting at hour 15, as shown in Fig. 18. In contrast, the Pelton Dam hits its lower water volume limit at hour 16 but remains operationally viable as the surge in water discharge from the Round Butte Dam replenishes its reservoir and supports its power generation needs. For Case-30, the optimal objective value is $98,854.26, with a final MILP gap of 0.41%. The results in Table 5 indicate that including more

Table 3 Parameters of hydropower units

Dispatchable ranges	Round butte dam			Pelton dam		
	Unit #1	Unit #2	Unit #3	Unit #1	Unit #2	Unit #3
Lower power bound (MW)	1	1	1	10	10	10
Upper power bound (MW)	112	130	130	38	36	36

Table 4 Forebay levels and water volume limits

	Round butte dam	Pelton dam
Forebay level limits (Feet)	[1,944, 1,945]	[1,576.5, 1,578.5]
Dam water volume limits (SFD)	[136,250, 138,260]	[940, 1,494]
Initial water volume (SFD)	137,000	1,000

scenarios has limited impact due to the relatively smooth daily variation in water inflows, allowing a smaller number of scenarios to effectively capture the trend while balancing computational efficiency.

The average water spillage rate (WSR) is calculated as the ratio of the average total water spillage across all scenarios to the average total water inflow across all scenarios. Similarly, for individual scenarios, the WSR is defined as the total water spillage divided by the total water inflow. For Case-20 and Case-30, the average WSRs are 7.27% and 6.75%, respectively. In both cases, the majority of water spillage occurs at the Pelton Dam, contributing 97.8% and 97.2% of the average spillage for Case-20 and Case-30, respectively. This spillage primarily results from the need to align the scenario-independent forebay levels across different scenarios. For example, in a scenario with higher water inflow, the dam water volume might rise to a certain level at specific hours. However, in a scenario with lower water inflow, achieving the same water volume level may be impractical or uneconomical. To reconcile this

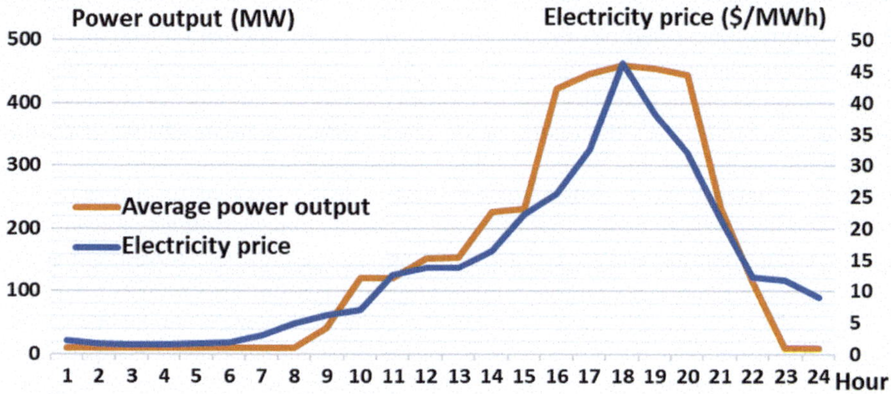

Fig. 18 Electricity price and average power output

Table 5 Summary of the scheduling result

Case-20	SA water inflow (SFD)		3,952.06
	Objective ($)		99,163
	SA water spillage (SFD)		287.66
	Water volume (SFD)	Round Butte	136,562–138,260
		Pelton	940–1,417
Case-30	SA water inflow (SFD)		3913.70
	Objective ($)		98,854
	SA water spillage (SFD)		264.17
	Water volume (SFD)	Round Butte	136,565–138,260
		Pelton	940–1,414

difference, the first scenario may spill water to prevent raising the dam water volume excessively, ensuring consistency in forebay levels across all scenarios.

Figure 18 illustrates the electricity prices and scenario average (SA) power outputs of the Round Butte-Pelton CHS in Case-20. It is evident that the CHS's power outputs closely follow the electricity price trend, demonstrating the system's ability to respond effectively to market signals. During peak electricity price hours, the CHS swiftly ramps up its output to reach maximum values, optimizing operational efficiency and maximizing economic benefits.

PGE employs a so-called "matched flow" policy as the operational strategy for managing forebay levels in its Round Butte-Pelton CHS project. Under this approach, the power output ratio between the Round Butte Dam and the Pelton Dam is consistently maintained at 7:3. The "matched flow" policy treats the forebay levels (dam water volumes) of the two dams as variables, constrained within their respective lower and upper bounds throughout the operating day. In contrast, the stochastic optimization model generates scenario-independent dam water volume schedules. On the operating day, operators adhere closely to these schedules by adjusting the unit commitment and power outputs of hydropower units and managing water spillage as necessary to maximize the total daily profit. This ensures that the dam water volumes remain fixed in accordance with the scenario-independent schedules, aligning operations with the optimization results while accommodating real-time conditions.

PGE's "matched flow" policy is compared with the stochastic optimal policy, which follows the scenario-independent dam water volume schedules derived from the stochastic optimization model. This comparison uses a set of generated out-of-sample scenarios and an actual water inflow case. In this setup, each scenario or actual inflow case is individually input into the stochastic optimization model (68)–(70), causing it to degenerate into a deterministic model that simulates the actual operations for the given operating day. To simulate the "matched flow" policy, the constraint, constraint $\sum_{i\in\mathcal{I}^{RB}} p_{i,t,s}^{RB} = \frac{7}{3}\sum_{i\in\mathcal{I}^{PL}} p_{i,t,s}^{PL}$ is added, enforcing the fixed 7:3 power output ratio between the Round Butte and Pelton Dams. In contrast, for the stochastic optimal policy, the dam water volume variables v_t^{RB} and v_t^{PL} are fixed according to the scenario-independent schedules from the stochastic optimization model. By solving these two modified deterministic models, the simulated operational outcomes for the day under both policies can be compared, providing insights into their respective effectiveness in managing operations and maximizing profits.

The comparison of results is summarized in Table 6. Both the "matched flow" policy and the stochastic optimal policy demonstrate the ability to produce feasible operations for the physical CHS. However, the stochastic optimal policy consistently achieves higher profits in the actual water inflow case and across four generated out-of-sample scenarios. It is noteworthy that the stochastic optimal policy results in a slightly higher WSR in the actual water inflow case and Scenario #3. This occurs because, as discussed, to accommodate low water inflow scenarios, water spillage is sometimes necessary during high water inflow periods to adhere to the prescribed dam water volumes dictated by the stochastic optimal policy. The trade-off is essential for managing future inflow uncertainties effectively. In summary, the day-ahead

Table 6 Matched flow policy and the stochastic optimal policy comparison with case-20

Scenario	Real-time operation strategy	Objective ($)	WSR (%)	Feasibility
Actual	Stochastic optimal policy	101,778	13.61	√
	"Matched flow" policy	97,312	7.39	√
Scenario 1	Stochastic optimal policy	97,543	4.09	√
	"Matched flow" policy	91,925	7.51	√
Scenario 2	Stochastic optimal policy	98,582	5.18	√
	"Matched flow" policy	92,845	7.53	√
Scenario 3	Stochastic optimal policy	101,440	11.3	√
	"Matched flow" policy	96,391	7.2	√
Scenario 4	Stochastic optimal policy	99,859	7.35	√
	"Matched flow" policy	94,280	7.58	√

operation schedules derived from the stochastic optimization model deliver greater economic benefits compared to PGE's current "matched flow" policy, highlighting the value of the stochastic approach in addressing variability and maximizing CHS performance.

4 Pumped Storage Hydropower Plants

4.1 Introduction to Pumped Storage Hydropower Plants

A pumped storage hydropower plant, as shown in Fig. 19, consists of a powerhouse that hosts turbines and generators, and water pipes, i.e., penstocks, that connect the two reservoirs through the powerhouse. Unlike hydropower plants, turbined water will be temporally stored in the lower reservoir and when the system has surplus electrical power, the turbines will switch to the pumping mode, transporting water from the lower reservoir back to the upper reservoir. The reservoirs could be constructed based on the terrain or be artificially excavated. The construction of pumped storage plants relies heavily on geographical conditions, requiring a relatively significant difference in elevation between the upper and lower reservoirs as well as storage capacities.

Based on the technologies used in installed generators, pumped storage hydropower (PSH) plants are categorized into single-speed and adjustable-speed types. The single-speed category also includes a specialized variant known as ternary PSH plants, which have limited current applications.

- **Single-Speed PSH Plants**: This type utilizes a synchronous generator. In generating mode, power output can be adjusted by regulating water inflow through the wicket gate, enabling flexible operation with a reaction turbine. However, in

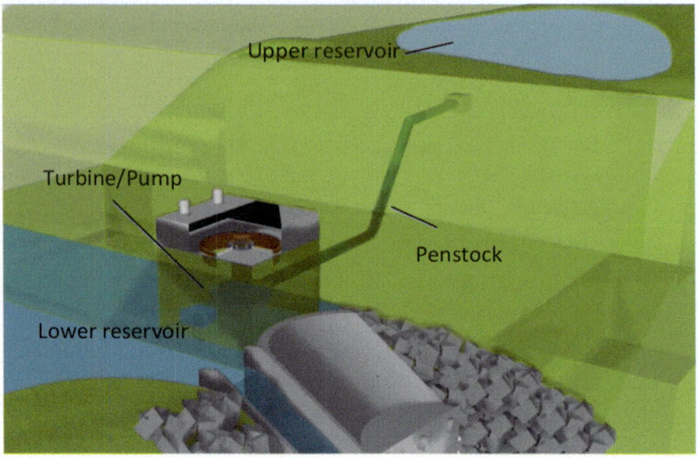

Fig. 19 A pumped storage hydropower plant

pumping mode, the wicket gate remains fixed, and the generator's rotor speed is synchronized to the system frequency, resulting in a fixed pumping power level. Consequently, single-speed PSH plants can only provide regulation and contingency reserves while operating in generating mode.

- **Ternary PSH Plants**: This variation of single-speed PSH also utilizes a synchronous generator but features separate turbine and pump components that can operate simultaneously. When the hydraulic bypass is activated during pumping mode, the turbine can generate power to partially offset the pump's energy consumption, enabling power dispatchability. As a result, ternary PSH plants can provide regulation and contingency reserves in both pumping and generating modes, offering greater operational flexibility.
- **Adjustable-Speed PSH Plants**: This type employs an asynchronous generator, specifically a doubly fed induction machine, with the rotor connected to the grid via an AC/DC/AC converter. In generating mode, their operation is similar to single-speed PSH plants, allowing for adjustable power output. However, in pumping mode, while the wicket gate position remains fixed, the pumping power can be varied by adjusting the generator's rotor speed. This is achieved by controlling the frequency of the rotor voltage and current through the converter. As a result, adjustable-speed PSH plants can provide regulation and contingency reserves in both generating and pumping modes, offering enhanced operational flexibility.

The main functions of PSH can be summarized in three aspects. First, PSH plants can help balance the electricity supply and demand by providing a means to store surplus energy generated during off-peak periods and release it during peak periods. Particularly, during periods of peak demand, when generation resources become scarce and the production cost goes high, PSHs are significant to peak shaving and operational cost reduction. Second, PSHs can enhance grid stability by providing quick responses to fluctuations in demand and unexpected changes in power generation from renewable sources, helping maintain a steady frequency and voltage of the grid as well as facilitating the accommodation of intermittent renewable energy sources, such as wind and solar energy, into the grid. Third, PSHs can serve as an emergency backup and black-start resources during grid failures.

With the increasing penetration of renewable energy, natural gas-fired units, combined-cycle units, and PSH plants have become critical providers of system flexibility. Among them, PSH plants stand out due to their fast response capabilities and high ramp rates, making them ideal for addressing fluctuations in other generation resources. In practice, PSH plants are playing an increasingly significant role in delivering regulation and contingency reserves, effectively supporting grid stability in the face of growing renewable energy integration.

4.2 Modeling of Pumped Storage Hydropower Plants

The dispatching of PSHs in electricity markets is predominantly governed by a state-of-charge (SOC)-based framework, which monitors and manages the dynamic SOC levels through evolution equations while adhering to operational boundaries and dispatchable range constraints. This formulation empowers system operators to integrate and optimize the dual roles of PSHs—pumping and generating—alongside other energy resources and demand-side activities. By doing so, it facilitates the achievement of key objectives, such as maximizing social welfare or minimizing system-wide operational costs. Over time, the SOC-based approach has emerged as the cornerstone of PSH dispatching strategies, inspiring a wide array of advanced variants tailored to address complex market and grid dynamics.

It is worth noting that the SOC-based formulation to be introduced is from a system perspective, with which the entire PHS plant is treated as a whole, with SOC representing its energy state and decision variables, including the overall power output as well as the indicators of corresponding modes, interfacing with the system. The SOC-based PSH formulation can be easily integrated into unit commitment problems under the framework robust or stochastic optimization [16, 17], which represents the advantage of its simplicity.

In fact, similar to hydropower plants, the power output of hydropower units within a PSH is jointly determined by water head, water tail, and water discharge. However, the SOC-based PSH model does not reflect these aspects due to the particular perspective. The SOC-based formulation of the market orientated dispatching model is shown in (83)–(94), where t is the index of hours. Set $\mathcal{T} \triangleq \{1,2,\ldots,T\}$ $\mathcal{T}$ collects the hours in the scheduling horizon, where T represents the last hour. If the model is embedded in a day-ahead unit commitment problem, T will be equal to 24 and the time interval will be hourly:

$$p_t^G + r_t^G + c_t^G \leq P^{G,UB} \cdot I_t^G; \forall t \in \mathcal{T} \tag{83}$$

$$P^{G,LB} \cdot I_t^G \leq p_t^G - r_t^G; \forall t \in \mathcal{T} \tag{84}$$

$$p_t^P + r_t^P \leq P^{P,UB} \cdot I_t^P; \forall t \in \mathcal{T} \tag{85}$$

$$P^{P,LB} \cdot I_t^P \leq p_t^P - r_t^P - c_t^P; \forall t \in \mathcal{T} \tag{86}$$

$$0 \leq r_t^G \leq R^{G,UB} \cdot I_t^G; \forall t \in \mathcal{T} \tag{87}$$

$$0 \leq r_t^P \leq R^{P,UB} \cdot I_t^P; \forall t \in \mathcal{T} \tag{88}$$

$$0 \leq c_t^G \leq C^{G,UB} \cdot I_t^G; \forall t \in \mathcal{T} \tag{89}$$

$$0 \leq c_t^P \leq C^{P,UB} \cdot I_t^P; \forall t \in \mathcal{T} \tag{90}$$

$$I_t^G + I_t^P \leq 1; \forall t \in \mathcal{T} \tag{91}$$

$$E \cdot soc_t = E \cdot soc_{t-1} - 1/\eta \cdot p_t^G + \eta \cdot p_t^P; \forall t \in \mathcal{T} \tag{92}$$

$$soc_0 = soc_T; \tag{93}$$

$$SOC^{LB} \leq soc_t \leq SOC^{UB}; \forall t \in \mathcal{T}. \tag{94}$$

The power dispatchable ranges for PSHs in generating and pumping modes are defined by Eqs. (83)–(84) and (85)–(86), respectively. These constraints accurately capture the operational capabilities of PSHs in both modes. Furthermore, the model incorporates bidirectional symmetrical regulation reserves and up-direction contingency reserves, ensuring enhanced operational flexibility and reliability to support system stability and dynamic requirements. In (83)–(86), p_t^G and p_t^P are power dispatches of the PSH at hour t respectively under generating and pumping modes; r_t^G and r_t^P are regulation reserve dispatches under the two modes at hour t; c_t^G and c_t^P are contingency reserve dispatches under the two modes at hour t; binary variables I_t^G and I_t^P are indicators of the generating and pumping modes; $P^{G,LB}$ and $P^{G,UB}$ are power lower and upper bounds under generating mode; $P^{P,LB}$ and $P^{P,UB}$ are power lower and upper bounds under pumping mode.

Reserves of the PSH under different operation modes are limited by constraints (87), (88). Ramping constraints can be ignored, considering the strong ramping ability of PSHs for both generating and pumping modes. In (87)–(90), $C^{G,UB}$ and $C^{P,UB}$ are contingency reserve capacity under generating and pumping modes; $R^{G,UB}$

and $R^{P,UB}$ respectively represent regulation reserve capacity under generating and pumping modes.

Constraint (91) ensures that the pumping and generating modes of PSHs are mutually exclusive, preventing simultaneous operation in both modes. The SOC evolution of PSHs is described by (92), while constraint (93) requires the terminal SOC to return to its initial value, disregarding any SOC changes resulting from reserve deployment. In practice, the generation and pumping efficiencies of PSHs depend on nonlinear factors such as the net water head and reservoir flow rate. Capturing this relationship accurately would necessitate more complex constraints, significantly increasing computational demands. To balance accuracy and efficiency, constant efficiency assumptions are commonly used as a practical trade-off in modeling. Therefore, η in (92) is a given constant. SOC of the PSH is further limited by capacity bounds as in (94). In (92)–(94), soc_t represents the scheduled SOC of the PSH at hour t. SOC^{LB} and SOC^{UB} are respectively SOC lower and upper bounds of the PSH.

4.3 *Modeling of Single-Speed Pumped Storage Hydropower and Water Head*

While adjustable-speed and ternary PSHs represent more advanced technologies, the majority of PSH projects worldwide, particularly in the U.S., are single-speed PSHs, which is largely because these facilities were constructed between the 1960s and 1980s, prior to the development of advanced PSH technologies. Additionally, although ternary PSHs achieve dispatchability in pumping mode through a different mechanism, their operational modeling aligns closely with that of adjustable-speed PSHs when internal operations are not explicitly accounted for.

For a single-speed PSH, the pump operates at a fixed speed, moving water at a constant rate regardless of varying hydraulic conditions, which, being reflected by the output power, means the power of the PSH is constant in pumping mode. Therefore, constraint (95) can be formulated to enforce that in pumping mode at hour t, the pumping power p_t^P is fixed to P^P:

$$p_t^P = P^P \cdot I_t^P; \forall t \in \mathcal{T}. \tag{95}$$

Constraint (95) ignores the influence of the water head of the upper reservoir on pumping. In fact, the pumping rate varies with different water heads, which can be expressed by (96), (97). In (96), (97), $I_{j,t}^H$ is the indicator of water head selection. When $I_{j,t}^H$ that indicates a certain water head is 1 while the PSH is in pumping mode, the pumping rate is fixed to a given value corresponding to the water head. With the pumping water q_t^P, the pumping power can be calculated as in (98). As q_t^P are discrete values, p_t^P are discrete as well. Constraints (96) introduce the nonlinear term $I_{j,t}^H \cdot I_t^P$, which can be equivalently linearized as in (99a–d). Because of the feature of the single-speed PSH, reserves under the pumping mode are unavailable:

$$q_{j,t}^{P} = Q_j^{P} \cdot I_{j,t}^{H} \cdot I_t^{P}; \forall j \in \mathcal{J}, \forall t \in \mathcal{T} \tag{96}$$

$$q_t^{P} = \sum\nolimits_j q_{j,t}^{P}; \forall j \in \mathcal{J}, \forall t \in \mathcal{T} \tag{97}$$

$$p_t^{G} = \sum\nolimits_j B_j \cdot I_t^{P} + C_j \cdot q_t^{P}; \forall j \in \mathcal{J}, \forall t \in \mathcal{T} \tag{98}$$

$$I_{j,t}^{H,P} = I_{j,t}^{H} \cdot I_t^{P}; \tag{99.a}$$

$$I_{j,t}^{H,P} \leq I_{j,t}^{H} + I_t^{P} - 1; \tag{99.b}$$

$$I_{j,t}^{H,P} \leq I_{j,t}^{H}; \tag{99.c}$$

$$I_{j,t}^{H,P} \leq I_t^{P}. \tag{99.d}$$

The modeling of the generating mode can be extended to incorporate the influence of the water head as shown in (100)–(105). There are two main differences compared to (96)–(98): first, constraints (100) and (101) consider the scheduling of regulation reserve and contingency reserve; second, in $\sum_k z_{k,j,t} = I_{j,t}^{H} \cdot I_t^{G}$, only when $I_{j,t}^{H}$ that indicates a specific water head is 1, while the PSH is in generating mode, the corresponding $z_{k,j,t}$ become active:

$$p_t^{G} + r_t^{G} + c_t^{G} \leq \sum\nolimits_j P_j^{G,UB} \cdot I_{j,t}^{H} \cdot I_t^{G}; \forall t \in \mathcal{T} \tag{100}$$

$$\sum\nolimits_j P_j^{G,LB} \cdot I_{j,t}^{H} \cdot I_t^{G} \leq p_t^{G} - r_t^{G}; \forall t \in \mathcal{T} \tag{101}$$

$$p_t^{G} = \sum\nolimits_j \left(\sum_k B_{k,j} \cdot z_{k,j,t} + \sum\nolimits_k C_{k,j} \cdot q_{k,j,t}^{G} \right); \forall t \in \mathcal{T} \tag{102}$$

$$q_t^{G} = \sum\nolimits_j \sum\nolimits_k q_{k,j,t}^{G}; \forall t \in \mathcal{T} \tag{103}$$

$$\sum_k z_{k,j,t} = I_{j,t}^{H} \cdot I_t^{G}; \forall j \in \mathcal{J}, \forall t \in \mathcal{T} \tag{104}$$

$$0 \leq q_{k,j,t} \leq Q^{G} \cdot z_{k,j,t}; \forall t \in \mathcal{T}. \tag{105}$$

It is worth mentioning that (96)–(98) and (100)–(105) only consider a single unit or an equivalent unit. However, as discussed in the first section, it is straightforward to extend (96)–(98) and (100)–(105) to consider multiple units.

The full formulation is given below. Most constraints are self-explained based on (96)–(98) and (100)–(105). Constraints (109) and (114) respectively calculate the integrated power of all units under pumping and generating modes. Constraints (117)–(119) together make pumping and generating exclusive of the entire PSH plant and avoid simultaneous pumping and generating of units. Constraint (120) is the reservoir water volume evolution constraint:

$$q^{P}_{i,j,t} = Q^{P}_{i,j} \cdot I^{H}_{j,t} \cdot I^{P}_{i,t}; \forall i \in \mathcal{I}, \forall j \in \mathcal{J}, \forall t \in \mathcal{T} \tag{106}$$

$$q^{P}_{i,t} = \sum_{j} q^{P}_{i,j,t}; \forall i \in \mathcal{I}, \forall t \in \mathcal{T} \tag{107}$$

$$p^{P}_{i,t} = \sum_{j} B_{i,j} \cdot I^{P}_{i,t} + C_{i,j} \cdot q^{P}_{i,t}; \forall i \in \mathcal{I}, \forall t \in \mathcal{T} \tag{108}$$

$$p^{P}_{t} = \sum_{i} p^{P}_{i,t}; \forall t \in \mathcal{T} \tag{109}$$

$$p^{G}_{i,t} + r^{G}_{i,t} + c^{G}_{i,t} \leq \sum_{j} P^{G,UB}_{i,j} \cdot I^{H}_{j,t} \cdot I^{G}_{i,t}; \forall i \in \mathcal{I}, \forall t \in \mathcal{T} \tag{110}$$

$$\sum_{j} P^{G,LB}_{i,j} \cdot I^{H}_{j,t} \cdot I^{G}_{i,t} \leq p^{G}_{i,t} - r^{G}_{i,t}; \forall i \in \mathcal{I}, \forall t \in \mathcal{T} \tag{111}$$

$$p^{G}_{i,t} = \sum_{j} \left(\sum_{k} B_{i,k,j} \cdot z_{k,j,t} + \sum_{k} C_{i,k,j} \cdot q^{G}_{i,k,j,t} \right); \forall i \in \mathcal{I}, \forall t \in \mathcal{T} \tag{112}$$

$$q^{G}_{i,t} = \sum_{j} \sum_{k} q^{G}_{i,k,j,t}; \forall i \in \mathcal{I}, \forall t \in \mathcal{T} \tag{113}$$

$$p^{G}_{t} = \sum_{i} p^{G}_{i,t}; \forall t \in \mathcal{T} \tag{114}$$

$$\sum_{k} z_{i,k,j,t} = I^{H}_{j,t} \cdot I^{G}_{i,t}; \forall i \in \mathcal{I}, \forall j \in \mathcal{J}, \forall t \in \mathcal{T} \tag{115}$$

$$0 \leq q_{i,k,j,t} \leq Q^{G} \cdot z_{i,k,j,t}; \forall i \in \mathcal{I}, \forall j \in \mathcal{J}, \forall k \in \mathcal{K}, \forall t \in \mathcal{T} \tag{116}$$

$$I^{P}_{i,t} \leq I^{P}_{t}; \forall i \in \mathcal{I}, \forall t \in \mathcal{T} \tag{117}$$

$$I^{G}_{i,t} \leq I^{G}_{t}; \forall i \in \mathcal{I}, \forall t \in \mathcal{T} \tag{118}$$

$$I^{G}_{t} + I^{P}_{t} \leq 1; \forall t \in \mathcal{T} \tag{119}$$

$$v_{t} = v_{t-1} - \sum_{i} q^{G}_{i,t} + \sum_{i} q^{P}_{i,t}; \forall t \in \mathcal{T}. \tag{120}$$

4.4 Scheduling of Pumped Storage Hydropower with Secured Reserve

Unlike thermal units, PSH plants and other energy storage systems possess a distinctive feature: their SOC. The SOC serves as a memory of cumulative power deviations over time and is constrained by specific upper and lower boundaries. This characteristic creates a notable difference in how reserves are delivered. When reserves are deployed, the actual SOC may diverge from the planned schedule. If there are repeated or substantial reserve deployments over time, these deviations can accumulate, resulting in either insufficient or excessive stored energy in later hours. Such imbalances can prevent PSH plants from executing their planned generation or pumping operations, as they risk breaching SOC capacity limits. Additionally, reserve commitments scheduled for those hours may become unattainable, potentially compromising system operational security. As a result, system operators must exercise caution and precision in scheduling reserves for PSH plants to maintain system reliability and stability.

Given the energy-limited nature of PSH plants and the accelerating expansion of renewable energy sources, it becomes increasingly crucial to improve the reserve deliverability of PSH plants and other energy-constrained resources. Addressing this challenge within the unit commitment problem is essential. This problem is typically modeled as a Mixed-Integer Linear Programming (MILP) formulation, which balances accuracy with computational efficiency to ensure practical solutions.

Constraints (121) and (122) treat the reserve scheduling of energy-limited PSHs in a manner similar to that of thermal units. These constraints either disregard SOC limits altogether or ensure that the stored energy at each time interval is sufficient to meet the scheduled power and reserve requirements for that specific interval only. However, these approaches may result in reserve schedules that are not fully deployable due to potential SOC violations. This issue arises because the SOC evolution Eq. (92) does not account for the cumulative effects of reserve deployment over time, which can lead to deviations that compromise the feasibility of the reserve schedule.

To address this, regulation reserve secure constraints (121) and (122) are introduced to ensure that the SOC of PSHs remains within allowable limits, even under the worst-case reserve deployment scenarios. Constraint (121) ensures that the scheduled SOC at hour t-1, soc_{t-1}, is sufficient to support the full deployment of regulation up reserves across all prior hours {1, …, t-1}, as well as contingency reserve and scheduled power generation at hour t. Similarly, constraint (122) guarantees that soc_{t-1} is low enough to accommodate the full deployment of regulation down reserves across all prior hours {1, …, t-1}, in addition to scheduled pumping power and contingency reserve at hour t.

It is important to note that the required continuous deployment duration for regulation reserves, denoted by τ, can vary depending on market rules. In many cases, PSHs may be permitted to release their regulation reserve obligation by substituting it with generation from other thermal units after maintaining reserve deployment for a specified time period. The flexibility allows PSHs to manage their SOC more effectively while ensuring compliance with operational and market requirements:

$$E \cdot soc_{t-1} - \frac{1}{\eta} \cdot \sum_{t'=1}^{t-1} r_{t'}^{G} \cdot \tau - \eta \cdot \sum_{t'=1}^{t-1} r_{t'}^{P} \cdot \tau - \frac{1}{\eta} \cdot (p_t^G + r_t^G \cdot \tau + c_t^G) \geq E \cdot SOC^{LB};\ \forall t \in \mathcal{J} \tag{121}$$

$$E \cdot soc_{t-1} + \frac{1}{\eta} \cdot \sum_{t'=1}^{t-1} r_{t'}^{G} \cdot \tau + \eta \cdot \sum_{t'=1}^{t-1} r_{t'}^{P} \cdot \tau + \eta \cdot (p_t^P + r_t^P \cdot \tau) \leq E \cdot SOC^{UB};\ \forall t \in \mathcal{J}. \tag{122}$$

Constraints (121) and (122) may be excessively conservative because they assume the worst-case scenario in which the regulation reserve scheduled at every hour is fully deployed. In practice, such a situation is highly unlikely, as reserve deployment is typically partial and varies dynamically based on system needs. Intuitively over-conservatism can lead to underutilization of PSHs' operational flexibility and unnecessarily restrict their participation in system optimization. A less conservative approach that considers probabilistic or expected reserve deployment levels could potentially improve efficiency without compromising system reliability. To this end, four regulation reserve usage factors $\zeta_t^{G,UP}$, $\zeta_t^{G,DN}$, $\zeta_t^{P,UP}$, and $\zeta_t^{P,DN}$ can be introduced:

$\zeta_t^{G,UP}/\zeta_t^{G,DN}$: ratio of the deployed regulation up/down reserve of a PSH at hour t, respectively, divided by r_t^G, under generating mode.

$\zeta_t^{P,UP}/\zeta_t^{P,DN}$: ratio of the deployed regulation up/down reserve of a PSH at hour t, respectively, divided by r_t^P, under pumping mode.

To address the limitation of constraints (121) and (122), which assume full reserve deployment at all hours, chance-constrained reserve secure constraints are introduced as in (123) and (124). These constraints leverage the above four random parameters to characterize the uncertainties in reserve deployment over multiple hours. By incorporating a confidence level, constraints (123) and (124) ensure that SOC capacity boundaries are satisfied under uncertain reserve deployment scenarios, reducing the over-conservativeness inherent in (121) and (122). However, constraints (123) and (124) adopt the same assumption as (121) and (122) on contingency reserve deployment. This is because, even contingency reserves are deployed only when contingencies occur, which are considered low-probability events; if a contingency does occur, full reserve deliverability must be guaranteed.

While joint chance constraints could theoretically provide stronger security guarantees by addressing multiple uncertainties simultaneously, they are computationally intensive due to their nonlinear nature. Reformulating them into a linear form suitable for wholesale market clearing within practical timeframes is challenging. As a result, individual chance constraints are adopted as a practical compromise. These constraints, combined with the distributionally robust chance-constrained formulation and the specific characteristics of PSH reserve deployment, strike a balance between operational security and computational feasibility, offering a solution with a controlled level of conservativeness:

$$\mathbb{P}\begin{pmatrix} E \cdot soc_{t-1} - \frac{1}{\eta} \cdot \sum_{t'=1}^{t-1} \zeta_{t'}^{G,UP} \cdot r_{t'}^{G} \cdot \tau \\ -\eta \cdot \sum_{t'=1}^{t-1} \zeta_{t'}^{P,UP} \cdot r_{t'}^{P} \cdot \tau \\ -\frac{1}{\eta} \cdot (p_t^G + r_t^G \cdot \tau + c_t^G) \geq E \cdot SOC^{LB} \end{pmatrix} \geq 1 - \epsilon; \forall t \in \mathcal{T} \tag{123}$$

$$\mathbb{P}\begin{pmatrix} E \cdot soc_{t-1} + \frac{1}{\eta} \cdot \sum_{t'=1}^{t-1} \zeta_{t'}^{G,DN} \cdot r_{t'}^{G} \cdot \tau \\ +\eta \cdot \sum_{t'=1}^{t-1} \zeta_{t'}^{P,DN} \cdot r_{t'}^{P} \cdot \tau \\ +\eta \cdot (p_t^P + r_t^P \cdot \tau) \leq E \cdot SOC^{UB} \end{pmatrix} \geq 1 - \epsilon; \forall t \in \mathcal{T}. \tag{124}$$

Constraints (123) and (124) are reformulated to a common compact form (125), where $\boldsymbol{\zeta}_t$, $\boldsymbol{X}_t$, and B_t for (123) are given as in (126a–c) and those for (124) are shown as in (127a–c):

$$\mathbb{P}\left((\boldsymbol{\zeta}_t)^T \cdot \boldsymbol{X}_t \leq B_t\right) \geq 1 - \epsilon; \forall t \in \mathcal{T} \tag{125}$$

$$\boldsymbol{X}_t = \tau \cdot \left[\frac{1}{\eta} \cdot r_1^G, \cdots, \frac{1}{\eta} \cdot r_t^G, \eta \cdot r_1^P, \cdots, \eta \cdot r_{t-1}^P\right]; \tag{126.a}$$

$$\boldsymbol{\zeta}_t = \left[\zeta_1^{G,UP}, \cdots, \zeta_t^{G,UP}, \zeta_1^{P,UP}, \cdots, \zeta_{t-1}^{P,UP}\right]; \tag{126.b}$$

$$B_t = E \cdot \left(-SOC^{LB} + soc_{t-1}\right) - \frac{1}{\eta} \cdot \left(p_t^G + c_t^G\right); \tag{126.c}$$

$$\boldsymbol{X}_t = \tau \cdot \left[\frac{1}{\eta} \cdot r_1^G, \cdots, \frac{1}{\eta} \cdot r_{t-1}^G, \eta \cdot r_1^P, \cdots, \eta \cdot r_t^P\right]; \tag{127.a}$$

$$\boldsymbol{\zeta}_t = \left[\zeta_1^{G,DN}, \cdots, \zeta_{t-1}^{G,DN}, \zeta_1^{P,DN}, \cdots, \zeta_t^{P,DN}\right]; \tag{127.b}$$

$$B_t = E \cdot \left(SOC^{UB} - soc_{t-1}\right) - \eta \cdot p_t^P. \tag{127.c}$$

A significant computational challenge associated with constraints (123) and (124) lies in the fact that the probability distributions of the four random parameters are unknown and may not conform to standard probability density functions, such as Normal or Weibull distributions. The unknown uncertainty makes it difficult to apply traditional probabilistic modeling techniques, as these methods typically rely on well-defined distributional assumptions. As a result, alternative approaches, such as distributionally robust optimization, are often utilized to handle these unknown distributions while maintaining computational tractability and ensuring the reliability of the constraints.

To this end, constraints (123) and (124) are transformed into the distributionally robust formulation (128), in which, $\mathcal{D}$ represents a family of probability distributions characterized by shared attributes, such as specified mean, covariance, and support, while $\lceil(\cdot)$ denotes the distribution of $\boldsymbol{\zeta}_t$. The essence of the reformulation lies in ensuring that constraint (125) is satisfied under the worst-case distribution of $\boldsymbol{\zeta}_t$ within the defined distribution family $\mathcal{D}$.

By adopting the distributionally robust optimization, the requirement for accurately estimating the exact distribution of $\boldsymbol{\zeta}_t$ is significantly relaxed, enabling the model to account for uncertainties in the underlying probability distribution. However, this enhanced flexibility introduces a trade-off, as the model becomes inherently more conservative to safeguard against the most adverse distribution scenarios in $\mathcal{D}$. The conservativeness ensures operational security while embracing the uncertainty inherent in real-world conditions:

$$inf_{\lceil(\zeta_t)\in\mathcal{D}}\mathbb{P}\left((\boldsymbol{\zeta}_t)^T\cdot\boldsymbol{X}_t\leq B_t\right)\geq 1-\epsilon;\forall t\in\mathcal{T}. \tag{128}$$

The mean and covariance of $\boldsymbol{\zeta}_t$ are employed as key characteristics to define the family of distributions $\mathcal{D}$. Using the specified mean $\boldsymbol{\mu}_t$ and the positive semi-definite covariance matrix $\boldsymbol{\Gamma}_t$ of $\boldsymbol{\zeta}_t$, constraint (129) can be equivalently reformulated to account for these distributional features. Constraint (129) ensures that the constraint incorporates the statistical properties of $\boldsymbol{\zeta}_t$, enabling a robust representation of the uncertainty within the distribution family $\mathcal{D}$ while maintaining mathematical equivalence with the original constraint:

$$\sqrt{(1-\epsilon)/\epsilon}\cdot\sqrt{(\boldsymbol{X}_t)^T\cdot\boldsymbol{\Gamma}_t\cdot\boldsymbol{X}_t}+(\boldsymbol{\mu}_t)^T\cdot\boldsymbol{X}_t\leq B_t;\forall t\in\mathcal{T}. \tag{129}$$

Constraint (129) is a second-order cone constraint, which, while convex, is inherently nonlinear. To maintain the unit commitment problem as a MILP model, constraint (129) can be linearized using the polyhedral approximation method. This approach approximates the nonlinear constraint with a set of linear inequalities, enabling the problem to remain within the MILP framework and ensuring compatibility with commercial optimization solvers while preserving computational efficiency.

The second-order cone constraint (129) can be expressed in a generalized form, as in (130), and further decomposed into two parts as illustrated in (131) by introducing an auxiliary variable y_1. Through repeated application of the splitting process, a second-order cone constraint involving multiple variables can be reformulated into several second-order cone constraints, each containing only three variables. This reformulation simplifies the structure of the constraint, facilitating its handling in optimization models and enabling more efficient solution methods.

Thereafter, without loss of generality, $\sqrt{x_1^2+x_2^2}\leq y_1$ is used as an example. With ancillary variables v_k and u_k, $\sqrt{x_1^2+x_2^2}\leq y_1$ can be linearized as in (132)–(134), where $cos(\pi/2^k)$ and $sin(\pi/2^k)$ are constant coefficients. Importantly, K is a preset parameter to control linearization accuracy.

To illustrate the linearization process, considering the example $\sqrt{x_1^2 + x_2^2} \le y_1$, it can be reformulated as a set of linear inequalities, as in (132)–(134), where v_k and u_k are ancillary variables introduced to approximate the relationship; $cos(\pi/2^k)$ and $sin(\pi/2^k)$ are constant coefficients derived from trigonometric properties; K is a predefined parameter that controls the accuracy of the linearization by determining the number of approximating segments. This linearization approach transforms the nonlinear constraint into a series of linear constraints, making it compatible with MILP solvers while allowing for adjustable precision based on the choice of K:

$$\sqrt{x_1^2 + x_2^2 + x_3^2 + \cdots + x_n^2} \le z; \tag{130}$$

$$\sqrt{x_1^2 + x_2^2} \le y_1; \sqrt{y_1^2 + x_3^2 + \cdots + x_n^2} \le z; \tag{131}$$

$$v_1 + x_1 = 0; -u_1 \le x_2 \le u_1; \tag{132}$$

$$v_{k+1} - v_k \cdot cos(\pi/2^k) - u_k \cdot sin(\pi/2^k) = 0; k \in \{1, \ldots, K-1\} \tag{133.a}$$

$$u_{k+1} - u_k \cdot cos(\pi/2^k) + v_k \cdot sin(\pi/2^k) \ge 0; k \in \{1, \ldots, K-1\} \tag{133.b}$$

$$u_{k+1} + u_k \cdot cos(\pi/2^k) - v_k \cdot sin(\pi/2^k) \ge 0;\ \mathrm{k} \in \{1, \ldots, K-1\} \tag{133.c}$$

$$v_K \cdot cos(\pi/2^k) + u_K \cdot sin(\pi/2^k) - y_1 = 0;\ \mathrm{k} \in \{1, \ldots, K-1\}. \tag{134}$$

Equation (135) indicates the linearization error of a second-order cone constraint, where $\tilde{x}_1$, ..., $\tilde{x}_n$, and $\tilde{z}$ represent the optimal solution derived from the linearized constraints (131)–(133). Across the entire optimization model, the maximum linearization error is defined as the highest error observed among all linearized second-order cone constraints. The parameter K, which governs the granularity of the linearization, plays a crucial role in controlling accuracy—larger values of K yield more precise approximations. For further insights, the theoretical relationship between K and the resulting linearization error is explored comprehensively in reference [18]:

$$max\left(\sqrt{\tilde{x}_1^2 + \tilde{x}_2^2 + \tilde{x}_3^2 + \cdots + \tilde{x}_n^2} - \tilde{z}, 0\right). \tag{135}$$

One common method for generating samples of $\boldsymbol{\zeta}_t$ involves using historical data on scheduled reserves and their actual deployment. However, this approach has inherent limitations, as each historical day yields only a single sample of $\boldsymbol{\zeta}_t$, requiring an exceptionally long observation period to accumulate a meaningful sample size. Furthermore, significant shifts in system operating conditions over time—such as seasonal variations in load demand, changes to network topology, and periods of generator maintenance—can introduce non-stationarity into the distribution of ζ_t. These changes undermine the validity of the historical data, making it less representative of current or future system dynamics and challenging the robustness of the sampling approach.

To address the limitations of relying solely on historical data, a simulation-based method can be employed to generate multiple samples of the random parameters ζ_t. This approach begins by selecting a historical day and creating multiple load scenarios based on variations around its historical load profile, which serve as reserve deployment scenarios. For each scenario, the regulation reserve deployment problem (136)–(144) is solved using the known day-ahead schedule from the historical day, including unit commitments, power dispatches, and reserve allocations, which produces one regulation reserve deployment sample, corresponding to a single ζ_t. By repeating this procedure across all scenarios, multiple samples can be generated for each historical day. Consequently, by focusing on a limited number of recent historical days with comparable system operating conditions, a sufficient set of ζ_t samples can be obtained without the need for an extensive timespan, ensuring both relevance and efficiency.

Usually, the regulation reserve is actually allocated among all resources for deployment, proportional to their awarded quantities. These proportional distribution factors, $K_{g,t}$ and $K_{h,t}$, can be pre-calculated as $\tilde{r}_{g,t}/R_t^{RQ}$ and $\left(\tilde{r}_{h,t}^G + \tilde{r}_{h,t}^P\right)/R_t^{RQ}$. Subscripts g and h are respectively used to index generation units and PSHs.

In newly generated scenarios, strictly adhering to the proportional distribution factors $K_{g,t}$ and $K_{h,t}$ may not fully balance net load fluctuations while satisfying the line flow constraint (138). To this end, slack variables $r_{g,t}^{D,+}$, $r_{g,t}^{D,-}$, $r_{h,t}^{D,+}$, and $r_{h,t}^{D,-}$ are introduced. These variables allow deviations from the pre-calculated proportional allocation strategy when necessary. The objective function (136) aims to minimize the total deviation from the original proportional strategy, maintaining the solution as close as possible to the planned allocation. In addition to constraint (138), the model incorporates the power balance constraint (137) and regulation reserve deployment boundaries (139), (140). Constraints (141), (142) calculate the actual regulation reserve deployment, while constraints (143), (144) ensure that all slack variables remain non-negative. Together, these constraints ensure a feasible and minimally deviated reserve deployment strategy that respects system requirements and operational constraints:

$$min \sum_{t\in\mathcal{T}} \left[\sum_{g\in\mathcal{G}} \left(r_{g,t}^{D,+} + r_{g,t}^{D,-} \right) + \sum_{h\in\mathcal{H}} \left(r_{h,t}^{D,+} + r_{h,t}^{D,-} \right) \right]; \tag{136}$$

$$\sum_{f\in\mathcal{F}} \Delta P_{f,t} = \sum\nolimits_{g\in\mathcal{G}} r_{g,t}^{D} + \sum\nolimits_{h\in\mathcal{H}} r_{h,t}^{D}; \forall t \in \mathcal{T} \tag{137}$$

$$-F_l^{UB} \leq F_{l,t} + \sum_{i \in \mathcal{I}} SF_{l,i} \cdot \left(\sum_{g \in \mathcal{G}_i} r_{g,t}^D + \sum_{h \in \mathcal{H}_i} r_{h,t}^D - \sum_{f \in \mathcal{F}_i} \Delta P_{f,t} \right) \leq F_l^{UB}; \forall l \in \mathcal{L}, \forall t \in \mathcal{T} \quad (138)$$

$$-\tilde{r}_{g,t} \leq r_{g,t}^D \leq \tilde{r}_{g,t}; \forall g \in \mathcal{G}, \forall t \in \mathcal{T} \quad (139)$$

$$-\left(\tilde{r}_{h,t}^G + \tilde{r}_{h,t}^P\right) \leq r_{h,t}^D \leq \left(\tilde{r}_{h,t}^G + \tilde{r}_{h,t}^P\right); \forall h \in \mathcal{H}, \forall t \in \mathcal{T} \quad (140)$$

$$r_{g,t}^D = K_{g,t} \cdot \sum_{f \in \mathcal{F}} \Delta P_{f,t} + r_{g,t}^{D,+} - r_{g,t}^{D,-}; \forall g \in \mathcal{G}, \forall t \in \mathcal{T} \quad (141)$$

$$r_{h,t}^D = K_{h,t} \cdot \sum_{f \in \mathcal{F}} \Delta P_{f,t} + r_{h,t}^{D,+} - r_{h,t}^{D,-}; \forall h \in \mathcal{H}, \forall t \in \mathcal{T} \quad (142)$$

$$r_{g,t}^{D,+} \geq 0; r_{g,t}^{D,-} \geq 0; \forall g \in \mathcal{G}, \forall t \in \mathcal{T} \quad (143)$$

$$r_{h,t}^{D,+} \geq 0; r_{h,t}^{D,-} \geq 0; \forall h \in \mathcal{H}, \forall t \in \mathcal{T}. \quad (144)$$

With solutions to $r_{h,t}^D$, samples of $\zeta_{h,t}^{G,UP}$, $\zeta_{h,t}^{G,DN}$, $\zeta_{h,t}^{P,UP}$, and $\zeta_{h,t}^{P,DN}$ can be obtained together with known generating or pumping modes of individual hours. Specifically, under the generating mode, a sample of $\zeta_{h,t}^{G,UP}/\zeta_{h,t}^{G,DN}$ can be generated as $\left|r_{h,t}^D\right|/\tilde{r}_{h,t}^G$, if $r_{h,t}^D$ is positive/negative; and the same logic can be applied to obtain samples of $\zeta_{h,t}^{P,UP}$ and $\zeta_{h,t}^{P,DN}$ under the pumping mode.

Considering (136)–(144), the proportional deployment strategy among all units is applied without strictly enforcing compliance with SOC boundaries. This is based on the assumption that SOC boundary violations will not occur if the day-ahead reserve and SOC schedules are robust enough to fully support the proportional reserve deployment strategy in real-time, which eliminates the need for complex real-time mitigation strategies to address undesired SOC boundary violations. However, reserve deployment scenarios in the simulation that lead to SOC boundary violations are significant. These scenarios highlight instances of reserve over-scheduling in the day-ahead market, where additional headroom is required to ensure secure real-time operations. Such scenarios are crucial for identifying gaps in the day-ahead scheduling process and will be used to estimate the ratios of deployed regulation reserves. These ratios will then inform the reserve secure constraints (129), enhancing their ability to capture the necessary reserve levels to prevent SOC boundary violations during real-time operations.

To incorporate the mutual exclusivity of generating and pumping modes alongside the regulation reserve schedules determined in the day-ahead market, valid data points from $\boldsymbol{\zeta}_t$ samples are carefully filtered to estimate the mean values of the four random parameters independently. Similarly, valid sample pairs corresponding to each pair of random parameters are used to estimate the entries of the covariance matrix $\boldsymbol{\Gamma}_t$. However, the estimated $\boldsymbol{\Gamma}_t$ matrix may not inherently satisfy the positive semi-definiteness requirement, which is essential for the reformulation of (128) into (129). To address this, a positive semi-definite matrix that closely approximates the estimated $\boldsymbol{\Gamma}_t$ is constructed by minimizing the Frobenius norm between the original and adjusted matrices, ensuring both computational stability and compliance with mathematical requirements. Additionally, when estimating the covariance of two random parameters, if the number of valid sample pairs is insufficient—falling below a specified threshold—it is assumed that these parameters are uncorrelated, and their covariance is set to zero. This approach mitigates the risk of unreliable estimations due to sparse data, ensuring that the resulting covariance matrix accurately reflects the underlying system dynamics while maintaining robustness and feasibility.

4.5 Numerical Simulation of Secured Reserve

A modified 118-bus transmission system featuring three identical PSHs connected at buses 4, 46, and 60 is used to assess the performance of reserve secure constraints for PSHs. To facilitate the evaluation, the energy and reserve bidding prices of the PSHs are adjusted to make them more competitive and likely to be dispatched. The regulation reserve deployment duration τ is set at 1 h, with a scheduling horizon T of 24 h. The confidence level factor ϵ in constraints (123), (124) is set to 5%, and the threshold S for valid sample pairs is set to 10. The SOC lower and upper bounds SOC^{LB} and SOC^{UB} are defined as 25% and 100%, respectively.

Three models are compared for validation:

- The Base Model (B-Model) excludes reserve secure constraints entirely.
- The Reserve Secured Model (R-Model) includes the reserve secure constraints but does not linearize the second-order cone constraint (129). The parameter K for linearization is set to 6.
- The Conservatively Reserve Secured Model (RC-Model) incorporates constraints (121), (122) to provide a more conservative approach.

The three to be compared models allow for a comprehensive comparison of the constraints' impact on PSH operations under different conditions.

Since the IEEE 118-bus system lacks historical operation data, synthetic historical load profiles are created by introducing significant load deviations to the original load curve. These variations generate multiple load curves that simulate historical load patterns. A unit commitment model without energy reserve constraints is then solved for these synthetic load profiles, producing unit commitments, power dispatches, and reserve schedules to form the synthetic historical operation dataset. Additionally, for each synthetic historical day, multiple reserve deployment scenarios are created by applying randomly generated load fluctuations. These fluctuations are modeled as a normal distribution with a mean of zero and a standard deviation equal to 1% of the hourly load level. Using these scenarios, 10 synthetic historical days with known unit commitment and dispatch results are selected for further analysis. For each historical day, 200 reserve deployment scenarios are simulated by solving the problem defined in (136)–(144), generating samples for $\zeta_t^{G,UP}$, $\zeta_t^{G,DN}$, $\zeta_t^{P,UP}$, and $\zeta_t^{P,DN}$. The samples are then used to estimate the mean $\boldsymbol{\mu}_t$ and covariance $\boldsymbol{\Gamma}_t$ of $\boldsymbol{\zeta}_t$, which are required inputs for the R-Model and RC-Model. The introduced approach ensures that $\boldsymbol{\mu}_t$ and $\boldsymbol{\Gamma}_t$ reflect the statistical properties of the generated scenarios, supporting accurate constraint formulation in the models.

An excessively conservative reserve scheduling model can undermine the economic efficiency of PSHs, whereas an overly aggressive approach increases the risk of SOC violations. To evaluate the balance between these two extremes, the R-Model is analyzed in terms of reserve schedule conservativeness and reserve deployment security. Its performance is compared to the B-Model, which excludes reserve secure constraints, and the RC-Model, which adopts a more conservative approach by incorporating stricter constraints. This comparison highlights the trade-offs between economic optimization and operational security across the models.

Table 7 summaries the objectives and reserve scheduling results of the three models. The B-Model demonstrates the most aggressive approach, with PSHs providing the largest share of total regulation reserves, resulting in the lowest objective value due to minimal reliance on thermal units. In contrast, the RC-Model adopts the most conservative strategy, with PSHs contributing the smallest amount of regulation reserve. The shortfall is compensated by less economical thermal units, leading to the highest objective value.

The R-Model strikes a balance between these two extremes in terms of both the objective value and the scheduled regulation reserve. Compared to the B-Model, the market clearing price for regulation reserves under the R-Model increases in 17 of the 24 h, with an average increase of \$2.03 over the entire day, corresponding to an average relative rise of 26.25%. This highlights the trade-off in the R-Model, which improves reserve deployment security while maintaining better economic efficiency than the RC-Model.

For contingency reserve scheduling, similar to regulation reserve, the B-Model schedules significantly more from PSHs compared to the other two models. However, unlike regulation reserve, the RC-Model schedules an amount of contingency reserve nearly identical to the R-Model. This difference arises because, in constraints (121), (122), contingency reserve is treated as non-cumulative. As a result, the RC-Model prioritizes maintaining contingency reserve scheduling by reducing the amount of

Table 7 Objective and reserve scheduling result

Model	Objective (\$)	All-hour scheduled reserve from all PSHs	
		Regulation reserve (%)	Contingency reserve (%)
B-Model	3,181,637	19.06	6.19
R-Model	3,183,282	11.19	3.65
RC-Model	3,185,336	4.67	3.50

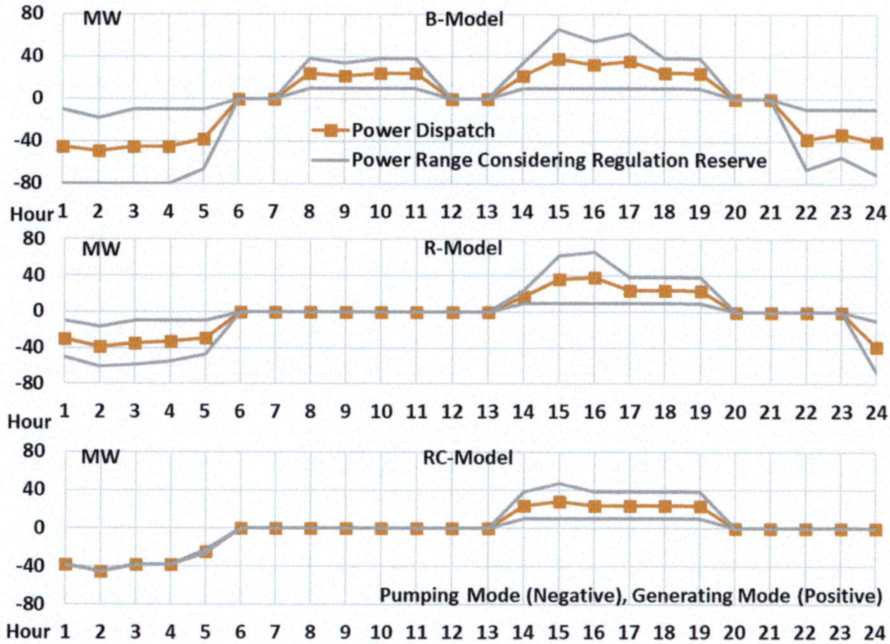

Fig. 20 Scheduling results of PSH #1 with the three models

regulation reserve it assigns to PSHs, which ensures compliance with contingency reserve requirements without overly compromising SOC boundaries.

Figure 20 illustrates the power dispatches and regulation reserve schedules for PSH #1 across the three models. The power dispatch pattern for PSH #1 follows the typical operational cycle: pumping during the early morning and late night hours, and generating during peak load periods. However, notable differences emerge in the regulation reserve schedules across the models.

The B-Model schedules the highest amount of regulation reserve for PSH #1, particularly during the pumping period in the early hours (e.g., hours 1–5). In comparison, the R-Model schedules less regulation reserve during these hours, reflecting its consideration of deployment probabilities rather than assuming worst-case scenarios. Meanwhile, the RC-Model restricts regulation reserve scheduling to generating hours only, which is because constraints (121), (122) in the RC-Model apply a cumulative tightening effect on SOC bounds; scheduling reserve in early hours reduces the flexibility for reserve scheduling in subsequent hours. In contrast, constraint (129) in the R-Model mitigates this effect by accounting for the likelihood of reserve deployment, making it less restrictive.

It is worth noting that the conservativeness of constraints (121), (122) in the RC-Model can be adjusted by scaling the second and third terms with tuning coefficients. However, these coefficients must be calibrated carefully to avoid unintended impacts on system operations. Additionally, both the R-Model and RC-Model result in PSH #1 being less active compared to the B-Model, spending more time in idle mode.

While the reserve schedule conservativeness analysis demonstrates that the B-Model achieves better economic performance, this section highlights that these economic gains come at the expense of compromised reserve deployment security. By prioritizing aggressive reserve scheduling, the B-Model exposes PSHs to higher risks of SOC violations and operational instability, undermining the reliability of reserve deployment in real-time operations.

The regulation reserve deployment problem serves as a critical tool for evaluating the reserve deployment security of the scheduling models. To conduct this analysis, a new set of reserve deployment scenarios is generated for the scheduling day. For each scenario, the regulation reserve deployment problem is solved using the commitment and dispatch schedules derived from the studied models.

By solving for $r_{h,t}^{D}$ and accounting for the known generating or pumping statuses of the PSHs, the actual SOC of each PSH is determined using Eq. (145). At every hour, the calculated SOC $\widehat{E}_{h,t}$ is assessed against the predefined SOC limits E_h^{LB} and E_h^{UB}. Any instance where the SOC of a PSH breaches these limits is flagged as a violation, indicating that the schedule fails to ensure security for that specific scenario and thereby the ability of the scheduling models to maintain SOC boundaries under various reserve deployment conditions is tested, offering a comprehensive evaluation of their effectiveness in handling operational uncertainties while safeguarding system reliability:

$$E_h^{CAP} \cdot \hat{soc}_{h,t} = E_h^{CAP} \cdot \hat{soc}_{h,t-1} - 1/\eta_h \cdot (\tilde{p}_{h,t}^{G} + \tilde{r}_{h,t}^{D} \cdot \tilde{I}_{h,t}^{G} \cdot \tau) + \eta_h \cdot (\tilde{p}_{h,t}^{P} - \tilde{r}_{h,t}^{D} \cdot \tilde{I}_{h,t}^{P} \cdot \tau); \forall h \in \mathcal{H}, \forall t \in \mathcal{T}. \quad (145)$$

Table 8 summarizes the test results of the three models across 500 scenarios. The B-Model fails to ensure reserve deployment security in all scenarios, highlighting its vulnerability under uncertain conditions. In contrast, both the R-Model and RC-Model maintain reserve deployment security across all realizations, demonstrating their robustness.

Figure 21 illustrates the scheduled SOC for PSH #1 under the three models. The B-Model operates at the SOC boundaries, hitting the upper limit of 100% at hour 5 and the lower limit of 25% at hour 19. This leaves no flexibility to handle adverse SOC deviations accumulated earlier, increasing the risk of operational failures. The R-Model mitigates this risk by reducing the maximum SOC to 86.61% and raising the minimum SOC to 41.42%, creating sufficient buffer room to address uncertainties

Table 8 Security test result of reserve scheduling

Model	Insecure scenarios	Violation type
B-Model	466	SOC lower and upper bounds
R-Model	None	–
RC-Model	None	–

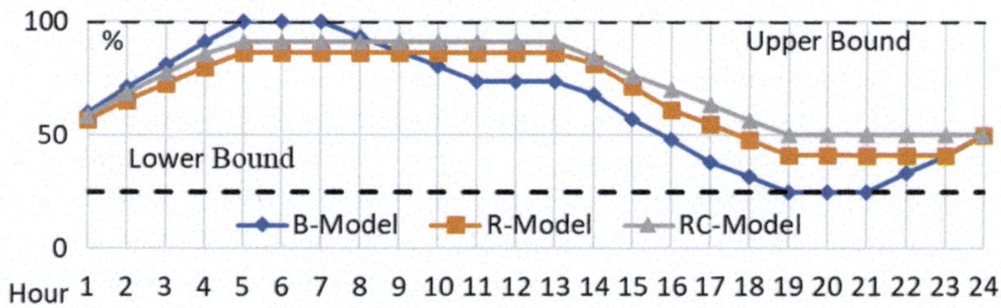

Fig. 21 SOC schedules of PSH #1 with the three models

in reserve deployment. The RC-Model further expands these buffers, providing even greater security.

During hours 1–6, the SOC of the RC-Model is higher than that of the R-Model, as no reserve is scheduled in this period, eliminating the need for additional SOC room. Figure 21 also reveals that PSH #1 remains in idle mode for longer durations in the R-Model and RC-Model compared to the B-Model. This behavior results from the reserve secure constraints tightening the SOC boundaries, effectively reducing the total energy that can be cycled. As a result, PSH #1 reaches the reduced SOC bounds more quickly in both generating and pumping modes, after which it enters idle mode and waits for the next mode transition in subsequent hours.

Although the reserve secure constraint (129) is designed as an individual chance constraint for each PSH at each hour, rather than a more robust joint chance constraint spanning multiple PSHs and/or hours, the R-Model still demonstrates strong security performance, which indicates that the individual chance constraint formulation effectively balances computational feasibility with reserve deployment reliability, ensuring secure operations without the added complexity of joint chance constraints.

To rigorously assess the reserve deployment security of the R-Model under extreme conditions and to explore the influence of the confidence level ϵ, 500 intentionally designed reserve deployment scenarios are generated. These scenarios represent consistent upward load fluctuations across all 24 h, mimicking a situation where the load forecast systematically underestimates the actual demand. This forces the deployment of regulation up reserves, making SOC violations at the lower bounds the critical concern. By varying ϵ, the results are analyzed and compared in Table 9, providing insights into how different confidence levels affect the R-Model's ability to handle worst-case scenarios, secure reserve deployment, and maintain SOC boundaries under such adverse conditions.

Table 9 Reserve security test result with biased scenarios

ϵ setting of R-model	Objective ($)	Insecure scenarios
3%	3,183,562	None
5%	3,183,282	2
8%	3,183,134	8
10%	3,183,106	14

Table 9 reveals a clear trade-off between economic efficiency and reserve deployment security as the confidence level ϵ increases from 3 to 10%. As ϵ rises, the objective value decreases, indicating improved economic efficiency, but the reserve deployment security worsens. Notably, the decrease in the objective value is uneven across the range; for instance, the reduction from ϵ=3% to ϵ=5% is $280.14, whereas the reduction from ϵ=8% to ϵ=10% is merely $27.43. Intuitively, the uneven trend highlights the importance of choosing an appropriate ϵ to strike a balance between economic gains and security. Given that the occurrence probability of such extreme upward load fluctuation scenarios is relatively low, setting ϵ to a very small value could unnecessarily compromise economic efficiency. Therefore, it is advisable to lean toward settings that prioritize economic efficiency while maintaining an acceptable level of security.

References

1. G.L.M. Fredo, E.C. Finardi, V.L. de Matos, Assessing solution quality and computational performance in the long-term generation scheduling problem considering different hydro production function approaches. Renew. Energy **131**, 45–54 (2019)
2. A.L. Diniz, M.E.P. Maceira, A four-dimensional model of hydro generation for the short-term hydrothermal dispatch problem considering head and spillage effects. IEEE Trans. Power Syst. **23**(3), 1298–1308 (2008)
3. Z. Zhao, C. Cheng, S. Liao, Y. Li, Q. Lü, A MILP based framework for the hydro unit commitment considering irregular forbidden zone related constraints. IEEE Trans. Power Syst. **36**(3), 1819–1832 (2021)
4. A. Arce, T. Ohishi, S. Soares, Optimal dispatch of generating units of the Itaipu hydroelectric plant. IEEE Trans. Power Syst. **17**(1), 154–158 (2002)
5. T. Kim, J. Shin, H. Kim et al., Ensemble-based neural network modeling for hydrologic forecasts: addressing uncertainty in the model structure and input variable selection. Water Resour Research **56**(6), 1–19 (2020)
6. H. Apaydin, H. Feizi, M.T. Sattari et al., Comparative analysis of recurrent neural network architectures for reservoir inflow forecasting. Water **12**(5), 1–18 (2020)
7. D. Liu, W. Jiang, L. Mu et al., Streamflow prediction using deep learning neural network: case study of Yangtze river. IEEE Access **8**, 90069–90086 (2020)
8. S.P. Vasseur, J.L. Aznarte, Comparing quantile regression methods for probabilistic forecasting of NO2 pollution levels. Sci. Rep. **1**, 1–8 (2021)
9. P. Lauret, M. David, H.T.C. Pedro, Probabilistic solar forecasting using quantile regression models. Energies **10**(10), 1–17 (2017)
10. S. Camal, F. Teng, A. Michiorri et al., Scenario generation of aggregated wind, photovoltaics and small hydro production for power systems applications. Appl. Energy **242**, 1396–1406 (2019)
11. W. Yuan, S. Zhang, C. Su et al., Optimal scheduling of cascade hydropower plants in a portfolio electricity market considering the dynamic water delay. Energy **252**, 124025 (2022)
12. J.P.S. Catalao, S.J.P.S. Mariano, V.M.F. Mendes et al., Scheduling of head-sensitive cascaded hydro systems: a nonlinear approach. IEEE Trans. Power Syst. **24**(1), 337–346 (2009)
13. X. Lei, Q. Tan, X. Wang et al., Stochastic optimal operation of reservoirs based on copula functions. J. Hydrol. **557**, 265–275 (2018)
14. Q. Tan, X. Wen, G. Fang et al., Long-term optimal operation of cascade hydropower stations based on the utility function of the carryover potential energy. J. Hydrol. **580**, 124359 (2020)

15. Z. Ding, X. Wen, Q. Tan et al., A forecast-driven Fecision-making model for long-term operation of a hydro-wind-photovoltaic hybrid system. Appl. Energy **291**, 116820 (2021)
16. L. Wu, M. Shahidehpour, T. Li, Stochastic security-constrained unit commitment. IEEE Trans. Power Syst. **22**(2), 800–811 (2007)
17. H. Wu, M. Shahidehpour, Stochastic SCUC solution with variable wind energy using constrained ordinal optimization. IEEE Trans. Sustain. Energy **5**(2), 379–388 (2014)
18. W.V. Ackooij, R. Henrion, A. Möller et al., On probabilistic constraints induced by rectangular sets and multivariate normal distributions. Math. Methods Oper. Res. **71**, 535–549 (2010)

Economic Dispatch of Wind and Thermal Bundled Power Systems

Chao Lei and Qianggang Wang

Abstract The chapter introduces a look-ahead rolling economic dispatch approach of wind and thermal bundled power systems. The chapter focuses on technical factors such as the variable ramp rate of retrofitted coal-fired units and flexible voltage-constrained load transfer strategy via high-voltage distribution networks. The chapter presents second-order cone ramping constraints for retrofitted coal-fired units, which can be validated with acceptable inner-approximated error, and employs for multiple wind and thermal bundled power system agents to accommodate wind power fluctuations for minimizing the production, purchasing, and switch-over operation costs. Moreover, the chapter introduces the multi-cut generalized Benders decomposition as a decentralized method to solve the sub-problems in parallel, which facilitates solving a large-scale rolling economic dispatch model. Through numerical case studies, the chapter shows the computing performance is satisfactory and has high efficiency.

Keywords Look-ahead rolling economic dispatch · Wind-thermal-bundled power system · Distribution-level topology optimization

1 Introduction

Growing penetration of renewable energy in power generation areas provides a green solution to the decarbonization of power systems [1–4]. To accommodate more renewable energy integration, the wind-thermal-bundled power system (WTBPS) is a suitable option to increase power system flexibility [5]. To explore an efficient economic dispatch of WTBPS, distribution-level topology optimization [6] and energy storage [7] can be used to increase the flexibility of WTBPS in which this generation system instantly accommodates the rapid growth of wind farms.

C. Lei (✉)
Department of Electrical and Electronic Engineering, The Hong Kong Polytechnic University, Hong Kong, China
e-mail: 21118924r@connect.polyu.hk

Q. Wang
School of Electrical Engineering, Chongqing University, Chongqing, China

F. Shahnia and F. Wen (eds.), *Power and Energy Resources*, Power Systems,
https://doi.org/10.1007/978-981-96-2612-0_3

The transition to multi-energy generation systems is crucial for improving power system reliability and efficiency, particularly through the use of Wind-Turbine-Based Power Systems (WTBPS) to manage renewable resource variability. Current research has explored economic dispatch in hybrid AC/DC systems [8], integrated renewable sources with energy storage [9], multi-fuel strategies [10], and integrated electricity and natural gas system [11], while also incorporating demand-side resources like energy hubs [12], electric vehicles [13], and controllable loads [14]. However, many studies overlook the potential of retrofitting flexibility measures on coal-fired power plants and the integration of High-Voltage Distribution Networks (HVDNs). Addressing these gaps could lead to more robust and flexible economic dispatch solutions, enhancing the overall sustainability of energy systems.

On the one hand, retrofitting coal-fired plants is gaining global attention, especially in middle-income countries [15] where these facilities continue to operate. Without retrofits, coal plants typically run at a minimum load of 50% and have limited ramp rates of 0.6–2% per minute of rated power [16]. However, retrofitted plants can significantly enhance their flexibility [13, 17], reducing minimum load levels to 15–30% and increasing ramp rates to 2–6%. For example, countries like China, Poland, and Germany [17] have successfully achieved these improvements, with plants like Bexbach and Wes Weiler demonstrating reduced minimum loads and increased ramp rates. Reference [18] suggests that ramping limits should be dynamic, depending on the unit's generating output, and existing studies categorize these rates into piecewise linear [19] and stepwise linear functions [20]. An advanced model incorporating a status transition approach further refines this dynamic ramping [20], while retrofitted units with molten-salt thermal energy storage [13] can achieve faster ramp rates, helping to ease peak regulation challenges.

On the other hand, utilizing energy storage systems [21] and distribution-side load transfer actions [22] can serve as alternatives to address the insufficient ramping margins of coal-fired units in Wind-Turbine-Based Power Systems (WTBPS). For instance, Ref. [23] involves using adjustable heat storage strategies from solar power stations to shift excess wind power in combined wind-thermal generation systems. However, the operational costs of stored energy remain prohibitively high compared to solar resources, limiting its widespread adoption. Distribution network load resources inherently possess flexibility and hold significant potential for integration with WTBPS, thereby minimizing wind curtailment. As effective load providers, distribution-level topology actions can be performed through reconfigurable High-Voltage Distribution Networks (HVDNs) that are integrated with renewable resources under stability-constrained conditions [24]. These HVDNs, typically operating at a 110 kV sub-transmission level [22], are constructed in a meshed topology but function in radial structures, as seen in countries like China [22], Spain [24], and Finland [25]. Comprising specific topological units, HVDNs can streamline the computational complexity of distribution-level topology optimization compared to conventional network reconfiguration methods. While existing optimization models in [22] utilize simplified *DistFlow* to incorporate voltage security constraints, they can inadvertently increase computational complexity and reduce the efficiency of power flow variables. Therefore, our goal is to enhance this distribution-level topology optimization model by implementing more streamlined yet effective voltage security constraints.

This chapter introduces a distribution-level topology optimization contributing to the flexibility enhancement of a look-ahead rolling economic dispatch of WTBPS. This distribution-level topology optimization offsets the insufficient ramping margins of retrofitted coal-fired units. Since WTBPS connects to HVDNs, graph characterization of typical HVDNs is summarized, and a simplified voltage-constrained load transfer strategy can be developed using topological structures. This approach can be integrated into a look-ahead economic dispatch model, which is formulated as a mixed-integer second-order cone programming (MISOCP) problem. The model can be further optimized through multi-cut generalized Benders decomposition (MGBD), a technique that allows for the parallel solving of multiple sub-problems, thereby accelerating the overall solution process. This combination of strategies enhances the efficiency and effectiveness of managing power flows within the system while ensuring voltage security and operational flexibility.

2 Modeling of WTBPS and Associated Constraints

2.1 *Static Modeling of WTBPS*

The 110 kV dispatchable load resources from substations within HVDNs typically range from 0 to 480 MW, or even higher, representing about 5–30% of the capacity of WTBPS. This flexibility provides a superior quick-response capability compared to coal-fired units. As a result, integrating HVDNs into a wind-thermal generation system—complete with connected AC/DC tie-lines to other grids—creates a WTBPS that delivers more stable output power. This flexibility allows multiple independent agents within the WTBPS to connect with the same large-scale HVDNs, with each agent linked to tie-lines for inter-grid connections, as illustrated in Fig. 1.

For the uth agent of WTBPS, wind power $P^t_{\mathrm{W},u}$ is bundled with several retrofitted coal-fired units of total output power $P^t_{\mathrm{G},u} = \sum_{i=1}^{N_G} r^t_{i,u} P_{Gi,u,\max}$, where $r^t_{i,u} = P^t_{Gi,u}/P_{Gi,u,\max}$ and N_{G} is the total number of retrofitted coal-fired units in the uth agent of WTBPS. Meanwhile, $P^t_{\mathrm{G},u}$ can simultaneously coordinate with dispatchable loads $P^t_{D,u}$ with reconfigurable HVDNs. For every $t \in T_s$, constraint (1) can be cast at bus i_h subject to transmission power $P_{T,u} \in [(1-\delta\%)P_{\mathrm{base}}, (1+\delta\%)P_{\mathrm{base}}]$ on tie-lines:

$$P^t_{\mathrm{W},u} + P^t_{\mathrm{G},u} - P^t_{D,u} = P^t_{\mathrm{T},u},\ (1-\delta\%)P_{\mathrm{base}} \le P^t_{\mathrm{T},u} \le (1+\delta\%)P_{\mathrm{base}} \tag{1}$$

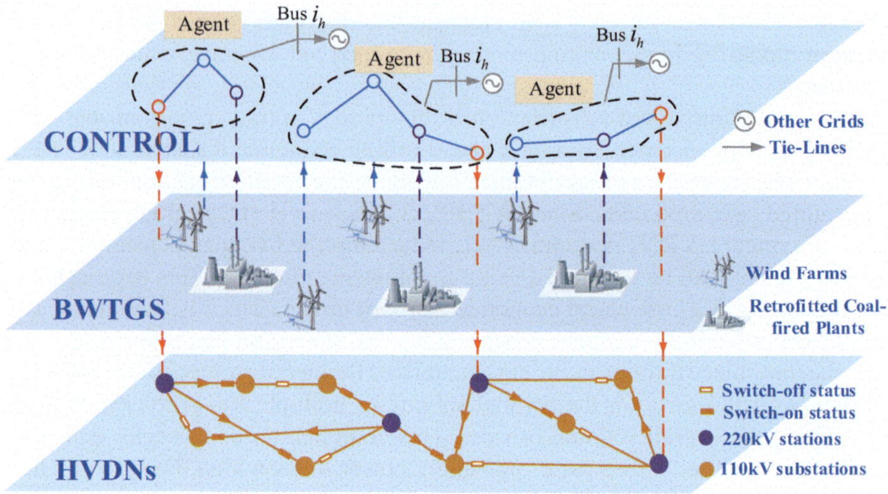

Fig. 1 An illustration of WTBPS, comprising three distinct layers: the HVDNs layer at the bottom, the wind-thermal-bundled layer in the middle, and the control layer at the top. In the HVDNs layer, load flows are represented by arrows, indicating the direction of energy flow. Vertical dashed lines connect the nodes in the HVDN layer to nodes in the wind-thermal-bundled layer, and further to nodes in the control layer. These connections aggregate at bus i_h, which serves as a critical point for different agents. Additionally, bus i_h for each agent acts as the starting node for tie-lines that link to other grids

2.2 Operational Constraints of Retrofitted Coal-Fired Units

Retrofitting flexibility measures involve upgrading control systems, reducing the wall thickness of key components, and using auxiliary firing with dried lignite ignition burners during booster operations, among other strategies [17]. These enhancements have been increasingly implemented in coal-fired plants in recent years. Table 1 highlights the key differences between conventional coal-fired plants without retrofits and their retrofitted counterparts. Notably, the ramp rate is calculated by dividing the ramping margin per minute by the corresponding rated capacity. In this context, we specify that ramp rates operate on a minute level, while operational ramping margins—whether untightened or tightened—are assessed over a dispatch period ΔT.

As shown in Table 1, conventional coal-fired plants are inherently less flexible than retrofitted plants in terms of minimum load levels, ramp rate utilization, and hot and cold start-up times. Specifically, retrofitted coal-fired plants can reduce their minimum load levels to 15–35% of rated power and enhance ramp rates to 2–6% per minute of rated power. These improvements contribute to a more flexible rolling dispatch of Wind-Turbine-Based Power Systems (WTBPS), with ramping limits being modeled as a linear function of the unit's generating output [18].

Table 1 Differences between conventional and retrofitted coal-fired plants

Items	Conventional power plants (without retrofits)	Retrofitted coal-fired power plants
Minimum load level (%)	50–60	15–35
Ramp rate (min/%)	0.6–2	2–6
Hot start-up time (h)	3–5	1.5–4
Cold start-up time (h)	5–8	5–6

Measuring ramping limits in relation to each output power point poses significant practical challenges. To address this, we utilize the upper envelope approach to approximate the ramping limits per minute, allowing us to capture the dynamic ramp rates across various output levels. Figure 2 illustrates the historical ramp rates (%/min) of a retrofitted coal-fired plant with a rated capacity of 600 MW located in northern China. On the x-axis, we represent the percentage of the output power of this coal-fired unit, while the y-axis indicates the operational ramp rates observed between two consecutive periods (%/min). Prior to retrofitting, the ramp rate of this 600 MW coal-fired plant was designed at 1.1%/min, with a minimum load level set at 60% of its rated capacity.

Figure 2 presents 2,500 samples of operational ramp rates corresponding to different output power points, where the minimum load level has been reduced to 30% and the ramp rate improved to 2%. The red line segment in the figure represents the upper envelope of these dynamic ramp rates as a linear function. It is observed that the operational ramp rate remains fixed at 2% when the output power exceeds 60%. For output power levels between 30 and 60%, the ramping limit can be approximated by a linear function of the unit's generating output. Therefore, the dynamic ramp rates $v_{i,u}^{t}$ (%/min) at time t can be reasonably treated as a linear function of output generation power, which yields

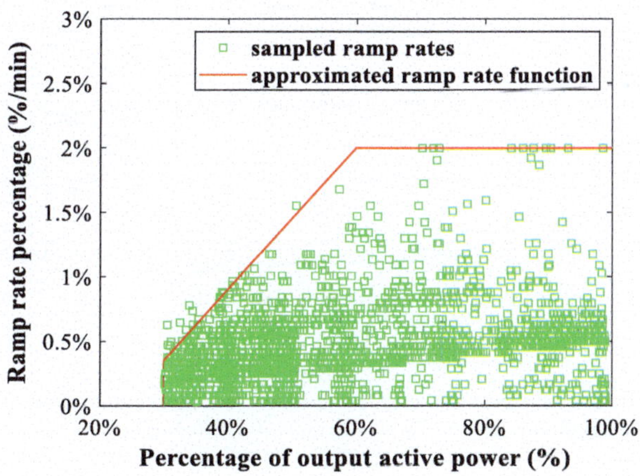

Fig. 2 Historical ramp rates for the retrofitted coal-fired unit

$$v_{i,u}^t = \begin{cases} a \cdot r_{i,u}^t - b, & r_{i,u}^t \le d \\ c, & r_{i,u}^t > d \end{cases} \tag{2}$$

where parameter c is determined by the enhanced ramp rate of retrofitted coal-fired units, while parameters a, b, and d are estimated from historical ramping data, minimum load level, and $d = (c + b)/a$; and $r_{i,u}^t$ refers to the percentage of $P_{Gi,u}^t / P_{Gi,u,\max}$.

Three types of models can be utilized to describe dynamic ramp rates: the approximated linear model (as shown in Eq. (2)), the piecewise linear model based on specified thresholds [18], and the stepwise linear model also defined by fixed thresholds [18]. The primary distinction among these models is that the first model provides a continuous function, while the latter two represent discrete functions with specific thresholds. Given that retrofitted coal-fired units are capable of delivering more stable and precise power outputs within minutes, the linear model of dynamic ramp rates effectively captures the available operational ramp rates. In contrast, the piecewise and stepwise linear models lack the fast-tracking capability due to their reliance on fixed thresholds. For example, to illustrate the differences between the proposed linear model and the piecewise linear model, Fig. 3a shows the ramp rate $v_{i,u}^t$ using Eq. (2) with parameters $a = 0.055$, $b = 0.013$, $c = 0.02$, and $d = 0.6$ and using the piecewise linear model [18] in the dashed green line for a retrofitted coal-fired unit, respectively. The minimum load level is set to 30% of $P_{Gi,u,\max}$. The piecewise linear model is represented by three fixed thresholds, depicted as green line segments in Fig. 3a. These segments are defined by the slopes of each segment and the interval ranges between consecutive periods. This structure allows the model to capture the ramp rates at specific output power levels. Each segment reflects a distinct operational regime, with ramp rates varying based on the defined thresholds, thus illustrating the discrete nature of this model.

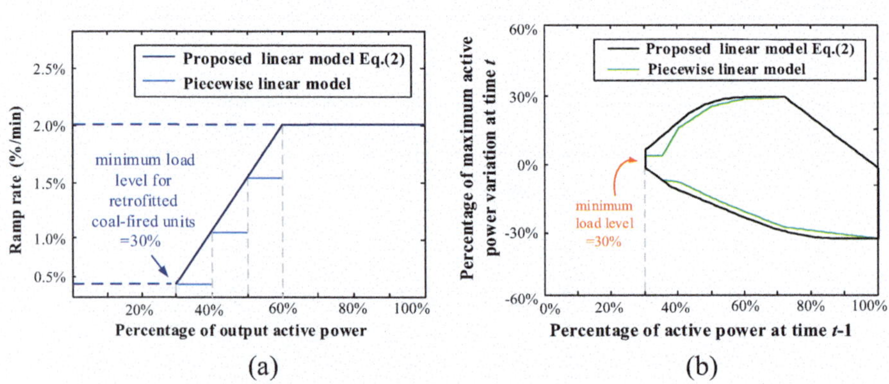

Fig. 3 **a** Proposed and piecewise linear models of dynamic ramp rates per minute, and **b** percentage of maximum active power variations for fifteen minutes using the proposed and piecewise linear models of dynamic ramp rates

Figure 3b illustrates the maximum active power variations of a retrofitted coal-fired unit over a dispatch period of $\Delta T = 15$ min, using both the proposed and piecewise linear models of dynamic ramp rates, where x-axis refers to $y_{i,u}^{t-1}$ and y-axis indicates the percentage of maximum active power variations $y_{i,u}^{t}$ at time t. And $\Delta y_{i,u}^{t}$ can be sketched out by blue curves and several green line segments, which correspond to the proposed and piecewise linear models shown in Fig. 3a, respectively. In this figure, the variations $\Delta y_{i,u}^{t}$ are depicted by blue curves and several green line segments, corresponding to the proposed and piecewise linear models shown in Fig. 3a. The maximum active power variations indicated by the blue line segments encompass an area that is larger than that enclosed by the green segments. This discrepancy is particularly pronounced when $r_{i,u}^{t} \leq 60\%$. In other words, Eq. (2) provides a more adaptable solution to the substantial fluctuations in wind power, especially during peak shaving operations. The gaps between the two areas in Fig. 3b clearly demonstrate that utilizing the piecewise linear model can lead to suboptimal dispatch solutions. Consequently, it can be concluded that the piecewise linear model is not suitable for accurately determining the ramp-up and ramp-down margins of retrofitted coal-fired units during a dispatch period of $\Delta T =$ 15 min.

Without loss of generality, $r_{i,u}^{t}$ can be analytically deduced as (3a)–(3d) for upper boundaries as ramp-up margins and (4a)–(4d) for lower boundaries as ramp-down margins under a dispatchable period ΔT in the $(r_{i,u}^{t-1}, r_{i,u}^{t})$-space. Therefore, the operational ramping constraints are expressed as

$$r_{i,u}^{t} \leq r_{i,u}^{t-1} + c \cdot (\Delta T - 1) \quad \text{if } r_{i,u}^{t-1} \geq d \tag{3a}$$

$$r_{i,u}^{t} \leq (r_{i,u}^{t-1} - b/a)(a+1)^{\Delta T - 1} + b/a \quad \text{if } r_{i,u}^{t-1} \leq \xi_{i,u} \tag{3b}$$

$$r_{i,u}^{t} \leq 1 \tag{3c}$$

$$r_{i,u}^{t} \leq (r_{i,u}^{t-1} - b/a)(a+1)^{\eta - 1} + b/a + c \cdot (\Delta T - \eta) \text{ if } \xi_{i,u} < r_{i,u}^{t-1} < d \tag{3d}$$

$$r_{i,u}^{t} \geq r_{i,u}^{t-1} - c \cdot (\Delta T - 1) \quad \text{if } r_{i,u}^{t-1} \geq \chi_{i,u} \tag{4a}$$

$$r_{i,u}^{t} \geq (r_{i,u}^{t-1} - b/a)(1-a)^{\Delta T - 1} + b/a \quad \text{if } r_{i,u}^{t-1} \leq d \tag{4b}$$

$$r_{i,u}^{t} \geq 0.3 \tag{4c}$$

$$r_{i,u}^{t} \geq \left(r_{i,u}^{t-1} - c \cdot (\omega_{i,u} - 1) - b/a\right)(1-a)^{\Delta T - \omega_{i,u}} + b/a \text{ if } d < r_{i,u}^{t-1} < \chi_{i,u} \tag{4d}$$

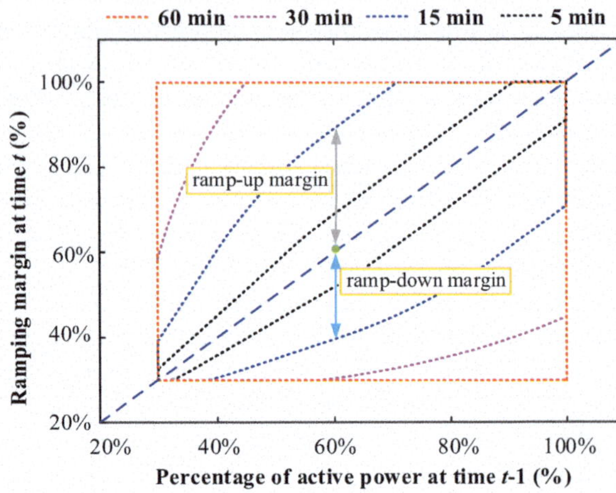

Fig. 4 Two-dimensional ramping boundaries under different dispatch periods

where $\xi_{i,u} = \frac{d-b/a}{(a+1)^{\Delta T-1}} + \frac{b}{a}$, $\chi_{i,u} = d + c(\Delta T - 1)$, and $d = (c+b)/a$, $\eta_{i,u} = round\left(\ln\left(c/\left(a \cdot r_{i,u}^{t-1} - b\right)\right)\ln(a+1) + 1\right)$, and $\omega_{i,u} = \text{round}\left(\left(r_{i,u}^{t-1} - d\right)/c + 1\right)$, in which $round(\cdot)$ refers to a function of round toward negative infinity.

An intuitive explanation of ramping margins is illustrated in Fig. 4, which examines different dispatch periods ΔT scaled at 60, 30, 15, and 5 min. The figure demonstrates that longer dispatch periods result in larger ramp-up and ramp-down margins compared to shorter ones. For instance, when $\Delta T = 15$ min and $r_{i,u}^{t-1} = 60\%$, the ramp-up band spans from 60 to 88%, while the ramp-down range extends from 60 to 37%. In contrast, with $\Delta T = 60$ min, the ramp-up and ramp-down bands can reach 100% and 30% for any $r_{i,u}^{t-1}$, respectively. This observation reveals that conventional ramping constraints—defined as the allowable changes in generation levels between two consecutive dispatch periods—become approximate when $\Delta T <$ 60 min. Conversely, these constraints become redundant when $\Delta T = 60$ min, highlighting the need for adaptive strategies in managing ramping margins based on the duration of the dispatch period.

Since some retrofitted coal-fired plants exhibit ramp rates exceeding 2%, we illustrate the ramp-up and ramp-down boundaries under $\Delta T = 15$ min using fixed parameters $a = 0.055$ and $d = 0.6$, while varying parameters $b = a{\cdot}d - c$ in Fig. 5. The results shown in Fig. 5 indicate that a larger value of parameter c leads to increased ramp-up and ramp-down margins under the $\Delta T = 15$ min setting. This enhancement in margins reflects the improved quick-response capability of retrofitted coal-fired plants, allowing them to better adapt to fluctuations in demand and integrate more effectively with variable renewable energy sources.

Without loss of generality, we illustrate the boundaries represented by Eqs. (3a)–(4d) in Fig. 6. In this figure, the line segments $\overline{CD}$, $\overline{AB}$, and $\overline{DE}$ refer to three linear constraints (3a)–(3c) $\overline{FG}$, $\overline{HI}$, and $\overline{IJ}$ refer to another three linear constraints (4a)–(4c), and two green curves $\overset{\frown}{BC}$ and $\overset{\frown}{HG}$ indicate nonlinear constraints indicated in (3d) and (4d).

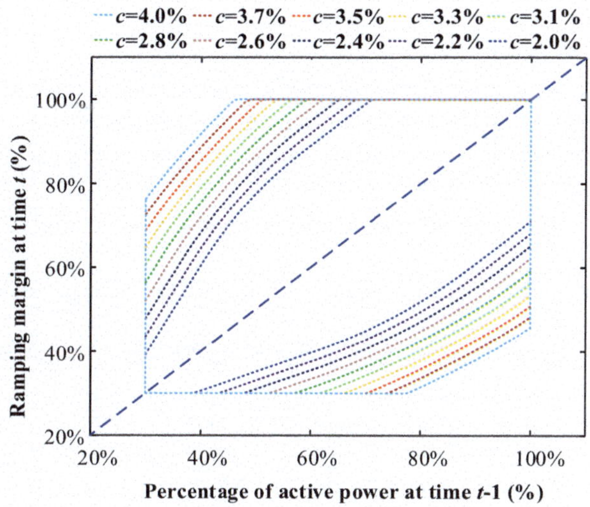

Fig. 5 Ramping boundaries under $\Delta T = 15$ min with different sets of parameters c and b

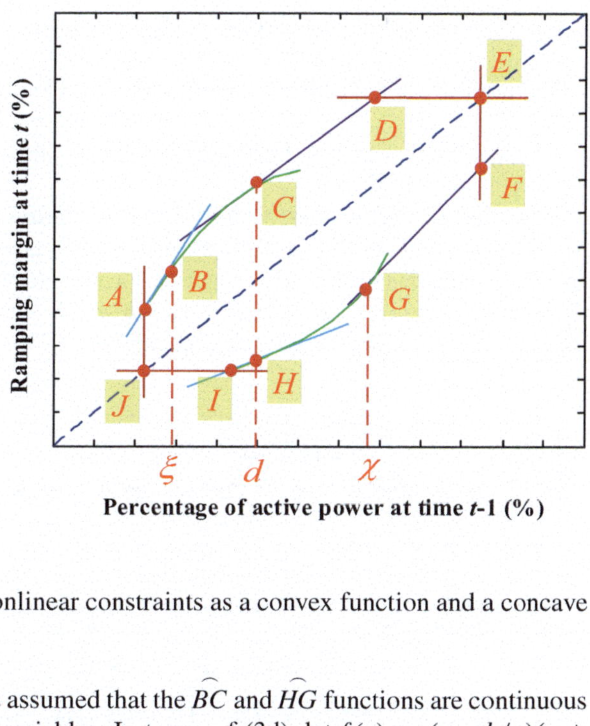

Fig. 6 General ramping boundaries of retrofitted coal-fired units

We can show these two nonlinear constraints as a convex function and a concave function, respectively.

Proof For (3d) and (4d), it is assumed that the $\overset{\frown}{BC}$ and $\overset{\frown}{HG}$ functions are continuous and η *and* ω are continuous variables. In terms of (3d), let $f(x) = (x - b/a)(a + 1)^{\eta - 1} + b/a + c \cdot (\Delta T - \eta)$ where $x = r_{i,u}^{t-1}$; we can deduce the following equations:

$$\begin{cases} f' = (a+1)^{\eta-1} + (x - b/a)\left[(a+1)^{\eta-1}\right]' - c\cdot\eta' \\ \left[(a+1)^{\eta-1}\right]' = (a+1)^{\eta-1}\ln(a+1)\cdot\eta' \\ \left[(a+1)^{\eta-1}\right]'' = (a+1)^{\eta-1}\ln(a+1)(\ln(a+1)\cdot(\eta')^2 + \eta'') \\ \eta' = -a/[(ax-b)\ln(a+1)], \quad \eta'' = a^2/\left[(ax-b)^2\ln(a+1)\right] \end{cases} \tag{5}$$

Based on the above, the resultant second derivative of $f(x)$ can be given by

$$f'' = 2\left[(a+1)^{\eta-1}\right]' + (x - b/a)\left[(a+1)^{\eta-1}\right]'' - c\cdot\eta'' \tag{6}$$

Notably, $\eta' < 0, \eta'' > 0$ hold for $\forall x \in (b/a, d)$. With (5), we obtain

$$\left[(a+1)^{\eta-1}\right]' < 0,\ \left[(a+1)^{\eta-1}\right]'' > 0 \tag{7}$$

From Eq. (6), we calculate $\Delta h = \left[(a+1)^{\eta-1}\right]' + \varepsilon\cdot\left[(a+1)^{\eta-1}\right]''$, where $\varepsilon = x - b/a > 0$:

$$\Delta h = (a+1)^{\eta-1}\ln(a+1)\{\eta' - \varepsilon\ln(a+1)(\eta')^2 - \varepsilon\eta'')\} \tag{8}$$

We substitute $\eta'' = c\ln(a+1)(\eta')^2$ from Eq. (5) into Eq. (8), and obtain

$$\Delta h = \eta'(a+1)^{\eta-1}\ln(a+1)\{1 - \varepsilon\ln(a+1)\eta'(1-c)\} \tag{9}$$

It is clear that $\Delta h < 0$ due to $\eta' < 0$, $\{1 - \varepsilon\ln(a+1)\eta'(1-c)\} > 0$. Consider $\Delta h < 0$ and $\left[(a+1)^{\eta-1}\right]' < 0$, thus causing $f'' = \Delta h + \left[(a+1)^{\eta-1}\right]' - c\cdot\eta'' < 0$. This shows that $f(x)$ is a convex function for $\forall x \in (b/a, d)$.

For (4d), let $g(x) = (x - c\cdot(\omega - 1) - b/a)(1-a)^{\Delta T - \omega} + b/a$ where $x = y_{i,u}^{t-1}$, and we similarly yield the following equations:

$$\begin{cases} g' = (1-a)^{\Delta T-\omega}(1 - c\cdot\omega') + (x - c\cdot(\omega-1) - b/a)\left[(1-a)^{\Delta T-\omega}\right]' \\ \left[(1-a)^{\Delta T-\omega}\right]' = -(1-a)^{\Delta T-\omega}\ln(1-a)\cdot\omega' \\ \left[(1-a)^{\Delta T-\omega}\right]'' = -(1-a)^{\Delta T-\omega}\ln(1-a)\cdot(\omega'' - \ln(1-a)\cdot(\omega')^2) \\ \omega' = 1/c, \quad \omega'' = 0 \end{cases} \tag{10}$$

Based on the above, we further achieve the second derivative of $g(x)$

$$\begin{aligned} g'' = &-\omega'(1-a)^{\Delta T-\omega}\ln(1-a)(1 - c\cdot\omega') \\ &+ (x - c\cdot(\omega-1) - b/a)\left[(1-a)^{\Delta T-\omega}\right]'' \\ &+ \left[(1-a)^{\Delta T-\omega}\right]' \end{aligned} \tag{11}$$

Notably, $\omega' > 0$, $\omega'' = 0$ hold for $\forall x \in (d, d + c \cdot (\Delta T - 1))$. Regarding $c \cdot (\omega - 1) = x - d, d = (b + c)/a$, and $\ln(1\text{–}a) < 0$, we obtain $g'' > 0 > 0$. This shows that $g(x)$ is a concave function for $\forall x \in (d, d + c \cdot (\Delta T - 1))$.

Moreover, the spinning reserve requirement of WTBPS must be taken into account prior to this economic dispatch model. The spinning reserve capacity supplied by retrofitted coal-fired units should not exceed the overall system spinning reserve capacity. Properly managing these reserves is crucial for effectively integrating renewable energy sources and ensuring that the power system can respond promptly to unexpected events:

$$\sum_{i=1}^{N_G \sum_u^{\text{spin}}} (1 - r_{i,u}^t) P_{Gi,u,\max} \tag{12}$$

where P_u^{spin} refers to system spinning reserve capacity for the uth agent of WTBPS and $u = 1, 2, \ldots, N_u$. Considering wind turbines have no contribution to the spinning reserve, P_u^{spin} is assumed to be 10% of the transmission power minus the wind power rated capacity.

2.3 *Distribution-Level Topology Optimization for HVDNs*

The load switch-over operations implemented on HVDNs can effectively mitigate wind curtailment and reduce generation costs for WTBPS during a look-ahead rolling economic dispatch task. Two typical grid-scale configurations for 110 kV HVDNs are the Direct Supply Connection (DSC) and the Serial Supply Connection (SSC), as shown in the following Fig. 7, where P_c^t and P_d^t are active loads at 110 kV substations C and D, respectively; P_{A1}^t and $P_{"A2"}^t$ refer to the total active dispatchable loads in 220 kV stations A1 and A2, respectively.

Thanks to the straightforward topological units of HVDNs, which typically consist of just one or two buses, the well-known simplified *DistFlow* equations can achieve acceptable accuracy with negligible power loss. As reported in Ref. [26], the nonlinear terms in these equations are minor compared to the linear terms, usually introducing an error of about 1%. Additionally, distribution systems tend to exhibit relatively flat voltage profiles, aided by voltage automatic control devices, resulting in a small approximation error of approximately ±1%. In light of this, we propose simplified power balance equalities represented as linear matrix equalities [22]. This transformation significantly reduces the complexity associated with HVDN reconfiguration, which is often treated as a non-convex and nonlinear problem [27]. By leveraging these simplified equations, we can enhance the efficiency of network optimization processes while maintaining accurate representations of power flows.

As indicated in Fig. 8, the general forms of Eqs. (13a) and (13b) can be reformulated as $\boldsymbol{P}_{S,j}^t = \boldsymbol{L}_j^t \boldsymbol{S}^t + \boldsymbol{b}_j^t$, $\forall j \in N_T, \forall t \in T_s$, where $\boldsymbol{P}_{S,j}^t$ denotes total active power

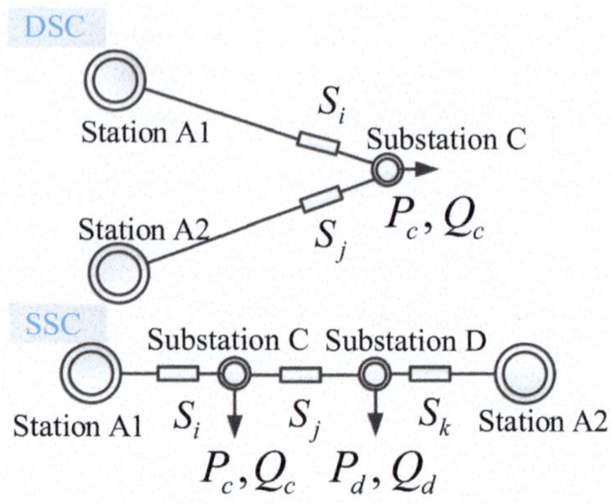

Fig. 7 Two typical HVDNs of DSC and SSC

on transformers in the jth station, and $\boldsymbol{L}_j^t$ and $\boldsymbol{b}_j^t$ refer to vectors of load coefficients at time t whose jth element indicates the j-th station j. Specifically, when the connected bus for the uth agent of the WTBPS is associated with the 220 kV station, then $\boldsymbol{P}_{D,u}^t = \boldsymbol{P}_{S,j}^t$ holds at time t; for other stations, $\boldsymbol{P}_{S,j}^t$ should satisfy (13a). Additionally, branch capacity constraints (14b) for cables and overhead lines within the HVDNs are also incorporated into the least trimmed squares (LTS) model, ensuring that the operational limits of the network are adhered to while optimizing power flows:

$$\begin{bmatrix} P_{A1}^t \\ P_{A2}^t \end{bmatrix} = \mathbf{L}^t \begin{bmatrix} S_1^t \\ S_2^t \end{bmatrix} + \mathbf{b}^t = \begin{bmatrix} P_c^t & 0 \\ 0 & P_c^t \end{bmatrix} \begin{bmatrix} S_1^t \\ S_2^t \end{bmatrix} + \mathbf{0} \tag{13a}$$

$$\begin{bmatrix} P_{A1}^t \\ P_{A2}^t \end{bmatrix} = \boldsymbol{L}^t \begin{bmatrix} S_3^t \\ S_4^t \\ S_5^t \end{bmatrix} + \boldsymbol{b}^t = \begin{bmatrix} P_{\mathrm{c}}^t + P_{\mathrm{d}}^t & P_{\mathrm{d}}^t & 0 \\ 0 & P_{\mathrm{c}}^t & P_{\mathrm{c}}^t + P_{\mathrm{d}}^t \end{bmatrix} \begin{bmatrix} S_3^t \\ S_4^t \\ S_5^t \end{bmatrix} + \begin{bmatrix} -P_{\mathrm{d}}^t \\ -P_{\mathrm{c}}^t \end{bmatrix} \tag{13b}$$

$$0 \le P_{S,j}^t \le \left(n_{s,j} \cdot S_{N,j}\right), \ \forall j \in N_T \tag{14a}$$

$$0 \le P_{S,j}^t \le S_{L,j}, \ \forall j \in N_T \tag{14b}$$

where $n_{s,j}$ refers to the number of transformers in the jth station; $S_{N,j}$ denotes the single transformer rated capacity in the jth station; $S_{L,j}$ indicates the branch capacity whose branch connects to the jth station.

Moreover, the approximate voltage drop for a branch section $\{m, k\}$ can be summarized based on simplified *DistFlow* equations. According to $v_m^2 - v_k^2 = 2\left(R_{mk}P_{mk}^t + X_{mk}Q_{mk}^t\right)$ in simplified *DistFlow* equations, we rewrite this equation:

$$\begin{aligned} R_{mk}P_{mk}^t + X_{mk}Q_{mk}^t &= \left(v_m^2 - v_n^2\right)/2 = \frac{(v_m - v_n)^2}{2} + v_n(v_m - v_n) \\ &= \frac{\left(\Delta v_{mk}^t\right)^2}{2} + v_n \Delta v_{mk}^t \le \frac{(\Delta \overline{v})^2}{2} + \overline{v} \cdot \Delta \overline{v} \end{aligned} \tag{15}$$

Fig. 8 **a** Ramping boundaries of a retrofitted coal-fired unit; **b** quadratic functions for $f(x)$ and $g(x)$

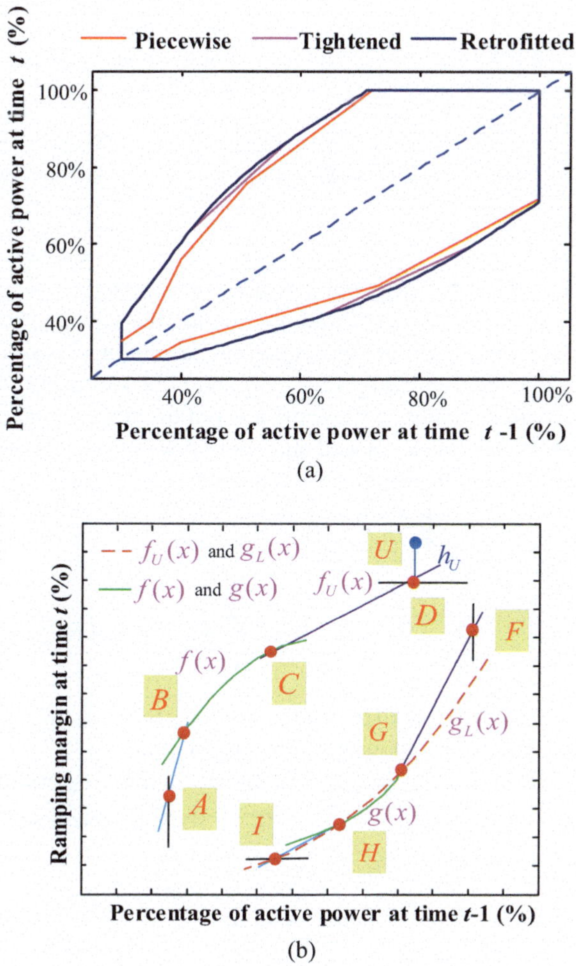

where R_{mk} and X_{mk} refer to resistance, reactance of branch (m, k), and P^t_{mk} and Q^t_{mk} denote active and reactive power flow on branch (m, k), $\overline{v}$ is the maximum voltage profile, and $\Delta\overline{v}$ refers to the allowable voltage drop, e.g. $\Delta\overline{v} = 0.1$ p.u. for DNs [22].

Let allowable squared voltage drop be $\Delta\overline{V} = \frac{(\Delta\overline{v})^2}{2} + \overline{v} \cdot \Delta\overline{v}$, and since DSC has a fixed directional power flow, we can construct linear voltage security constraints with the allowable ranges of voltage profiles:

$$R_{\text{A1-C}}P^t_c + X_{\text{A1-C}}Q^t_c R_{\text{A2-C}}P^t_c + X_{\text{A2-C}}Q^t_c \begin{bmatrix} S^t_1 \\ S^t_2 \end{bmatrix} \leq \Delta\bar{V} \tag{16}$$

The power flow between substations C and D in SSC depends on the circuit switching status of circuit breakers among S^t_3, S^t_4, and S^t_5. Given that this bi-directional power flow for different circuit switching variables among S_3^t, S_4^t, and S_5^t, we can establish two linear equivalent voltage security constraints:

$$\begin{bmatrix} \lambda_1 & 0 & R_{\text{C-D}}P_d^t + X_{\text{C-D}}Q_d^t \\ 0 & \lambda_2 & R_{\text{C-D}}P_c^t + X_{\text{C-D}}Q_c^t \end{bmatrix}\begin{bmatrix} S_3^t \\ S_4^t \\ S_5^t \end{bmatrix} - \begin{bmatrix} R_{\text{C-D}}P_d^t + X_{\text{C-D}}Q_d^t \\ R_{\text{C-D}}P_c^t + X_{\text{C-D}}Q_c^t \end{bmatrix} \le \begin{bmatrix} \Delta\bar{V} \\ \Delta\bar{V} \end{bmatrix} \quad (17)$$

where $\lambda_1 = R_{\text{A1-C}}P_c^t + X_{\text{A1-C}}Q_c^t + R_{\text{C-D}}P_{\text{d}}^t + X_{\text{C-D}}Q_{\text{d}}^t$, $\lambda_2 = R_{\text{A2-D}}P_{\text{d}}^t + X_{\text{A2-D}}Q_{\text{d}}^t + R_{\text{C-D}}P_{\text{c}}^t + X_{\text{C-D}}Q_{\text{c}}^t$; other symbols in (16) and (17) can be found in *Nomenclature.*

It can be observed that rearranging the voltage profile expressions derived from the simplified *DistFlow* equations results in a more compact matrix form, specifically represented by Eqs. (16) and (17). Compared to the approach in [22], these equations maintain the same level of accuracy in the optimal circuit switch scheme while utilizing a reduced number of power flow variables. Moreover, frequent circuit switching can diminish the remaining useful life of circuit breakers, making it essential to limit the switching frequency of each circuit breaker over a rolling window [28]. This constraint can be expressed as

$$\sum_{t=1}^{N_{Ts}} \left(S_j^t - S_j^{t-1}\right)^2 = \sum_{t=1}^{N_{Ts}} \left(\left(S_j^t\right)^2 + \left(S_j^{t-1}\right)^2 - 2 \cdot S_j^t S_j^{t-1}\right) \le N_w \quad (18)$$

where N_{Ts} refers to the number of time-horizons in a rolling window and N_w denotes the number of allowable switching actions.

Consider the property of binary number operations $(S_j^t)^2 = S_j^t$, and let $z_j^t = S_j^t + S_j^{t-1}$ that satisfy $z_j^t \in \{0, 1\}$. By using piecewise McCormick envelopes [29], we exactly replace $z_j^t = S_j^t + S_j^{t-1}$ with

$$z_j^t \ge S_j^t + S_j^{t-1} - 1, \quad z_j^t \le S_j^t, \quad z_j^t \le S_j^{t-1} \quad (19)$$

Due to $z_j^t = S_j^t + S_j^{t-1}$, constraint (18) can be subsequently rearranged as

$$\sum_{t=1}^{N_{Ts}} \left(S_j^t + S_j^{t-1} - 2 \cdot z_j^t\right) \le \tau \quad (20)$$

As mentioned in Sect. 1.1, dispatchable loads account for only about 5–30% of the capacity of WTBPS. This highlights the primary role of retrofitted coal-fired units in effectively tracking wind power, while load transfer operations serve a secondary function. Consequently, the substantial switching times of circuit breakers may not be feasible within the context of economic dispatch problems.

Alternatively, maintaining a radial operation requirement for HVDNs is crucial for supporting mesh topological units. As observed in Fig. 7, these radial configurations should satisfy

$$\text{DSC:}\quad S_1^t + S_2^t = 1 \qquad \text{SSC:}\quad S_3^t + S_4^t + S_5^t = 2 \tag{21}$$

Hereby, a more simplified LTS model (13)–(21) can be constructed for the following rolling economic dispatch model in Sect. 3.

3 Tightened Ramping Constraints

3.1 Linear Ramping Constraints

As shown in Fig. 7a, to simplify the convex constraints (3d) and (4d), we replace the arcs $\overset{\frown}{BC}$ and $\overset{\frown}{HG}$ with straight lines connecting points B and C, and H and G, respectively. It is clear that $\overset{\frown}{BC}$ and $\overset{\frown}{HG}$ are strictly tighter than arcs $\overset{\frown}{BC}$ and $\overset{\frown}{HG}$, resulting in a ramping area represented as a decagon with vertices labeled A, B, C, D, E, F, G, H, I, and J. We can further approximate the convex and concave constraints in Eqs. (3d) and (4d) as expressed in Eq. (22). This formulation leads to the establishment of linear ramp-up and ramp-down constraints for retrofitted coal-fired units:

$$\begin{bmatrix} 1 & -A_{\text{upper}} \\ -1 & A_{\text{lower}} \end{bmatrix} \begin{bmatrix} r_{i,u}^t \\ r_{i,u}^{t-1} \end{bmatrix} + \begin{bmatrix} -B_{\text{upper}} \\ B_{\text{lower}} \end{bmatrix} \leq 0 \tag{22}$$

where A_{upper}, B_{upper}, A_{lower}, and B_{lower} are obtained constants calculating from linear equations $\overline{BC}$ and $\overline{HG}$, given as

$$\begin{cases} A_{\text{upper}} = \frac{a(\Delta T-1)\,(a+1)^{\Delta T-1}}{(a+1)^{\Delta T-1}-1} \\ A_{\text{lower}} = \frac{1-(1-a)^{\Delta T-1}}{a(\Delta T-1)} \end{cases},$$
$$\begin{cases} B_{\text{upper}} = d - \frac{\Delta T-1}{(a+1)^{\Delta T-1}-1} \cdot \left(c + b\left(a+1\right)^{\Delta T-1}\right) \\ B_{\text{lower}} = d - \frac{\left(1-(1-a)^{\Delta T-1}\right)d}{a(\Delta T-1)} - \frac{c\left(1-(1-a)^{\Delta T-1}\right)}{a} \end{cases} \tag{23}$$

Figure 8a compares the ramping boundaries using the piecewise and proposed linear models, and tightened ramp-up/down margins by linear ramping constraints (22) under $\Delta T = 15$ min, respectively. As suggested in Fig. 8a, the tightened ramping boundaries in the pink line are very close to ramping boundaries using the linear model in blue line, and also are more accurate than ramping boundaries using the piecewise linear model in the green line, especially during peak shaving stages.

Let us derive the maximum inner-approximated errors Δe_U and Δe_L between linear and untightened ramp-up/down margins. We can write the maximum error of the linear retrofitted ramp-up constraint (22) as

$$\Delta f_{\text{upper}}(x) = f(x) - A_{\text{upper}}x - B_{\text{upper}} \tag{24}$$

where A_{upper}, B_{upper} can be found in (23).

Herein, $\mathrm{d}\Delta f_{\text{upper}}(x)\mathrm{d}x = 0$, yielding

$$-c \cdot \eta' - A_{\text{upper}} = 0 \tag{25}$$

Substituting $\eta' = -a/[(ax - b)\ln(a+1)]$ from Eq. (26), we can obtain the optimal x^* as

$$x^* = \frac{c}{A_{\text{upper}} \ln(a+1)} + \frac{b}{a} \tag{26}$$

Consequently, the maximum error between tightened and untightened ramp-up margins can be calculated as

$$\begin{aligned}\Delta f_{\text{upper}}(x^*) =& f(x^*) - A_{\text{upper}} x^* - B_{\text{upper}} \\ =& \left(\frac{c}{A_{\text{upper}} \ln(a+1)}\right)(a+1)^{\eta^*-1} \\ &+ \frac{b}{a}(1 - A_{\text{upper}}) + c \cdot (\Delta T - \eta^*) \\ &- \frac{c}{\ln(a+1)} - B_{\text{upper}}\end{aligned} \tag{27}$$

where $\eta^* = \text{round}\left(-\ln\left(\frac{a}{A_{\text{upper}} \ln(a+1)}\right)\ln(a+1) + 1\right)$.

Moreover, the maximum error of the linear ramp-down constraint (22) can be summarized as

$$\Delta f_{lower}(x) = A_{\text{lower}} x + B_{\text{lower}} - g(x) \tag{28}$$

where A_{lower}, B_{lower} can be found in (23).

According to $\mathrm{d}\Delta f_{\text{lower}}(x)\mathrm{d}x = 0$ and $c(\omega - 1) = (x - d)$, we express

$$A_{\text{lower}} - (1-a)^{\Delta T - \omega} \frac{-\ln(1-a)}{a} = 0 \tag{29}$$

As a result, we can obtain the optimal x^* as

$$x^* = c\left(\Delta T - \frac{\ln(-\frac{aA_{\text{lower}}}{\ln(1-a)})}{\ln(1-a)} - 1\right) + d \tag{30}$$

The maximum error between linear and untightened ramp-down margins can be calculated as

$$\Delta f_{\text{lower}}(x^*) = A_{\text{lower}} x^* + B_{\text{lower}} - g(x^*) \tag{31}$$

where $\omega^* = \text{round}\left(\Delta T - \ln\left(-\frac{aA_{\text{lower}}}{\ln(1-a)}\right)/\ln(1-a)\right)$.

3.2 Second-Order Cone (SOCO) Ramping Constraints

As previously mentioned, we simplify the convex constraints (3d) and (4d) into linear constraints, though this may introduce approximations that limit the actual ramp-up and ramp-down margins of retrofitted coal-fired units. To enhance accuracy, we propose replacing the arcs $\overset{\frown}{BC}$ and $\overset{\frown}{HG}$ with tightened quadratic curves. For convenience, we define $x = r_{i,u}^{t-1}$ and then ramp-up/down margin functions $f(x)$ and $g(x)$ for arcs $\overset{\frown}{BC}$ and $\overset{\frown}{HG}$ are

$$\begin{cases} f(x) = (x - b/a)(a+1)^{\eta-1} + b/a + c \cdot (\Delta T - \eta) \\ g(x) = (x - c \cdot (\omega - 1) - b/a)(1-a)^{\Delta T - \omega} + b/a \end{cases} \tag{32}$$

where $f(x)$ and $g(x)$ are defined as Lipschitz continuous functions. Let $f_U(x) = a_{f,\text{upper}} \cdot x^2 + b_{f,\text{upper}} \cdot x + c_{f,\text{upper}}$ and $g_L(x) = a_{f,\text{lower}} \cdot x^2 + b_{f,\text{lower}} \cdot x + c_{f,\text{lower}}$ represent the quadratic functions to approximate $f(x)$ and $g(x)$, where $a_{f,\text{upper}}$, $b_{f,\text{upper}}$, $c_{f,\text{upper}}$ and $a_{f,\text{lower}}$, $b_{f,\text{lower}}$, $c_{f,\text{lower}}$ are quadratic coefficients of $f_U(x)$ and $g_L(x)$, respectively.

As shown in Fig. 8b, the quadratic constraint $f_U(x)$ is represented by the red dashed lines, where the arc $\overset{\frown}{BCD}$ has the minimal inner-approximated error and can be utilized to approximate $f(x)$. We assume that our initial point is an arbitrary U on the vertical line of point D subject to $x_U = x_D$ and $v_U = v_D + h_U$. With this specified point U and fixed points B and C, we can determine a specific quadratic function to represent arc $\overset{\frown}{BCU}$:

$$\begin{bmatrix} x_B^2 & x_B & 1 \\ x_C^2 & x_C & 1 \\ x_U^2 & x_U & 1 \end{bmatrix} \begin{bmatrix} a_{f,\text{upper}} \\ b_{f,\text{upper}} \\ c_{f,\text{upper}} \end{bmatrix} = \Gamma \begin{bmatrix} a_{f,\text{upper}} \\ b_{f,\text{upper}} \\ c_{f,\text{upper}} \end{bmatrix} = \begin{bmatrix} y_B \\ y_C \\ y_D + h_U \end{bmatrix} \tag{33}$$

where Γ is the coefficient matrix of $f_U(x)$, and x_B, x_C, x_D, and x_U refer to x-coordinates of points B, C, D, and U, and y_B, y_C, y_D, and y_U refer to y-coordinates of points B, C, D, and U, respectively.

Since Γ is invertible, the quadratic coefficients $a_{f,\text{lower}}$, $b_{f,\text{lower}}$, and $c_{f,\text{lower}}$ can be expressed as

$$\begin{bmatrix} a_{f,\text{upper}} \\ b_{f,\text{upper}} \\ c_{f,\text{upper}} \end{bmatrix} = \Gamma^{-1} \begin{bmatrix} y_B \\ y_C \\ y_D \end{bmatrix} + \Gamma^{-1} \begin{bmatrix} 0 \\ 0 \\ h_U \end{bmatrix}$$

$$= \Gamma^{-1}\begin{bmatrix} y_B \\ y_C \\ y_D \end{bmatrix} - \frac{1}{|\Gamma|}\begin{bmatrix} x_C - x_B \\ x_B^2 - x_C^2 \\ x_C^2 x_B - x_B^2 x_C \end{bmatrix} \cdot h_U \tag{34}$$

where $|\Gamma|$ is the moduli of Γ and Γ is only with respect to the x-coordinates of points B, C, and U. By (34), we express $f_U(x)$ as

$$\begin{aligned} f_U(x) &= \left[x^2\ x\ 1\right]\begin{bmatrix} a_{f,\,\text{upper}} \\ b_{f,\,\text{upper}} \\ c_{f,\,\text{upper}} \end{bmatrix} \\ &= \left[x^2\ x\ 1\right]\left(\Gamma^{-1}\begin{bmatrix} y_B \\ y_C \\ y_D \end{bmatrix} - \frac{x_C - x_B}{|\Gamma|}\begin{bmatrix} 1 \\ -x_C - x_B \\ x_C x_B \end{bmatrix} \cdot h_U\right) \\ &= \left(\Gamma^{-1}\begin{bmatrix} y_B \\ y_C \\ y_D \end{bmatrix}\right)^T \begin{bmatrix} x^2 \\ x \\ 1 \end{bmatrix} + h_U \cdot \Upsilon^T \cdot \begin{bmatrix} x^2 \\ x \\ 1 \end{bmatrix} \end{aligned} \tag{35}$$

where $\Upsilon = -\frac{x_C - x_B}{|\Gamma|}\left[1\ -x_C - x_B\ x_C x_B\right]^T$.

Observing Eq. (35), $[x^2 \times 1]\cdot\Gamma^{-1}\cdot[y_B\ y_C\ y_D]^T$ is the quadratic function for arc $\overset{\frown}{BCD}$. If $h_U = 0$, arc $\overset{\frown}{BCD}$ overlaps arc $\overset{\frown}{BCD}$. Moreover, for any $\forall x \geq x_C$, we rearrange $\Upsilon^T \cdot \left[x^2\ x\ 1\right]^T$ as

$$\Upsilon^T \cdot \left[x^2\ x\ 1\right]^T = -\frac{x_C - x_B}{|\Gamma|}\left(x^2 - (x_C + x_B)x + x_C x_B\right) \tag{36}$$

where $x_C > x_B$ and $|\Gamma|<0$.

It is clear that $x^2 - (x_C + x_B)x + x_C x_B$ is a quadratic function where the parabola opens upward. The minimization achieves at $x^* = (x_B + x_C)/2$, but $x > x_C > x^*$, so the minimization must be at x_C. The minimal value of Eq. (36) yields $-\frac{x_C - x_B}{|\Gamma|}\left(x_C^2 - (x_C + x_B)x_C + x_C x_B\right) = 0$. Therefore, $\Upsilon^T \cdot \left[x^2\ x\ 1\right]^T \geq 0$.

With $\Upsilon^T \cdot \left[x^2\ x\ 1\right]^T \geq 0$, if $h_U \neq 0$ means that point U is higher than point D, then we can perceive that $f_U(x)$ can be a monotonically increasing function with respect to h_U for any fixed $\boldsymbol{x}$ from (35). If and only if $h_U = 0$, then the minimization of $f(x)$–$f_U(x)$ can be achieved, which indicates that the quadratic constraint $\overset{\frown}{BCD}$ has the minimal inner-approximated error for arc $\overset{\frown}{BC}$. It should be noted that $f_U(x_A)$ can be higher than $(0.3 - b/a)(a+1)^{\Delta T - 1} + b/a$ where $x_A = 0.3$. This indicates arc $\overset{\frown}{BCD}$ can envelop point A. Analogously, the quadratic constraint $f_L(x)$ can be justified as arc $\widehat{IHG}$, which can envelop the point F.

The maximum inner-approximated errors Δs_U and Δs_L between arcs $\widehat{BCD}$ and $\widehat{BC}$ and between arcs $\widehat{IHG}$ and $\widehat{HG}$ can be obtained by establishing $df(x)/dx = df_U(x)/dx$ and $dg(x)/dx = dg_L(x)/dx$, respectively. The solutions yield

$$\frac{a \cdot c}{\left(a \cdot x_+^* - b\right)\ln(a+1)} = 2a_{f,\,\text{upper}} x_+^* + b_{f,\,\text{upper}} \tag{37a}$$

$$2a_{f,lower} x_-^* + b_{f,lower} = (1-a)^{\Delta T - \frac{x_-^* - d}{c} - 1} \cdot \frac{-\ln(1-a)}{a} \tag{37b}$$

where x_+^* and x_-^* denote the x-coordinates of the maximum inner-approximated errors for arcs $\widehat{BCD}$ and $\widehat{IHG}$.

Equation (37a) is a quadratic equation and Eq. (37b) is a transcendental equation. Equation (37b) can be dealt with the Newton method [30]. With achieved roots x_+^* and x_-^*, the maximum inner-approximated errors Δs_U and Δs_L between quadratic and untightened ramp-up/down margins can be calculated as

$$\Delta s_U = f\left(x_+^*\right) - f_U\left(x_+^*\right),\ \Delta s_L = g_L\left(x_-^*\right) - g\left(x_-^*\right) \tag{38}$$

Based on the above, the quadratic ramping constraints with minimal inner-approximated errors are given as

$$a_{f,\,\text{upper}} \cdot (r_{i,u}^{t-1})^2 + b_{f,\,\text{upper}} \cdot r_{i,u}^{t-1} + c_{f,\,\text{upper}} \geq r_{i,u}^t \tag{39a}$$

$$a_{f,\,\text{lower}} \cdot (r_{i,u}^{t-1})^2 + b_{f,\,\text{lower}} \cdot r_{i,u}^{t-1} + c_{f,\,\text{lower}} \leq r_{i,u}^t \tag{39b}$$

where $a_{f,\text{upper}}$, $b_{f,\text{upper}}$, $c_{f,\text{upper}}$, $a_{f,\text{lower}}$, $b_{f,\text{lower}}$, and $c_{f,\text{lower}}$ can be obtained by arcs $\widehat{BCD}$ and $\widehat{IHG}$.

To convert this quadratic constraint (39a) and (39b) in the rotated SOCO form, we have

$$\begin{cases} 2 \cdot m_{\text{upper}} \cdot 1/2 \geq \left|\left|J_{\text{upper}}\right|\right|^2,\ m_{\text{upper}} = -r_{i,u}^t - \frac{b_{f,\,\text{upper}}^2}{4a_{f,\,\text{upper}}} + c_{f,\,\text{upper}} \\ J_{\text{upper}} = \sqrt{-a_{f,\,\text{upper}}} \cdot r_{i,u}^{t-1} - \frac{1}{2} b_{f,\,\text{upper}} \sqrt{-1/a_{f,\,\text{upper}}} \end{cases} \tag{40a}$$

$$\begin{cases} 2 \cdot m_{\text{lower}} \cdot 1/2 \geq \left|\left|J_{\text{lower}}\right|\right|^2,\ m_{\text{lower}} = r_{i,u}^t + \frac{b_{f,\,\text{lower}}^2}{4a_{f,\,\text{lower}}} - c_{f,\,\text{lower}} \\ J_{\text{lower}} = \sqrt{a_{f,\,\text{lower}}} r_{i,u}^{t-1} + \frac{1}{2} b_{f,\,\text{lower}} \sqrt{1/a_{f,\,\text{lower}}} \end{cases} \tag{40b}$$

Making use of the above, we can summarize that linear and SOCO operational constraints of retrofitted coal-fired units can be characterized by (3a)–(3c), (4a)–(4c), (12), (27), (28) and (3a)–(3c), (4a)–(4c), (12), (40a), (40b), respectively.

4 Distribution-Level Topology Optimization for Flexibility Enhancement in Look-Ahead Rolling Economic Dispatch Approach

4.1 Rolling Economic Dispatch Model

With the day-ahead unit commitment scheme and forecasted wind power data, we can develop the economic dispatch model over a rolling window, during which WTBPS maintain steady transmission power within specified limits. Each dispatch encompasses several look-ahead hours, and generally, wind power forecasts for the next 4 to 6 h exhibit relatively low forecast errors [13]. In this study, the look-ahead period is thus set at 5 h, including the current hour and the remaining hours in the intra-day stage. The objective of this rolling economic dispatch is to minimize both thermal production costs and switch-over operational costs while adhering to the operational constraints of WTBPS. Specifically, the thermal production cost, spanning a horizon of length N_{Ts}, is generated by retrofitted coal-fired units, while the switch-over operational cost pertains to the optimal load transfer solutions via HVDNs. In summary, we outline the following look-ahead rolling economic dispatch framework and the associated rolling process Fig. 9.

Since different y_i^t have similar production cost coefficients, we approximately express the coal-fired generation cost F_{o1} as

$$\min_{r'_{i=\sim}\in\mathbb{R}} F_{\text{o1}} = \sum_{t=1}^{N_{T_k}} \sum_{u=1}^{N_w} \sum_{i=1}^{N_{G=1}} \left[c_{2i,u} \left(r_{i,u}^t P_{\text{G}iu,\max} \right)^2 + c_{1i,u} r_{i,u}^t P_{\text{G}i,u,\max} + c_{0i,u} \right] \tag{41}$$

where $c_{2i,u}$, $c_{1i,u}$, and $c_{0i,u}$ refer to production cost coefficients for the ith retrofitted coal-fired unit in the uth agent of WTBPS; $N_{G,u}$ and N_u indicate the total number of retrofitted coal-fired units in the uth agent and the total number of agents, respectively.

Moreover, addressing the operational cost minimization issue in an LTS task is to seek the smallest number of total circuit switching operation times and the minimum amount of dispatchable loads during a rolling window. This chapter adopts circuit switching service cost $c_{H,u}$ to multiply switching operation times plus purchasing costs of dispatchable loads as $c_{D,u} \cdot (P_{D,u}^t)^2$. As a result, $F_{o2} = \sum_{t=1}^{N_{Ts}} \sum_{u=1}^{N_u} \left\{ c_{\text{D},u} \cdot (P_{D,u}^t)^2 + c_{\text{H},u} \sum_{j=1}^{N_s} \left(S_j^t - S_j^{t-1} \right)^2 \right\}$. After applying $z_j^t = S_j^t S_j^{t-1}$ to F_{o2}, F_{o2} can be rearranged as

$$F_{o2} = \sum_{t=1}^{N_{T_S}} \sum_{u=1}^{N_u} \left\{ c_{\text{D},u} \cdot (P_{D,u}^t)^2 + c_{\text{H},u} \sum_{j=1}^{N_s} (S_j^t - S_j^{t-1})^2 \right\} \tag{42}$$

In light of all these facts, we consider that two objectives are assumed to have equal priorities and weighting coefficients are selected as $\mu_1 = \mu_2 = 1/2$. Thus,

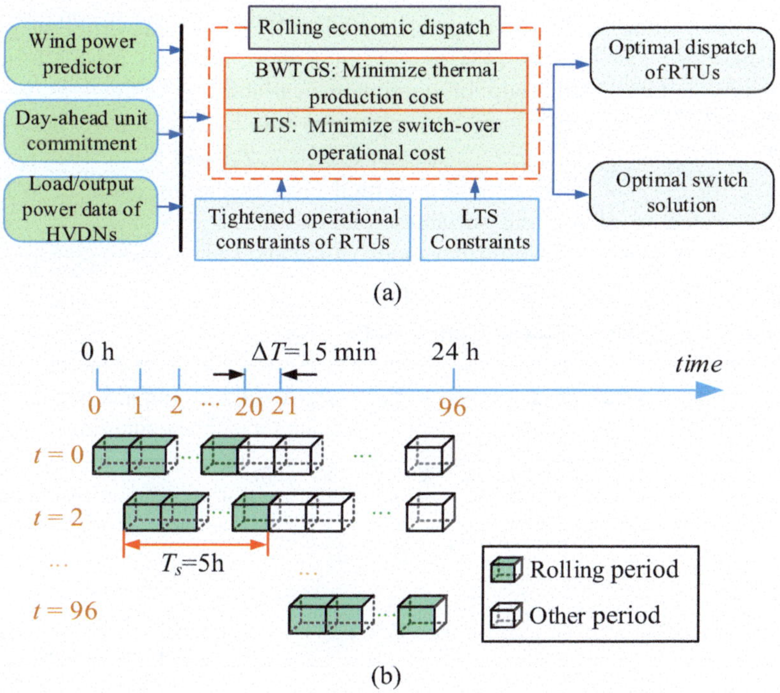

Fig. 9 **a** Look-ahead rolling economic dispatch framework; **b** rolling process

a look-ahead rolling economic dispatch model is the minimization of F_o subject to the power balance of tie-lines (1), operational limits of retrofitted coal-fired units (3a)–(3c), (4a)–(4c), and (12), and linear ramping constraints of retrofitted coal-fired units (22)–(23) or SOCO ramping constraints of retrofitted coal-fired units (41a) and (41b), plus simplified voltage-constrained LTS via HVDNs (13)–(21).

P1: Rolling economic dispatch model with linear ramping constraints

$$\begin{aligned} &\min_{r^t_{i,u},P^t_{D,u},P^t_{\mathrm{T},u}\in\mathbb{R},S^t_i,z^t_j\in\mathbb{N}} F_o = \mu_1 F_{o1} + \mu_2 F_{o2} \\ &\text{s.t.}\quad (1.1), (1.3a)-(1.3c), (1.4a)-(1.4c), (1.12)-(1.23),\ \ \forall t\in T_s \end{aligned} \tag{43}$$

P2: Rolling economic dispatch model with SOCO ramping constraints

$$\begin{aligned} &\min_{r^t_{i,u},P^t_{D,u},P^t_{\mathrm{T},u},m_{\mathrm{upper}},m_{\mathrm{lower}},J_{\mathrm{upper}},J_{\mathrm{lower}}\in\mathbb{R},S^t_i,z^t_j\in\mathbb{N}} F_o = \mu_1 F_{o1} + \mu_2 F_{o2} \\ &\text{s.t.}(1.1), (1.3a)-(1.3c), (1.4a)-(1.4c), (1.5) \\ &-(1.14), (1.25a), (1.25b)\ \ \forall t\in T_s \end{aligned} \tag{44}$$

Since this rolling economic dispatch model (43) or (44) has different groups of variables, we can express (43) or (44) in a more compact model form. To avoid heavy notion, one group of continuous variables is defined as $\boldsymbol{y}_{1,\boldsymbol{u}} := \left\{\left(\boldsymbol{r}_{\boldsymbol{u}}^{t}, \boldsymbol{P}_{\boldsymbol{D},\boldsymbol{u}}^{t}, \boldsymbol{P}_{\boldsymbol{T},\boldsymbol{u}}^{t}\right) \in \mathbb{R},\ t \in T_s\right\}$ and $\boldsymbol{Y}_1 =: \left\{\boldsymbol{y}_{1,\boldsymbol{u}}\right\}, u = 1, 2, \ldots, N_u$ for (43) and $\boldsymbol{y}_{2,\boldsymbol{u}} := \left\{\left(\boldsymbol{r}_{\boldsymbol{u}}^{t}, \boldsymbol{P}_{\boldsymbol{D},\boldsymbol{u}}^{t}, \boldsymbol{P}_{\boldsymbol{T},\boldsymbol{u}}^{t}, m_{\text{upper}}, m_{\text{lower}}, J_{\text{upper}}, J_{\text{lower}}\right) \in \mathbb{R},\ t \in T_s\right\}$ and $\boldsymbol{Y}_2 =: \left\{\boldsymbol{y}_{2,\boldsymbol{u}}\right\}, u = 1, 2, \ldots, N_u$ for (44) for all retrofitted coal-fired units in the uth agent of WTBPS. The other category of discrete variables refers to $\boldsymbol{X} := \left\{\left(\boldsymbol{S}^t, z^t\right) \in \mathbb{N},\ t \in T_s\right\}$. Without loss of generality, we further modify (43) and (44) in more general forms as (35a)–(45e) and (46a)–(46f), respectively.

Compact P1 Subject to Linear Constraints:

$$\min_{\boldsymbol{X}\in\mathbb{N},\ \boldsymbol{Y}_1\in\mathbb{R}} \mu_1(\boldsymbol{Y}_1^T D_{q_1}^y \boldsymbol{Y}_1 + (D_{l_1}^y)^T \boldsymbol{Y}_1) + \mu_2(\boldsymbol{X}^T D_q^x \boldsymbol{X} + (D_l^x)^T \boldsymbol{X}) \tag{45a}$$

$$\text{s.t.}\quad \boldsymbol{A}_l \cdot \boldsymbol{X} + \boldsymbol{B}_l \cdot \boldsymbol{Y}_1 \le \boldsymbol{b}_l \tag{45b}$$

$$\boldsymbol{E}_l \cdot \boldsymbol{Y}_1 \le \boldsymbol{h}_l \tag{45c}$$

$$\boldsymbol{F}_l \cdot \boldsymbol{X} \le \boldsymbol{r}_{le} \tag{45d}$$

$$\boldsymbol{G}_l \cdot \boldsymbol{X} = \boldsymbol{r}_{ls} \tag{45e}$$

where $\mathbf{Y}_1^{\mathrm{T}} D_{q_1}^y \mathbf{Y}_1 + \left(D_{l_1}^y\right)^T \mathbf{Y}_1 = \sum_{t=1}^{N_{T_s}} \sum_{u=1}^{N_u} \sum_{i=1}^{N_{G,u}} \left[c_{2i,u}(r_{i,u}^t P_{\mathrm{G}i,u,\max})^2 + c_{1i,u} r_{i,u}^t P_{\mathrm{G}i,u,\max}\right]$ and $D_{q_1}^y$, D_q^x, $D_{l_1}^y$ and D_l^x in (45a) are constant matrices summarizing from (41) and (42). Constraint (45b) refers to constraint (1) where $\boldsymbol{A}_l$, $\boldsymbol{B}_l$, and $\boldsymbol{b}_l$ are constant vectors drawn from constraint (1). Constraint (45c) includes constraints (3a)–(3c), (4a)–(4c) and (12), (22)–(23) where $\boldsymbol{E}_l$ and $\boldsymbol{h}_l$ can be deduced as constant matrices. Constraints (45d) and (45e) encompass inequalities and equalities for LTS-based network operational constraints (13)–(21) for the LTS-based HVDNs with deduced constant matrices $\boldsymbol{F}_l$, $\boldsymbol{r}_{le}$, $\boldsymbol{G}_l$, and $\boldsymbol{r}_{ls}$.

According to minor inner-approximated errors for rotated SOCO constraints (40a) and (40b), this compact **P2** is more accurate than **P1**, expressed as follows.

Compact P2 Subject to SOCO Constraints:

$$\min_{\boldsymbol{X}\in\mathbb{N},\boldsymbol{Y}_2\in\mathbb{R}} \mu_1(\boldsymbol{Y}_2^T D_{q_2}^y \boldsymbol{Y}_2 + (D_{l_2}^y)^T \boldsymbol{Y}_2) + \mu_2(\boldsymbol{X}^T D_q^x \boldsymbol{X} + (D_l^x)^T \boldsymbol{X}) \tag{46a}$$

$$\text{s.t.}\quad \boldsymbol{A}_l \cdot \boldsymbol{X} + \boldsymbol{B}_l \cdot \boldsymbol{Y}_2 \le b_l \tag{46b}$$

$$\boldsymbol{E}_l \cdot \boldsymbol{Y}_2 \le \boldsymbol{h}_l \tag{46c}$$

$$\boldsymbol{F}_l \cdot \boldsymbol{X} \le \boldsymbol{r}_{le} \tag{46d}$$

$$G_l \cdot X = r_{ls} \tag{46e}$$

$$Y_2^T Q_r Y_2 + l_r^T Y_2 \leq g_r \tag{46f}$$

where $\mathbf{Y}_2^{\mathrm{T}} D_{q2}^y \mathbf{Y}_2 + \left(D_{l_2}^y\right)^T \mathbf{Y}_2 = \sum_{t=1}^{N_{T_s}} \sum_{u=1}^{N_u} \sum_{i=1}^{N_{G,u}} \left[c_{2i,u}\left(r_{i,u}^t P_{Gi,u,\max}\right)^2 + c_{1i,u} r_{i,u}^t P_{Gi,u,\max}\right]$, and D_{q2}^y and $D_{l_2}^y$ in (46a) are constant matrices. Quadratic constraints in constraint (46f) can automatically convert to standard rotated SOCO constraints with the auxiliary variables (J_{upper}, m_{upper}, J_{lower}, m_{lower}). The constant matrices Q_r, l_r, and g_r are symbols of $a_{f,\text{upper}}$, $b_{f,\text{upper}}$, $c_{f,\text{upper}}$, $a_{f,\text{lower}}$, $b_{f,\text{lower}}$, and $c_{f,\text{ lower}}$.

4.2 *Multi-cut Generalized Benders Decomposition*

As observed in the compact formulations **P1 and P2**, this look-ahead economic dispatch model is classified as a MISOCP problem. This MISOCP-based optimization problem is designed for large-scale HVDNs and multiple agents of WTBPS, involving a significant number of continuous and integer variables throughout the look-ahead period. Running this extensive optimization model, combined with the LTS-based network operation model, can be time-consuming. Fortunately, the compact forms **P1** and **P2** can be decomposed into a relaxed master problem (**MP**) concerning variable X and multiple sub-problems related to Y_1 or Y_2. This decomposition aligns well with the GBD framework, facilitating more efficient computation by separating the optimization tasks into many smaller sub-problems.

Given the discrete variables $\hat{X}$ from ***MP***, we can develop the sub-problem focusing on the continuous variables $Y_1 =: \{y_{1,u}\}$ and $Y_2 =: \{y_{2,u}\}$, $u = 1, 2, \ldots, N_u$. By examining the block matrices $B_{l,u}$ and $E_{l,u}$ of constraints (45b) and (45c) for **P1**, as well as (46b) and (46c) for **P2**, we can observe that the structure of compact **P1** and **P2** exhibits a special "block-angular" form concerning the variables $y_{1,u}$ and $y_{2,u}$ in the sub-problem (SP_u) for the uth agent of WTBPS. Generally, y_u refers to $y_{1,u}$ for **P1** or $y_{2,u}$ for **P2**:

$$\begin{matrix}\text{Agent } 1\\ \text{Agent } 2\\ \vdots\\ \text{Agent } N_u\end{matrix}\begin{bmatrix} B_{l,1} & & & \\ & B_{l,2} & & \\ & & \ddots & \\ & & & B_{l,N_u}\end{bmatrix} \cdot \begin{bmatrix} y_1\\ y_2\\ \vdots\\ y_{N_u}\end{bmatrix} \leq \begin{bmatrix} b_{l,1}\\ b_{l,2}\\ \vdots\\ b_{l,N_u}\end{bmatrix} - \begin{bmatrix} A_{l,1}\\ A_{l,2}\\ \vdots\\ A_{l,N_u}\end{bmatrix} \cdot \hat{X} \tag{47}$$

We summarize the special "block-angular" decomposable structure in (47) through three key characteristics: convexity, linear separability, and linear independence.

(i) **Convexity**: **P2** is entirely convex with respect to the continuous variables $\boldsymbol{y_u}$ when the discrete variables $\hat{X}$ are given.
(ii) **Linear separability**: The constraints (45b)–(45e) and (46b)–(46e) exhibit linearity with respect to the discrete variables X while the continuous variables $\hat{y}_{\boldsymbol{u}}$ are fixed.
(iii) **Linear Independence**: Given the discrete variables $\hat{X}$, the different groups of continuous variables $\boldsymbol{y_u}$ are linearly independent. With these three characteristics, it is highly beneficial to employ the GBD method with multiple cuts, referred to as MGBD. MGBD allows the decomposition of the GBD sub-problem into multiple independent smaller optimization models, as expressed in Eq. (47). Each of these sub-problems generates multiple feasibility cuts and optimality cuts at every iteration. In contrast to classical GBD, which typically produces only one feasibility cut per iteration, MGBD generates a number of feasibility cuts and optimality cuts equal to N_u in each iteration. This capability enables MGBD to achieve faster convergence of the large-scale MP model, resulting in reduced computational time compared to the traditional single-cut GBD method.

Following the MGBD approach, we formulate the sub-problem ($\boldsymbol{SP_u}$) for the uth agent of WTBPS for **P2** at the kth iteration with given discrete variables $\hat{X}^{k-1}$:

$$\min_{\boldsymbol{y_{2,u}} \in \mathbb{R}} \mu_1(\boldsymbol{y}_{2,u}^T D_{q_2}^y \boldsymbol{y}_{2,u} + (D_{l_2}^y)^T \boldsymbol{y}_{2,u}) + \mu_2((\hat{X}^{k-1})^T D_q^x \hat{X}^{k-1} + (D_l^x)^T \hat{X}^{k-1}) \tag{48a}$$

$$\text{s.t.} \quad \boldsymbol{B}_{l,u} \cdot \boldsymbol{y}_{2,u} \le \boldsymbol{b}_{l,u} - \boldsymbol{A}_{l,u} \cdot \hat{X}^{k-1} \tag{48b}$$

$$\boldsymbol{E}_{l,u} \cdot \boldsymbol{y}_{2,u} \le \boldsymbol{h}_{l,u} \tag{48c}$$

$$\boldsymbol{y}_{2,u}^T \boldsymbol{Q}_{r,u} \boldsymbol{y}_{2,u} + \boldsymbol{l}_{r,u}^T \boldsymbol{y}_{2,u} \le \boldsymbol{g}_{r,u} \tag{48d}$$

where $\boldsymbol{A}_l = \{\boldsymbol{A}_{l,u}\}$, $\boldsymbol{l}_r = \{\boldsymbol{l}_{r,u}\}$, $\boldsymbol{B}_l = \{\boldsymbol{B}_{l,u}\}$, $\boldsymbol{b}_l = \{\boldsymbol{b}_{l,u}\}$, $\boldsymbol{E}_l = \{\boldsymbol{E}_{l,u}\}$, $\boldsymbol{h}_l = \{\boldsymbol{h}_{l,u}\}$, $\boldsymbol{Q}_r = \{\boldsymbol{Q}_{r,u}\}$ and $\boldsymbol{g}_r = \{\boldsymbol{g}_{r,u}\}$ for $u = 1, 2, \ldots, N_u$.

If this $\boldsymbol{SP_u}$ problem is optimal with $\hat{y}_{2,u}^k$, the upper bound is $UB_k = \left\{UB_{k-1}, \mu_1 \sum_{u=1}^{N_u} \left(\left(\hat{y}_{2,u}^k\right)^T D_{q_2}^y \hat{y}_{2,u}^k + \left(D_{l_2}^y\right)^T \hat{y}_{2,u}^k\right)\right.$ $\left.\mu_2\left(\left(\hat{X}^{k-1}\right)^T D_q^x \hat{X}^{k-1} + \left(D_l^x\right)^T \hat{X}^{k-1}\right)\right\}$; otherwise it is infeasible, and we turn to solve an l_1-minimization feasibility check problem with the relaxed variable $\sigma_u \ge 0$ for $u = 1, 2, \ldots, N_u$ is as follows:

$$\min_{\boldsymbol{y_u}, \sigma_u \in \mathbb{R}} \sigma_u \tag{49a}$$

$$\text{s.t.} \quad \boldsymbol{B}_{l,u} \cdot \boldsymbol{y}_{2,u} - \sigma_u \le \boldsymbol{b}_{l,u} - \boldsymbol{A}_{l,u} \cdot \hat{X}^{k-1} \tag{49b}$$

$$\boldsymbol{E}_{l,u} \cdot \boldsymbol{y}_{2,u} - \sigma_u \le \boldsymbol{h}_{l,u} \tag{49c}$$

$$\boldsymbol{y}_{2,u}^T \boldsymbol{Q}_{r,u} \boldsymbol{y}_{2,u} + \boldsymbol{l}_{r,u}^T \boldsymbol{y}_{2,u} - \sigma_u \le \boldsymbol{g}_{r,u} \tag{49d}$$

On the one hand, if the $\boldsymbol{SP_u}$ problem is feasible at kth iteration as indicated in (48a)–(48d), we substitute the obtained continuous variables $\hat{y}_{2,u}^k$ and Lagrange multiplier vectors $\hat{\lambda}_{u,1}^k$, $\hat{\lambda}_{u,2}^k$, and $\hat{\lambda}_{u,3}^k$ for constraints (48a)–(48d), which enforces the optimality cut as

$$\begin{aligned}(\hat{\lambda}_{u,1}^k)^T A_{l,u} X + &(\hat{\lambda}_{u,1}^k)^T (B_{l,u} \cdot \hat{y}_{2,u}^k - b_{l,u}) \\ &+ (\hat{\lambda}_{u,2}^k)^T (E_{l,u} \cdot \hat{y}_{2,u}^k - h_{l,u}) \\ &+ (\hat{\lambda}_{u,3}^k)^T ((\hat{y}_{2,u}^k)^T Q_{r,u} \hat{y}_{2,u}^k + l_{r,u}^T \hat{y}_{2,u}^k - g_{r,u}) \\ &+ \mu_1 \sum_{u=1}^{N_u} ((\hat{y}_{2,u}^k)^T D_{q_2}^y \hat{y}_{2,u}^k + (D_{l_2}^y)^T \hat{y}_{2,u}^k) \le q_u \end{aligned} \tag{50}$$

where q_u is an auxiliary variable from the $\boldsymbol{SP_u}$ for the $\boldsymbol{MP}$.

On the other hand, provided that the $\boldsymbol{SP_u}$ problem is infeasible at the kth iteration, by solving l_1-minimization feasibility check problem as indicated in (49a)–(49d), and then substituting continuous variables $\hat{y}_u^{f,k}$ and Lagrange multiplier vectors $\hat{\gamma}_{u,1}^k$, $\hat{\gamma}_{u,2}^k$, and $\hat{\gamma}_{u,3}^k$ for constraints (49b)–(49d) to form the feasibility cut:

$$\begin{aligned}(\hat{\gamma}_{u,1}^k)^T A_{l,u} X + &(\hat{\gamma}_{u,1}^k)^T \left(B_{l,u} \cdot \hat{y}_{2,u}^{f,k} - b_{l,u}\right) \\ &+ (\hat{\gamma}_{u,2}^k)^T \left(E_{l,u} \cdot \hat{y}_{2,u}^{f,k} - h_{l,u}\right) \\ &+ (\hat{\gamma}_{u,3}^k)^T \left(\left(\hat{y}_{2,u}^{f,k}\right)^T Q_{r,u} \hat{y}_{2,u}^{f,k} + l_{r,u}^T \hat{y}_{2,u}^{f,k} - g_{r,u}\right) \le 0 \end{aligned} \tag{51}$$

After imposing the optimality cuts (50) and the feasibility cuts (51) for $u = 1, 2, \ldots, N_u$, we establish the relaxed $\boldsymbol{MP}$ model at kth iteration for all agents:

$$\min_{X, q_u \in \mathbb{R}} \sum_{u=1}^{N_u} q_u + \mu_2 (\boldsymbol{X}^T D_q^x \boldsymbol{X} + (D_l^x)^T \boldsymbol{X}) \tag{52a}$$

$$\text{s.t.} \quad F_l \cdot \boldsymbol{X} \le r_{le}, \; G_l \cdot \boldsymbol{X} = r_{ls}, \text{ and } (1.50)-(1.51) \tag{52b}$$

The lower bound is obtained as $LB_k = \mu_2\left(\left(X^k\right)^T D_q^x X^k + \left(D_l^x\right)^T X^k\right) + \sum_{u=1}^{N_u} q_u^k$, where $\boldsymbol{X^k}$ and q_u^k are optimal solutions from (52a), (52b). Until $||LB_k - UB_k||$ is less than the given tolerance, this MGBD algorithm can be converged.

This MGBD framework can typically be executed by servers at the dispatch center, while each sub-problem $\boldsymbol{SP_u}$ can be addressed individually and simultaneously by

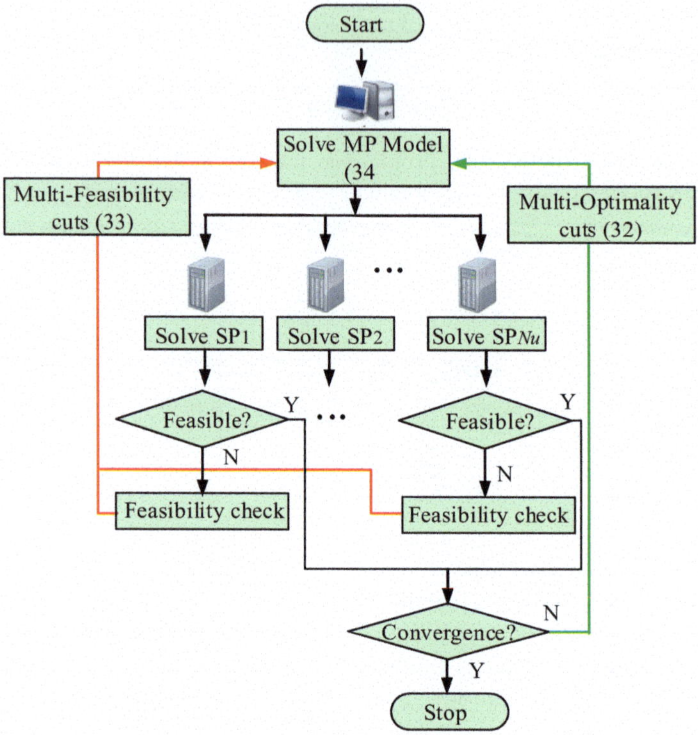

Fig. 10 Computation procedure of MGBD for **P1** and **P2**

each agent. Traditionally, in a centralized rolling economic dispatch system, each agent is required to upload sensitive local data to the dispatch center, including power outputs, power capacities, and utilization levels, often without adequate privacy protection [31]. However, with the MGBD computation framework, agents are able to share only the dispatchable loads and Lagrange multiplier vectors with the dispatch center, effectively preserving the privacy of sensitive information. In summary, the entire computation process, leveraging multiple parallel CPU cores, is illustrated in Fig. 10. This approach enhances both computational efficiency and data security in the dispatch operations.

5 Case Studies

A real modified HVDN system with 220 kV and 110 kV voltage levels in central China is shown in Fig. 11 [22]. Four agents A_1–A_4 of WTBPS are integrated in four different 220 kV stations, namely stations TH, ZJS, HTC, and SQZ, with tie-lines

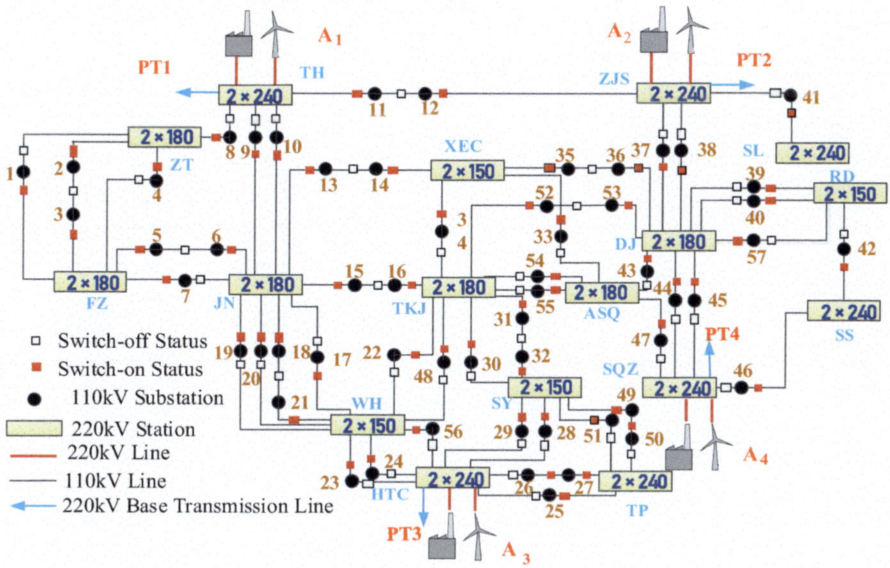

Fig. 11 A real modified HVDNs in central China

PT1–PT4. This integrated practical system is used for case studies to validate the approach.

Assuming that four agents have identical installed capacities for wind power and coal-fired power generation, each agent is equipped with five retrofitted coal-fired units that satisfy the condition $5 \cdot P_{\mathrm{G,max}} = P_{\mathrm{base}} = 1500$ MVA, $\Delta T = 15$ min, $T_{\mathrm{s}} = 5$ h, $N_{\mathrm{Ts}} = 20$, $P_u^{\mathrm{spin}} = 37.5$ MVA, $\delta\% = 5\%$, and $P_{\mathrm{wind}} = 1200$ MVA. The coefficients $c_{2i,u}$, $c_{1i,u}$, and $c_{0i,u}$ are set to 0.00049 ($/MW^{2}h), 18.23 ($/MWh), and 35.54 ($) for all agents. The costs for load transfer $c_{D,u}$ and $c_{H,u}$ are set to 4.17 ($/MWh) and 10 ($/p.u.), respectively. The HVDN system consists of 17 units of 220 kV stations, represented by long rectangular boxes with rated capacities in light yellow, and 57 units of 110 kV substations, indicated by black-filled circles. The status of the switch breakers is depicted in Fig. 11 for the initial period. The number of allowable switch actions during the rolling window is limited to 6 times, and load data for each substation, along with wind power data for the four agents, is forecasted for $T_{\mathrm{s}} = 5$ h ahead. The voltage profiles across all nodes are constrained within [0.95, 1.05] p.u., and the thermal capacities of branches are restricted to less than 0.126 p.u. Other network parameters can be found in Ref. [22], where the base power is set to 1000 MW. Additionally, the tolerance for the MGBD method is established at $\varepsilon = 10^{-3}$ for this study. The simulation experiments will compare conventional coal-fired units without retrofits and retrofitted coal-fired units with dynamic ramping, as detailed in Table 2.

In this study, we present comparative rolling ED models for agent A1, as outlined in Table 3. The models are categorized into two main factors: dispatchable loads and

Table 2 Comparative coal-fired plants

Types of coal-fired plants	Minimum load level (%)	Ramp rate (min/%)
Conventional coal-fired units with normal ramp rate	60	1.1%
Retrofitted coal-fired units with piecewise linear model of dynamic ramp rates [18]	30	$v_{i,u}^{t} = \begin{cases} 0.35\%, & r_{i,u}^{t} \in [0.3, 0.4] \\ 1.0\%, & r_{i,u}^{t} \in [0.4, 0.5] \\ 1.5\%, & r_{i,u}^{t} \in [0.5, 0.6] \\ 2\%, & r_{i,u}^{t} \in [0.6, 1.0] \end{cases}$
Retrofitted coal-fired units with dynamic ramp rates Eq. (2)	30	$v_{i,u}^{t} = \begin{cases} a \cdot r_{i,u}^{t} - b, & r_{i,u}^{t} \le d \\ c, & r_{i,u}^{t} > d \end{cases}$ where $a = 0.055$, $b = 0.013$, $c = 0.02$, $d = 0.6$

ramp rates. For dispatchable loads, Ref. [17] employs a load-shedding variable in the look-ahead rolling economic dispatch model, termed unrestricted dispatchable loads, which is indicated in the second column. In contrast, the LTS-based HVDN operation model only provides discrete dispatchable loads, as shown in the third column.

Regarding ramp rates, both conventional and retrofitted coal-fired units are incorporated into the ED models, with the types of coal-fired units detailed in Table 2. The fourth and fifth columns represent conventional coal-fired units with a normal ramp rate and retrofitted coal-fired units using a piecewise linear model of dynamic ramp rates, respectively. Lastly, the sixth and seventh columns denote the linear and SOCO-based tightened ramping constraints for retrofitted coal-fired units. Parameters for the normal ramp rate and dynamic ramp rates are sourced from Table 2, ensuring consistency across the models.

Table 3 Comparative rolling economic dispatch models for agent A1

ED models	Dispatchable loads		Normal ramp rate	Dynamic ramp rates		
	Unrestricted	HVDNs		Piecewise	linear	SOCO
ED1	√				√	
ED2	√					√
ED3	√		√			
ED4	√			√		
ED5	√					√
ED6	√					√
ED7		√				√
ED8		√			√	

5.1 Maximum Inner-Approximated Errors of Tightened Ramping Constraints

We discuss the accuracy of maximum inner-approximated errors between untightened, linear, and SOCO ramping constraints. Figure 13a–c illustrates these maximum inner-approximated errors Δe_U and Δe_L, for the comparison between untightened and linear ramping constraints, and Δs_U and Δs_L for the comparison between untightened and SOCO ramping constraints, across different ranges of experimental parameters *a, b,* and *c*. In our experiments, we define the parameters as follows, with *d* calculated as $d = (b + c)/a$.

In Fig. 12a, parameters *a* and *b* range from 0.0467 to 0.06 and from 0.001 to 0.015, respectively, while parameter c is held constant at 0.02.

In Fig. 12b, parameters *a* and *c* vary from 0.0467 to 0.06 and from 0.020 to 0.025, respectively, with parameter *b* fixed at 0.013.

In Fig. 12c, parameters *b* and *c* range from 0.011 to 0.015 and from 0.020 to 0.025, respectively, while parameter *a* is fixed at 0.055.

These settings allow for a comprehensive examination of how variations in parameters influence the accuracy of the ramping constraints.

From Fig. 12a–c, it is evident that the maximum inner-approximated errors—Δe_U, Δe_L, Δs_U, and Δs_L—are quantified at below 3.21, 2.45, 2.84, and 1.36% with varying parameter sets. The relationships $\Delta e_U > \Delta s_U$ and $\Delta e_L > \Delta s_L$ indicate that the SOCO ramping constraints offer greater accuracy compared to the corresponding linear ramping constraints. Specifically, for the SOCO ramping constraints, the maximum inner-approximated errors translate to ramp-up and ramp-down margins of 8.52 MW by $2.84\%{\cdot}P_{\text{base}}/5$ and 4.08 MW by $1.36\%{\cdot}P_{\text{base}}/5$, respectively, under a time interval of $\Delta T = 15$ min. This demonstrates the effectiveness of SOCO constraints in maintaining tighter control over generation variability.

5.2 Boundaries of Wind Power Fluctuations

The allowable ramping margin of wind power turbines (WPTs) can be quantified using three types of ramping constraints: the normal ramp rate, the piecewise linear model, and the proposed linear model of dynamic ramp rates, as outlined in Table 2. In Fig. 13, the corresponding allowable ramping margins for these three types of constraints are illustrated with black, pink, and blue lines, respectively. The ramping boundary for the installed wind power capacity is represented by green lines, and the figure also includes three forecasted sets of wind power data. This visual representation effectively highlights the differences in ramping margins and the implications of various modeling approaches on the operational flexibility of wind power generation.

Figure 13 illustrates that the allowable ramping margin of WPTs can be increased when coal-fired plants are retrofitted, as evidenced by the ramping areas delineated by the black, blue, and pink lines. The theoretical ramping area, represented by

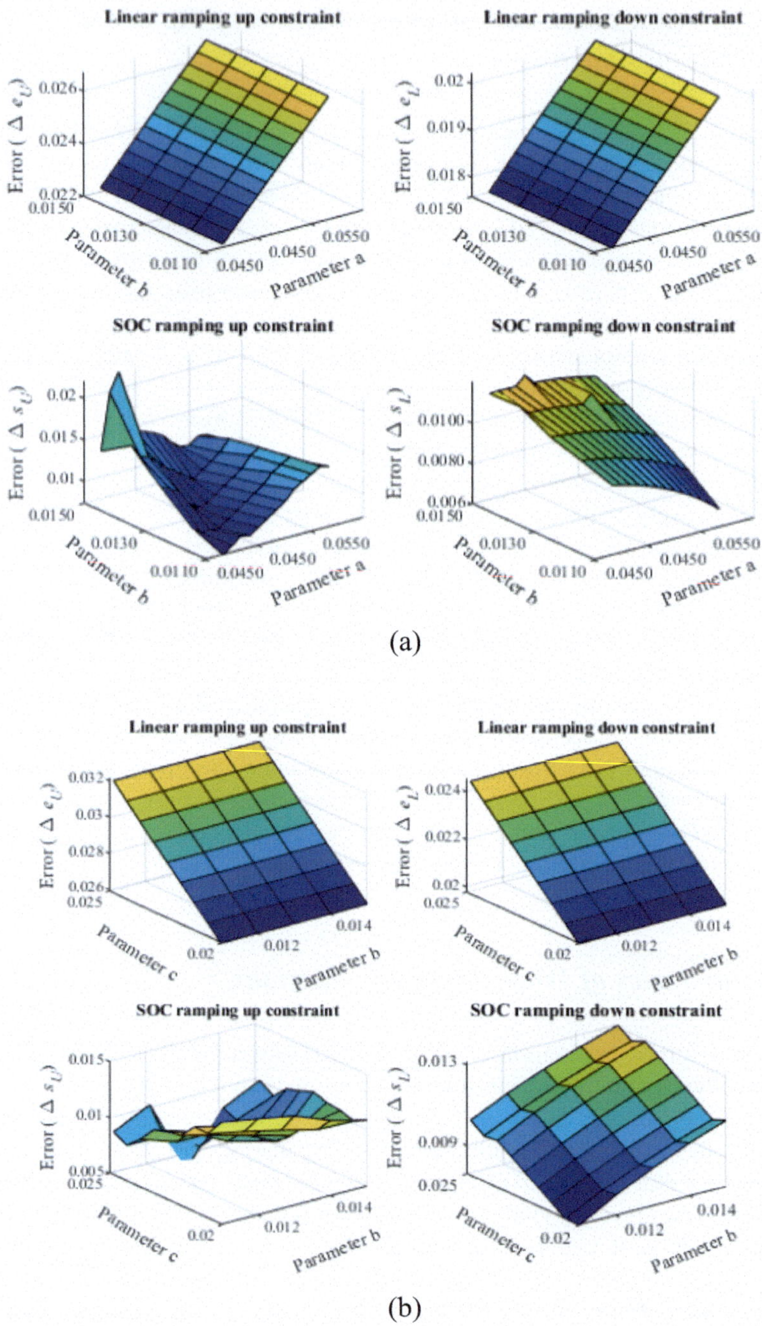

Fig. 12 Maximum inner-approximated errors using linear and SOCO ramping constraints with different sets of parameters

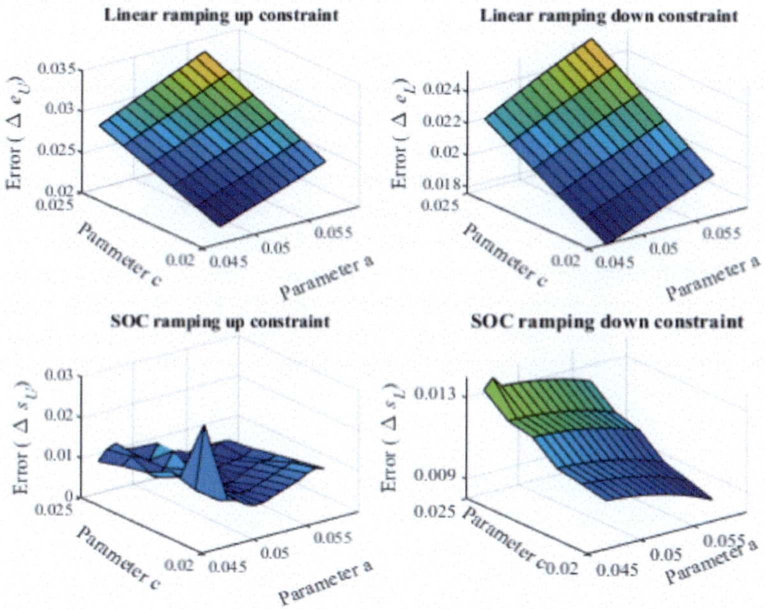

(c)

Fig. 12 (continued)

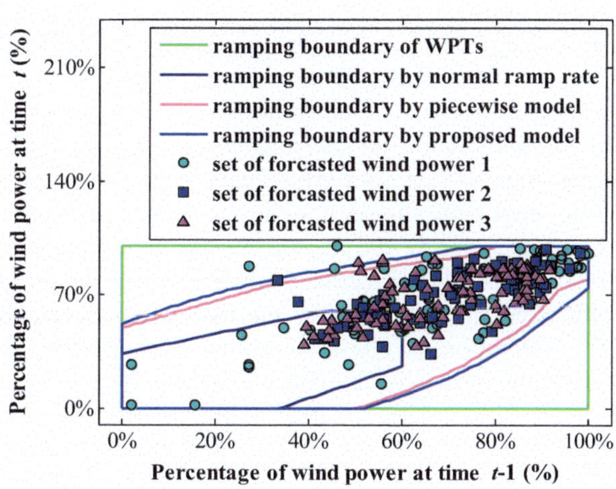

Fig. 13 Different allowable ramping margins of WPTs

the green lines, is determined by the installed wind power capacities. For WTBPS, the feasible wind power fluctuations of WPTs are bounded by these lines between adjacent periods. This explains why only the forecasted wind power set 2 is fully contained within the allowable ramping margin of WPTs, while sets 1 and 3 cannot be completely accommodated, leading to inevitable wind curtailment or load transfer for those datasets.

The figure also depicts both untightened and tightened allowable ramping margins of WPTs under $\Delta T = 15$ min. The blue line indicates the untightened boundary defined by operational ramping constraints, while the red and black lines represent the tightened ramping boundaries caused by linear and SOCO ramping constraints, respectively. Notably, one point from the forecasted wind power dataset 2, marked within a blue-filled square box, lies on the SOCO ramping constraint. If linear ramping constraints are applied, this results in the need to purchase dispatchable loads through a limited number of switch-over operations in HVDNs. In contrast, adopting tightened SOCO ramping constraints can effectively accommodate the variations in wind power output, enhancing operational flexibility.

5.3 Linear Versus SOCO Ramping Constraints

Following the feasibility-check procedure for wind power data, utilizing tightened ramping margins for WPTs illustrated in Fig. 14a, we proceed to conduct simulation experiments aimed at validating the accuracy of dispatch solutions for ED1 and ED2. To effectively differentiate between the two methods, we compare the optimal generation power solutions of ED1 and ED2 over a rolling window from 8:30 to 13:30, as depicted in Fig. 14b. This comparison considers both linear and SOCO ramping constraints, facilitating a clear assessment of how each approach responds to the wind power data and its associated challenges during the specified time frame.

Figure 15b reveals significant differences between the optimal solutions of ED1 and ED2, with ED2 demonstrating greater accuracy. The optimal solution of ED2 closely aligns with the untightened ramping margin, represented by the blue line in the graph. Maximum error analysis presented in Fig. 15a–c indicates that the rolling economic dispatch solution of ED2 remains reliable despite variations in wind power during the assessment periods. Consequently, the SOCO constraints employed in ED2 effectively replace nonlinear untightened constraints, yielding optimal economic dispatch solutions with acceptable errors. This suggests that ED2 offers a more robust framework for managing wind power fluctuations while maintaining efficiency in dispatch operations.

5.4 Normal, Piecewise Versus SOCO Ramping Constraints

To illustrate the quick-response capabilities, we compare the optimal generation power solutions and optimal dispatchable loads over the rolling window from 8:30 to 13:30 using models ED3, ED4, and ED5, as depicted in Fig. 15a, b. The optimal ED solutions for ED3, ED4, and ED5 are evaluated under typical load demands and the mean of wind power data across 100 random scenarios, as shown in Fig. 15a.

Fig. 14 **a** Untightened and tightened ramping margins of WPTs under $\Delta T = 15$ min; **b** Optimal rolling ED solutions between ED1 and ED2

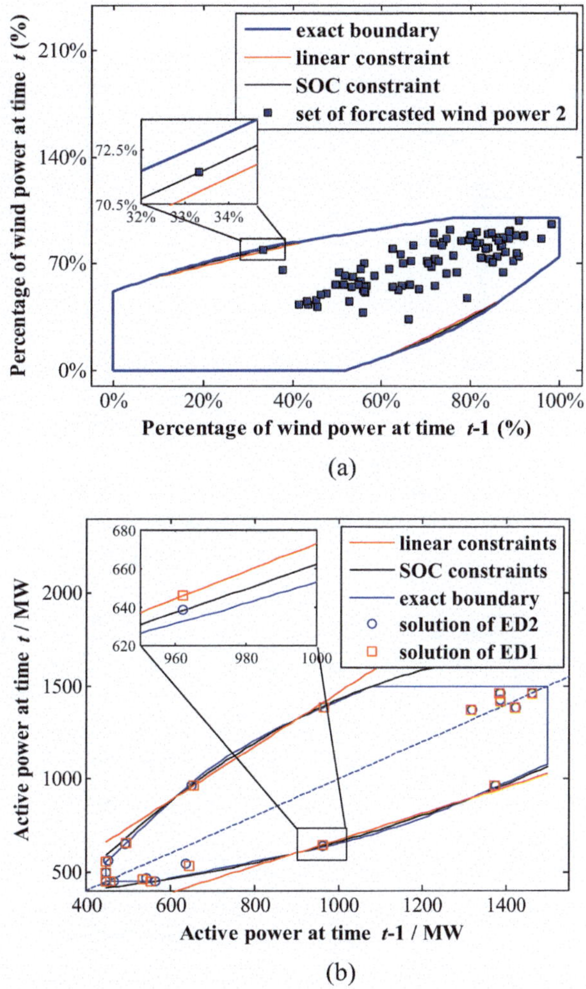

Meanwhile, the optimal dispatchable loads are gathered from 100 random wind power scenarios, which are represented in Fig. 15b.

Table 4 presents the average objective costs for ED3, ED4, and ED5 across these 100 wind power scenarios. The second column details the generation cost F_{o1}, while the third column indicates the operational cost F_{o2}. The final column F_o represents the total cost associated with this rolling economic dispatch. This comparison highlights the efficiency and responsiveness of each economic dispatch model in adapting to varying conditions and requirements.

Figure 15a demonstrates that models ED4 and ED5 can quickly respond to wind power fluctuations, with multiple output powers falling below 600 MW, while ED3 fails to accommodate these variations. This indicates that retrofitted coal-fired units offer greater flexibility compared to conventional coal-fired plants. Among the two,

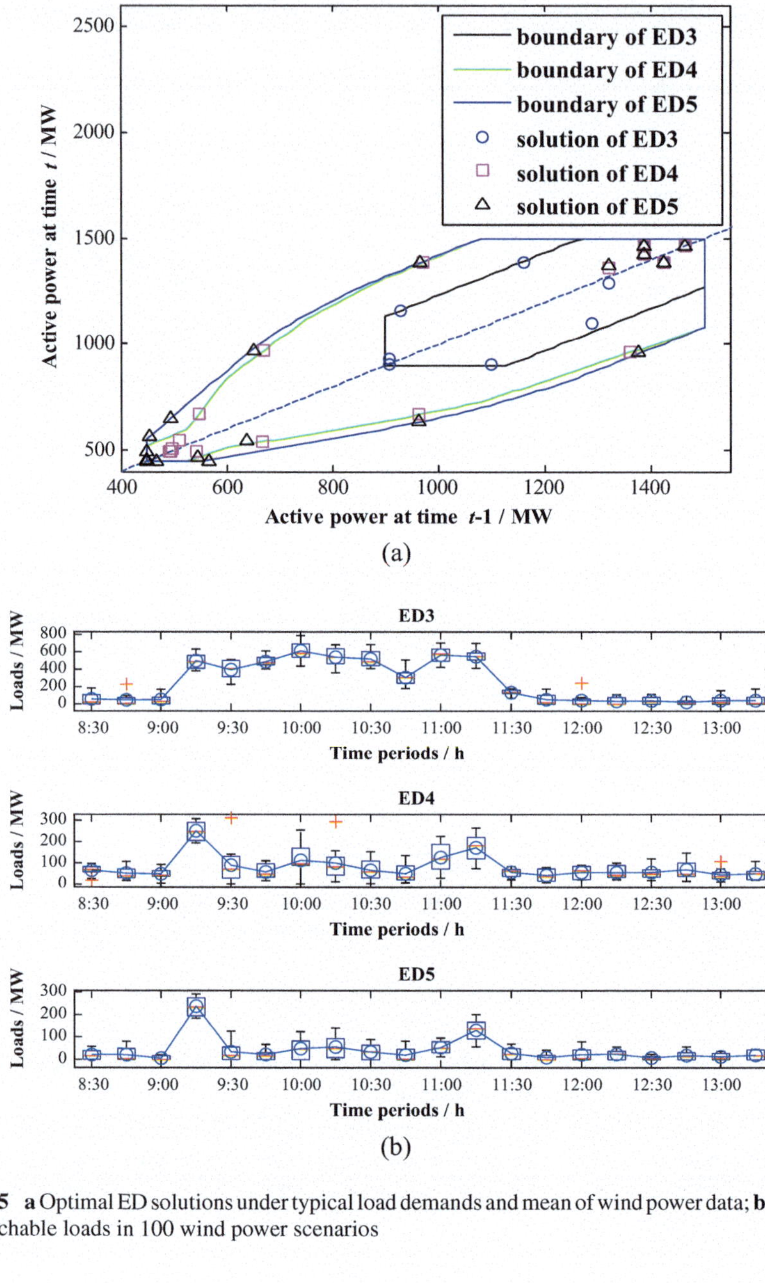

Fig. 15 **a** Optimal ED solutions under typical load demands and mean of wind power data; **b** optimal dispatchable loads in 100 wind power scenarios

Table 4 Average costs of 100 wind power scenarios

ED models	F_{o1} ($\times 10^5$\$)	F_{o2} ($\times 10^5$\$)	F_o ($\times 10^5$\$)
ED3	10.40	82.420	46.41
ED4	8.67	5.400	7.01
ED5	8.49	3.031	5.75

ED5 outperforms ED4, as evidenced by a larger number of ED5 solutions under 600 MW compared to ED4.

Additionally, Fig. 15b shows that ED5 requires fewer dispatchable loads from HVDNs than the quantities obtained by ED3 and ED4 during each dispatch period. These findings further confirm ED5's superiority in responding swiftly to unexpected wind power fluctuations, which is also reflected in Table 4, where ED5 exhibits the smallest total costs F_{o1}, F_{o2}, and F_o. This highlights the efficiency and effectiveness of the ED5 model in managing operational costs while maintaining responsiveness to dynamic conditions.

5.5 *Unrestricted Versus LTS-Based HVDNs*

Figure 16a,b illustrates the rolling ED differences between scenarios ED6 and ED7, utilizing wind power data from forecasted sets 2 and 3, respectively. In Fig. 16a, it is evident that ED7 and ED8 yield identical solutions, as there are no circuit switching actions involved. However, during periods of sharp wind power fluctuations between adjacent intervals, such as in Fig. 16b, ED6 and ED7 adopt different strategies for purchasing dispatchable loads from HVDNs. Specifically, at 09:30, the switched load of 213.27 MW becomes crucial as the wind power variation reaches 614.72 MW—exceeding the maximum ramp-down margin of 251.67 MW for retrofitted coal-fired units at that time. Without this switching, load transfer operations would be necessary. Notably, ED7 procures a greater amount of dispatchable loads from the LTS-based network compared to ED6 during specific times, such as 09:30, 11:15, and 11:30. This difference is attributed to the continuous circuit switching variable utilized in ED7, as opposed to the network-constrained discrete circuit switching variable in ED6.

5.6 *Centralized Versus Decentralized MBGD Methods*

In our study, we conduct simulation experiments to evaluate the computational efficiency of four distinct methods for economic dispatch (ED) in scenarios ED7 and ED8. The methods compared include a centralized MILP solver for ED8 (M1), a centralized MISOCP solver for ED7 (M2), a centralized MISOCP solver integrated within a GBD framework for ED7 (M3), and a decentralized MISOCP solver

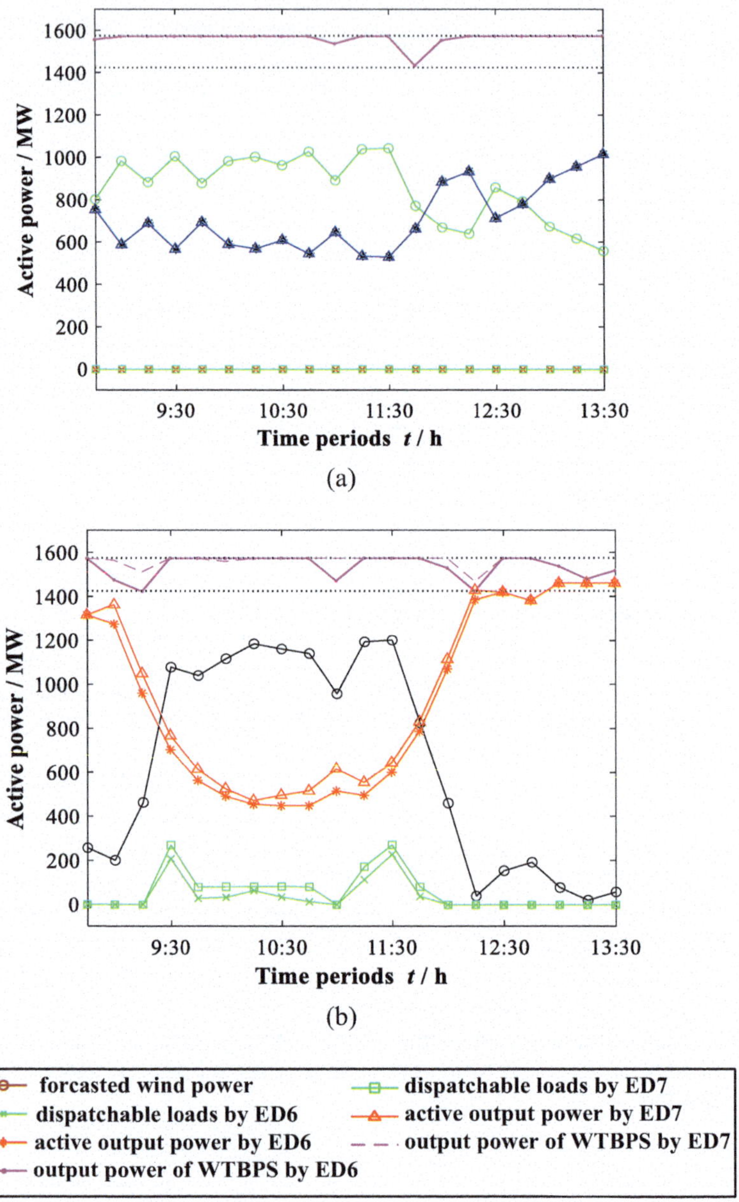

Fig. 16 Optimal ED solutions between ED6 and ED7: **a** set of forecasted wind power 2; **b** set of forecasted wind power 3; and **c** legends for **a** and **b**

embedded in a MGBD framework for ED7 (M4). These methods were implemented using CPLEX and MOSEK in a MATLAB environment on an AMD Ryzen 75800X 8-core CPU at 3.80 GHz. To simulate real-world conditions, we generated three types of stochastic wind power scenarios for four agents, with maximum power fluctuations relative to wind power rated capacity between periods set at ΔP values of 0.3, 0.5, and 0.8. Importantly, all forecasted wind power data adhered to a tightened ramping margin for WPTs under a ΔT of 15 min, while also respecting load transfer boundaries.

Table 5 presents the numerical results from case studies across different wind power scenarios. In Case II, we examine eight agents of WTBPS connected to 220 kV stations FZ, RD, XEC, and SY. Notably, the number of constraints and variables in this case is twice that of Case I. The first column lists the case numbers, reflecting varying ranges of wind power fluctuations, while the second column details four methods utilized in the analysis.

The third column indicates the total number of iterations needed to reach the optimum for the GBD and MGBD approaches. The fourth column presents the total CPU time (in seconds) required to solve the overall problem, and the fifth column includes the CPU running time (in seconds) for both the master problem and the sub-problems in methods M3 and M4, respectively.

For methods M1 and M2, M1 converges more rapidly than M2, requiring less CPU time across the two cases. In the larger system size of Case II, M1 continues to outperform M2, largely due to the use of the MILP-based ED8, while M2 addresses the MISOCP-based ED7, which provides a more accurate solution. Comparatively, M4 demonstrates significantly lower CPU time and fewer iterations than M3 in both cases. However, the CPU processing time for M3 increases substantially with larger system sizes.

Based on the above, M4 achieves effective results within less than 8 s of CPU processing time, resulting in over a 65% reduction in computation time compared to M3. Overall, M4 outperforms both M2 and M3 in terms of CPU time while also exceeding M1 in solution accuracy. When executed on multiple distributed computing servers, M4 is expected to enhance computational efficiency significantly within the decomposition-coordination computing framework, particularly for large-scale optimization problems.

Table 5 Computing performance of four methods

Cases	Methods	Iterations	Entire CPU time (s)	CPU time (s)	
				The master problem	The sub-problems
I $\Delta P = 0.3$ $N_u = 4$	M1	/	5.34	/	/
	M2	/	12.11	/	/
	M3	4	12.89	9.85	3.04
	M4	2	6.05	4.43	1.62
I $\Delta P = 0.5$ $N_u = 4$	M1	/	4.11	/	/
	M2	/	11.29	/	/
	M3	5	13.12	10.05	3.07
	M4	2	5.85	4.23	1.62
I $\Delta P = 0.8$ $N_u = 4$	M1	/	5.44	/	/
	M2	/	12.89	/	/
	M3	7	12.42	10.06	2.36
	M4	3	5.96	4.11	1.85
II $\Delta P = 0.3$ $N_u = 8$	M1	/	5.34	/	/
	M2	/	15.11	/	/
	M3	8	18.42	14.22	4.20
	M4	4	7.56	5.54	2.02
II $\Delta P = 0.5$ $N_u = 8$	M1	/	5.41	/	/
	M2	/	16.36	/	/
	M3	9	20.08	15.75	4.33
	M4	5	7.47	5.12	2.35
II $\Delta P = 0.8$ $N_u = 8$	M1	/	5.02	/	/
	M2	/	16.81	/	/
	M3	15	21.73	15.24	6.49
	M4	5	7.51	5.40	2.11

References

1. Q. Yang et al., Prospective contributions of biomass pyrolysis to China's 2050 carbon reduction and renewable energy goals. Nat. Commun. **12**, 1–12 (2021)
2. G. He, J. Lin, F. Sifuentes, X. Liu, N. Abhyankar, A. Phadke, Rapid cost decrease of renewables and storage accelerates the decarbonization of China's power system. Nat. Commun. **11**, 1–9 (2020)
3. S. Mallapaty, How China could be carbon neutral by mid-century. Nature **586**, 482–483 (2020)
4. C. Figueres et al., Emissions are still rising: ramp up the cuts. Nature **564**, 27–30 (2018)
5. S. Impram, S.V. Nese, B. Oral, Challenges of renewable energy penetration on power system flexibility: a survey. Energ. Strat. Rev. **31**, 1–12 (2020)
6. C. Lei, S. Bu, Q. Wang, Q. Chen, L. Yang, Y. Chi, Look-ahead rolling economic dispatch approach for wind-thermal-bundled power system considering dynamic ramping and flexible load transfer strategy. IEEE Trans. Power Syst. (2023) (Early Access)

7. D. J. Olsen, D. S. Kirschen, Profitable emissions-reducing energy storage. IEEE Trans. Power Syst. **35**(2), 1509–1519 (2020)
8. M. Zhou, J.Y. Zhai, G.Y. Li, J.W. Ren, Distributed dispatch approach for bulk AC/DC hybrid systems with high wind power penetration. IEEE Trans. Power Syst. **33**(3), 3325–3336 (2017)
9. Z. Chen, J. Zhu, H. Dong, W. Wu, H. Zhu, Optimal dispatch of WT/PV/ES combined generation system based on cyber-physical-social Integration. IEEE Trans. Smart Grid **13**(1), 342–354 (2022)
10. V.P. Sakthivel, P.D. Sathya, Single and multi-area multi-fuel economic dispatch using a fuzzified squirrel search algorithm. Prot. Control Mod. Power Syst. **6**(2), 147–159 (2021)
11. Z. Zhang, C. Wang, S. Chen, Y. Zhao, X. Dong, X. Han, Multi-time scale co-optimized dispatch for integrated electricity and natural gas system considering bidirectional interactions and renewable uncertainties. IEEE Trans. Ind. Appl. (2022) (Early Access)
12. C.Z. Shao, Y. Ding, P. Siano, Y.H. Song, Optimal scheduling of the integrated electricity and natural gas systems considering the integrated demand response of energy hubs. IEEE Syst. J. **15**(3), 4545–4553 (2020)
13. Y. Wang, S. Lou, Y. Wu, S. Wang, Flexible operation of retrofitted coal-fired power plants to reduce wind curtailment considering thermal energy storage. IEEE Trans. Power Syst. **35**(2), 1178–1187 (2020)
14. Z. Zhang, J. Shi, W. Yang, Z. Song, Z. Chen, D. Lin, Deep reinforcement learning based bi-layer optimal scheduling for microgrid considering flexible load control. CSEE J. Power Energy Syst. (2021) (Early Access)
15. United Nations Economic Commission For Europe. Opportunities for coupling wind and coal based generation in the electricity sector. https://unece.org/fileadmin/DAM/energy/se/pdfs/CES/ge12_WS/CEP-12_2016_INF.3v2.pdf
16. C. Na et al., Economic decision-making for coal power flexibility retrofitting and compensation in China. Sustainability **10**(348) (2018)
17. F. Ess, F. Peter, F. Klummp, *Flexibility in Thermal Power Plants-With a Focus on Existing Coal-Fired Power Plants*, (Agora Energiewende, Berlin, Germany, Tech. Rep. 115/04-S-2017/EN, 2017)
18. T. Li, M. Shahidehpour, Dynamic ramping in unit commitment. IEEE Trans. Power Syst. **22**(3), 1379–1381 (2007)
19. C.M. Correa-Posada, G. Morales-España, P. Dueñas, P. Sánchez-Martín, Dynamic ramping model including intraperiod ramp-rate changes in unit commitment. IEEE Trans. Sustain. Energy **8**(1), 43–50 (2017)
20. Z. Jin, K. Pan, L. Fan, T. Ding, Data-driven look-ahead unit commitment considering forbidden zones and dynamic ramp rate. IEEE Trans. Ind. Inf **15**(6), 3267–3276 (2019)
21. X. Li et al., Wind-Thermal-CSP bundling model with an adjustable heat storage strategy for CSP stations. IEEE Access **9**, 25984–25992 (2021)
22. C. Lei, S. Bu, Q. Wang, N. Zhou, L. Yang, X. Xiong, Load transfer optimization considering hot-spot and top-oil temperature limits of transformers. IEEE Trans. Power Deli. **37**(3), 2194–2208 (2022)
23. IEEE draft guide for overhead AC transmission line design. IEEE P1863/D10, pp. 1–96, 25 (2019)
24. *Distributed Generation in Europe—Physical Infrastructure and Distributed Generation Connection*, KEMA, Inc., pp. 26 (2011)
25. M. Laaja, General planning principles of high voltage distribution networks including wind power, Master of Science Thesis, Department of Computing and Electrical Engineering, Tampere University of Technology, Finland (2012). https://trepo.tuni.fi/handle/123456789/21124
26. H. Zhu, H. J. Liu, Fast local voltage control under limited reactive power: optimality and stability analysis. IEEE Trans. Power Syst. **31**(5), 3794–3803 (2016)
27. X. Zhang, Y. Liu, H. Gao, L. Wang, J. Liu, A bi-level corrective Line switching model for urban power grid congestion mitigation. IEEE Trans. Power Syst. **35**(4), 2959–2970 (2020)

28. J. Wang et al., Cooperative overload control strategy of power grid-transformer considering dynamic security margin of transformer in emergencies. Int. J. Electr. Power Energy Syst. **140**, 108098 (2022)
29. G. McCormick, Computability of global solutions to factorable nonconvex programs: part I convex underestimating problems. Math. Program. **10**, 146–175 (1976)
30. N.I. Muskhelishvili, *Singular Integral Equations* (Groningen, Netherlands, 1953)
31. B. Fan, X. Wang, Distributed privacy-preserving active power sharing and frequency regulation in microgrids. IEEE Trans. Smart Grid **12**(4), 3665–3668 (2021)

Modeling of Natural Gas Dynamics Impact on Power Systems with High Renewables

Reza Bayani and Saeed Manshadi

Abstract Natural gas-fired generation units help mitigate the volatility of renewable energy sources. The fast-ramping capability of gas-fired units has rendered them as one of the best solutions to possible fluctuations in the system. This chapter focuses on the modeling of natural gas system dynamics and discusses that accurate modeling is essential to manage the volatile demand of these units, particularly during short-term fluctuations. While steady-state models are commonly used to simplify these dynamics, they are argued to be unreliable for short-term operations. The chapter first discusses different approaches to modeling natural gas dynamics and present several convex relaxation methods. It also introduces techniques for the coordinated operation of interconnected electricity-gas systems. Through the case studies, the chapter presents the steady-state model of natural gas optimization and highlights its drawbacks. Finally, the chapter shows that the rank-minimization relaxation technique can accurately and efficiently solve the non-convex natural gas flow equations.

Keywords Optimal gas flow · Integrated electricity-gas system · Natural gas dynamics · Convex relaxation · Short-term operation

1 Introduction

Renewable energy sources (RESs) inherently produce variable and non-dispatchable power. In an electricity system integrated with RESs, net demand is calculated by subtracting the total RES generation from the total demand. As the penetration of RESs increases, so does the variability in net demand. Inability to meet this net

R. Bayani · S. Manshadi (✉)
Department of Electrical and Computer Engineering, San Diego State University, San Diego, USA
e-mail: smanshadi@sdsu.edu

R. Bayani
Department of Electrical and Computer Engineering, University of California San Diego, San Diego, USA

F. Shahnia and F. Wen (eds.), *Power and Energy Resources*, Power Systems,
https://doi.org/10.1007/978-981-96-2612-0_4

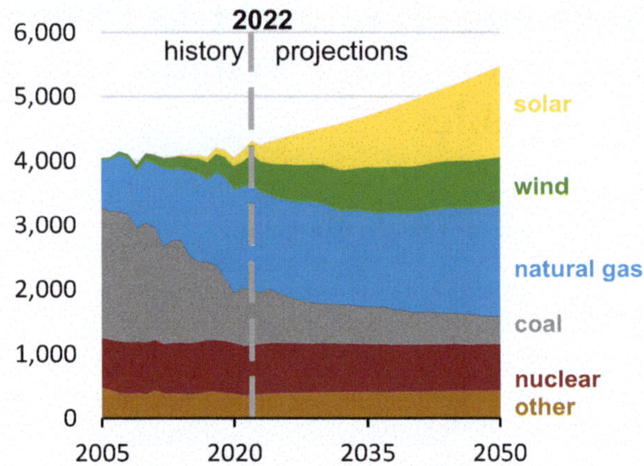

Fig. 1 U.S. net electricity generation by fuel [1]

demand can lead to significant economic and operational issues for system operators, including price volatility and load shedding. Consequently, it is essential to adopt strategies that address the challenges associated with the high penetration of non-dispatchable RESs. Natural gas-fired generation units are the primary source of flexibility in the electricity system, thanks to their fast-ramping capabilities. These units are mainly used during peak periods because of their rapid response capabilities, providing crucial support when renewable energy generation is insufficient. Figure 1 illustrates that gas-fired power accounted for 39% of the fuels contributing to electricity generation in 2022. At the same time, it is projected that the share of RESs' generation will reach 39% by 2050, highlighting the increasing significance of solutions that manage the uncertainties in the system.

Gas-fired units experience daily evening surges caused by the concurrent decline in solar generation and the increase in electricity consumption. Data displayed in Fig. 2 from the California Independent System Operator reveals daily evening surges in natural gas generation due to this pattern [2]. Because these variations affect electricity operations, accurate modeling of fluid dynamics is crucial for the integrated operation of electricity and natural gas systems. The coordinated operation of electricity and natural gas networks has been the focus of numerous studies, exploring areas such as demand response [3], renewable energy integration [4], expansion planning [5], and market operations [6]. The extent of coordination varies; some models [7] incorporate natural gas constraints as part of the electricity optimization process, while others [8] fully integrate the operational states of both networks.

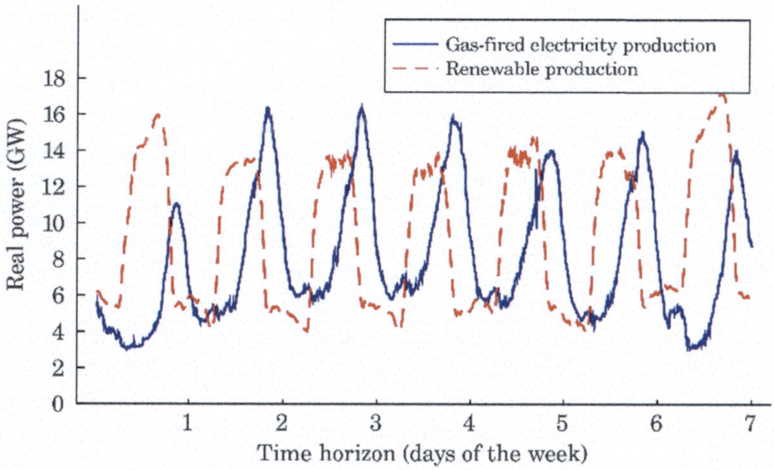

Fig. 2 Gas-fired and RES electricity generation during fourth week of June 20 in the network of California independent system operator [2]

1.1 Literature Review

Accurate modeling of natural gas dynamics is crucial, particularly for short-term planning applications where fluctuations in gas demand significantly impact electricity operations. Capturing natural gas dynamics entails adopting complex and computationally demanding models. In some studies, focused on mid-term and long-term planning, gas flow dynamics are not considered [9]. However, in short-term planning applications, variations are more pronounced, and fluid dynamics cannot be neglected. In such cases, choosing an appropriate optimal gas flow (OGF) model is critical due to the limited time available for model execution.

Simplified steady-state models are frequently employed in the literature to avoid the computational complexity of solving the differential equations in the OGF problem. The Weymouth model [10] is a widely used steady-state equation derived from approximating the flow equations governing fluid dynamics in the pipeline. To address the non-convexity of this equation, some studies have approximated the Weymouth model with linearization techniques [11], while others have employed convex relaxation methods such as second-order cone (SOC) programming [12] and semi-definite programming [13]. However, as demonstrated in [14], the Weymouth equation proves incorrect for short-term operation problems.

Steady-state approximations, although computationally efficient, result in inaccurate outcomes for short-term operation problems where natural gas consumption rates are highly variable. These methods fail to capture the dynamic relationships between natural gas pressure and mass flow rate due to gas compressibility and low velocity within pipelines and across different time steps. Therefore, for short-term planning periods, it is preferable to use dynamic models that accurately represent the intra-hour variations of natural gas. Additionally, steady-state models incorrectly assume an instantaneous supply–demand balance within the natural gas network. In reality, it can take several hours for natural gas to travel from the source to the demand junction, depending on the distance.

Dynamic models address the OGF problem by incorporating the governing partial differential equations of the natural gas network [15]. These equations are converted into time-difference equations using finite difference methods [16], enabling numerical solutions for the OGF problem. Despite this, solving these non-linear, non-convex, spatio-temporally discretized optimization problems remains challenging due to the extended computation time required, rendering them impractical for short-term planning periods. Also, these models often require non-linear solvers, which neither guarantee convergence to the optimal solution nor have a definitive gap. To address the aforementioned challenges, a rank-minimization relaxation method is presented in this chapter, which significantly reduces solver time compared to the original non-convex models.

2 Natural Gas Dynamics

The natural gas momentum equation, also known as the Navier–Stokes Eq. (1), describes the overall forces affecting natural gas particles, whereas (2) describes the natural gas mass conservation equation, also known as the material-balance equation [7]:

$$\frac{\partial}{\partial t}(\rho\omega) + \frac{\partial}{\partial x}\left(\rho\omega^2\right) + \frac{\partial \pi}{\partial x} + g\rho\frac{\partial H}{\partial x} + \frac{\lambda}{2D}\rho\omega^2 = 0 \tag{1}$$

$$\frac{\partial \rho}{\partial t} + \frac{\partial}{\partial x}(\rho\omega) = 0 \tag{2}$$

Here, ρ is the gas density (kg/m^3), ω is the gas velocity (m/s), π represents pressure (Pa), H is the altitude of natural gas pipeline (m), λ is the friction factor, and D is the diameter (m) of the pipelines. The first term in (1) describes the acceleration of natural gas particles, the third term represents the hydrostatic effects of the natural gas, while the last term reflects the second-order deviatoric stress tensor.

Next, we introduce some simplification assumptions to achieve a more tractable representation of (1). The second term, representing the kinetic energy (also known as the convective term), can be neglected for natural gas systems because the velocity of the fluid is negligible compared to the speed of sound [17]. The fourth term describes the force of gravity (g is the gravitational constant) and can be removed by assuming horizontal pipelines where the height variation of the pipes is zero. These assumptions together convert (1) to a more manageable form presented below

$$\frac{\partial}{\partial t}(\rho\omega) + \frac{\partial \pi}{\partial x} + \frac{\lambda}{2D}\rho\omega^2 = 0 \tag{3}$$

Next, we apply some conversions to express (3) in terms of mass flow (denoted by q in kg/s) and pressure, which are easier to measure. To achieve this, we utilize the thermodynamic state Eq. (4). Additionally, the relationship between pressure and density is defined in (5):

$$m = \rho\omega A \tag{4}$$

$$\pi = \rho c^2 \tag{5}$$

Here, A is the area of natural gas pipeline, and c is the speed of sound in the fluid (m/s), which can be considered constant under isothermal conditions. By integrating (4) and (5), Eqs. (2) and (3) are converted to the form presented in (6) and (7), respectively:

$$\frac{\partial \pi}{\partial t} + \frac{c^2}{A}\frac{\partial q}{\partial x} = 0 \tag{6}$$

$$\frac{1}{A}\frac{\partial q}{\partial t} + \frac{\partial \pi}{\partial x} + \frac{c^2\lambda}{2DA^2}\frac{q^2}{\pi} = 0 \tag{7}$$

2.1 Spatio-Temporal Representation of Natural Gas Dynamics

To implement (6) and (7) in an optimization model, we can apply the finite-difference method to avoid solving the partial differential equations directly. Subsequently, the spatio-temporally discretized forms of these equations for pipeline k are presented as

$$\frac{\pi^k_{(s+1,t+1)} - \pi^k_{(s+1,t)}}{2\Delta t} + \frac{\pi^k_{(s,t+1)} - \pi^k_{(s,t)}}{2\Delta t} + \frac{c^2}{A}\frac{q^k_{(s+1,t+1)} - q^k_{(s,t+1)}}{\Delta x} = 0 \tag{8}$$

$$\frac{1}{2A}\frac{q^k_{(s+1,t+1)} - q^k_{(s+1,t)}}{\Delta t} + \frac{1}{2A}\frac{q^k_{(s,t+1)} - q^k_{(s,t)}}{\Delta t} + \frac{\pi^k_{(s+1,t+1)} - \pi^k_{(s,t+1)}}{\Delta x} + \frac{c^2\lambda}{2DA^2}\frac{{q^k_{(s,t)}}^2}{\pi^k_{(s,t)}} = 0 \tag{9}$$

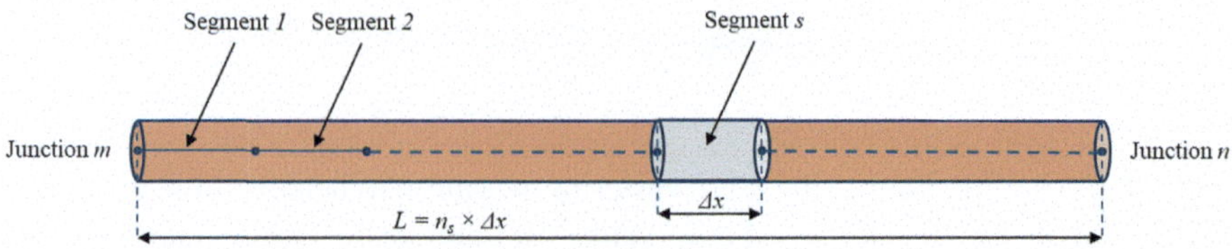

Fig. 3 Schematic of Pipe k in natural gas system

Here, $t \in T = \{1, 2, \dots, n_t\}$ is the set of time steps in the operation horizon, and $s \in S = \{1, 2, \dots, n_s\}$ represents the segmented sections of pipe k with total length L, where each segment has a length of Δx, as illustrated in Fig. 3.

2.2 *Steady-State (Weymouth) Model*

Often, research considers the equations presented in (8) and (9) in a steady-state mode. In this case, it is assumed that the temporal dynamics of natural gas particles have stabilized. To reflect this, the assumption $\Delta t \to \infty$ is applied to the spatio-temporal dynamics, resulting in the simplified equation known as the Weymouth equation:

$$q_{mn}|q_{mn}| = w_{mn}(\pi_m^2 - \pi_n^2) \tag{10}$$

As a result, it is observed that the mass flow rate throughout the pipeline mn (connecting junctions m and n) remains constant and is a function of the parameter $w_{mn} = \frac{2DA^2}{Lc^2\lambda}$ and the pressures at the two neighboring junctions.

3 Optimal Operation of Electricity and Natural Gas Systems

This section explores the optimal operation of RES-integrated electricity grids containing gas-fired generators, as well as the optimal operation of the natural gas system. In interconnected electricity-gas systems (IEGSs), the availability of natural gas can significantly impact the operation of gas-fired units within the electricity grid. We begin by presenting formulations for the independent operation of each system. The following section delves into mathematical solutions for managing natural gas dynamics, and the subsequent section addresses the joint operation of IEGSs.

3.1 Electricity System Operation Problem

The optimization problem for the daily operation of the electricity system is presented in (11a–11d). The objective function of this problem is shown in (11a). The first term considers the generation costs using a quadratic cost function, while the second term penalizes unserved energy by a factor of c_d:

$$\min \sum_{t \in T} \sum_{g \in G} (c_{2g} p_{g,t}^2 + c_{1g} p_{g,t} c_{0g}) + \sum_{t \in T} \sum_{d \in D} c_d (\overline{p}_{d,t} - p_{d,t}), \text{ subject to:} \tag{11a}$$

$$\underline{p}_{z,t} \leq p_{z,t} \leq \overline{p}_{z,t}, \forall z \in \{G, R, D, L\}, \forall t \in T \tag{11b}$$

$$\sum_{g \in G} p_{g,t} + \sum_{r \in R} p_{r,t} + \sum_{l \in L^{to}} p_{l,t} = \sum_{d \in D} p_{d,t} + \sum_{l \in L^{fr}} p_{l,t}, \forall b \in B, \forall t \in T \tag{11c}$$

$$p_{l,t} \cdot x_l = \theta_{i,t} - \theta_{j,t}, \forall l \in L, \forall t \in T \tag{11d}$$

Equation (11b) enforces the lower and upper bounds for the dispatched power of all elements in the system. For generation units ($g \in G$), the dispatched power at each time ($p_{g,t}$) is constrained by their operational limits. For RESs ($r \in R$), $p_{r,t}$ is subject to the availability of their sources (e.g., solar radiation and wind). The served demand ($d \in D$) at each demand bus ($p_{d,t}$) is limited between zero and the requested value ($\overline{p}_{d,t}$). For each line ($l \in L$), the power flow ($p_{l,t}$) is constrained by thermal limits. The nodal balance equation at each bus is shown in (11c), where $l \in L^{to}$ and $l \in L^{fr}$ denote the lines leading to and originating from a bus, respectively. The DC power flow model for each line is given in (11d), where x_l is the reactance of that line.

3.2 Natural Gas System OGF Problem

The equations for the OGF problem of the natural gas system are presented in their non-convex form in (12a–12e). The objective function calculates the cost of gas production based on the mass of the gas at each well $q_{w,t}$, with the linear cost term c_w:

$$\min \sum_{t \in T} \sum_{w \in W} c_w q_{w,t}, \text{ subject to:} \tag{12a}$$

$$\underline{\pi}_{j,t} \leq \pi_{j,t} \leq \overline{\pi}_{j,t}, \forall j \in J, \forall t \in T \tag{12b}$$

$$\underline{q}_{w,t} \leq q_{w,t} \leq \overline{q}_{w,t}, \forall w \in W, \forall t \in T \tag{12c}$$

$$q_{g,t} = h_g(p_{g,t}), \forall g \in G, \forall t \in T \tag{12d}$$

$$\sum\nolimits_{w \in W} q_{w,t} + \sum\nolimits_{k \in K^{to}} q_{k,t} = \sum\nolimits_{n \in N} q_{n,t} + \sum\nolimits_{g \in G} q_{g,t} + \sum\nolimits_{k \in K^{fr}} q_{k,t}, \forall j \in J, \forall t \in T \tag{12e}$$

The operational bounds for the pressure of natural gas at each junction ($j \in J$) and the flow of gas generated at each well ($w \in W$) are defined in (12b) and (12c), respectively. The natural gas demand of gas-fired generators is modeled in (12d), where h_g is a function that calculates the gas fuel required for a unit's operation given its scheduled dispatch. The supply–demand balance of natural gas at each junction is ensured in (12e). Additionally, the natural gas spatio-temporal dynamics must be accounted for, as represented by (8) and (9), which should also be considered in (12a)–(12e).

So far, we have formulated the operational models for the electricity (11a)–(11d) and natural gas (12a)–(12e) systems. These systems are interconnected through the fuel consumption of gas-fired units, which depends on their generation dispatch. Solving (12a)–(12e) along with (8) and (9) can become challenging for real-world applications due to the time-consuming nature of the OGF problem. Next, we discuss some relaxation approaches to transform the non-convex dynamic OGF formulation into a convex one.

4 Solving the OGF Problem

This section introduces a novel approach to solving the non-linear, non-convex OGF problem by leveraging a dynamic flow model and a rank-minimization technique through a bi-level optimization formulation. Traditional methods, such as the Weymouth equation, are criticized for their inadequacies in short-term scenarios due to their reliance on simplifying assumptions that fail to accurately capture the dynamic behavior of natural gas systems.

4.1 Convex Relaxation of OGF

We present a convex relaxation scheme that transforms the non-convex OGF problem into a more tractable form while preserving the integrity of the original dynamics. This relaxation method significantly reduces computational time compared to traditional non-linear solvers, which is essential for real-time applications. The first step in our relaxation process is to relax the non-convex term present in (9) by introducing the lifting operation $\frac{{q^k_{(s,t)}}^2}{\pi^k_{(s,t)}} \to \gamma^k_{(s,t)}$, resulting in a convex-relaxed form as shown in (13). Additionally, we introduce positive semi-definite (PSD) cones, as shown in

(14), to further tighten this relaxation:

$$\frac{1}{2A}\frac{q^k_{(s+1,t+1)} - q^k_{(s+1,t)}}{\Delta t} + \frac{1}{2A}\frac{q^k_{(s,t+1)} - q^k_{(s,t)}}{\Delta t} + \frac{\pi^k_{(s+1,t+1)} - \pi^k_{(s,t+1)}}{\Delta x} + \frac{c^2\lambda}{2DA^2}\gamma^k_{(s,t)} = 0 \tag{13}$$

$$\begin{bmatrix} \pi^k_{(s,t)} & q^k_{(s,t)} \\ q^k_{(s,t)} & \gamma^k_{(s,t)} \end{bmatrix} \succcurlyeq 0 \tag{14}$$

One can adopt the SOC relaxation method presented in (15) instead of (14). Applying either of these equations will increase the tightness of the lifting operation:

$$\left\| \begin{matrix} 2q^k_{(s,t)} \\ \gamma^k_{(s,t)} - \pi^k_{(s,t)} \end{matrix} \right\| \leq \gamma^k_{(s,t)} + \pi^k_{(s,t)} \tag{15}$$

4.2 *Standard Rank-Minimization*

Although replacing (9) with (13) and (14) addresses the non-convexity of the OGF problem, this relaxation is not very tight. If the tightness is not guaranteed, the solution to the relaxed OGF problem will not be physically meaningful. To enhance the tightness of the relaxation, we employ a rank-minimization technique, which minimizes the trace of the PSD matrix (14):

$$\min \sum\nolimits_{s\in S} \sum\nolimits_{k\in K} \sum\nolimits_{t\in T} (\pi^k_{(s,t)} + \gamma^k_{(s,t)}) \tag{16}$$

4.3 *Bi-Level Rank-Minimization*

The relaxation presented in (16) increases the tightness of the PSD cones; however, it may still result in solutions that are not necessarily feasible [14]. To ensure feasibility, a bi-level optimization problem is formulated. The upper-level problem consists of (16), minimizing the rank of the solution matrix to enforce tightness, while the lower-level problem (17a), (17b) optimizes the relaxed objective function subject to the natural gas flow constraints:

$$(16) \tag{17a}$$

$$(14), \gamma \in \text{argmin}\,\{(8), (12), (13)\} \tag{17b}$$

4.4 Single-Level Rank-Minimization

To solve the bi-level problem (17a), (17b), we leverage the convexity of (17b) and apply the Karush–Kuhn–Tucker (KKT) conditions to transform (17a), (17b) into its equivalent single-level convex form (18a)–(18d):

$$(16) \tag{18a}$$

$$(8), (12b) - (12e), (13), (14) \tag{18b}$$

$$\text{Dual constraints of } \{(8), (12b) - (12e), (13)\} \tag{18c}$$

$$\begin{aligned}
&\sum\nolimits_{t\in T}\Big(\sum\nolimits_{w\in W} c_w q_{w,t} + \sum\nolimits_{g\in G} c_g \left(h_g(p_{g,t}) - q_{g,t}\right)\Big) \\
&= \sum\nolimits_{t\in T}\Big(\sum\nolimits_{j\in J} \left(\mu^{lb}_{j,t} \cdot \pi_{j,t} - \mu^{ub}_{j,t} \cdot \bar{\pi}_{j,t}\right) \\
&+ \sum\nolimits_{w\in W} \left(\omega^{lb}_{w,t} \cdot \underline{q}_{w,t} - \omega^{ub}_{w,t} \cdot \bar{q}_{w,t}\right) \\
&- \sum\nolimits_{g\in G} \eta_{g,t} \cdot f_g\left(p_{g,t}\right) + \sum\nolimits_{n\in N} \psi_{n,t} \cdot q_{n,t}\Big)
\end{aligned} \tag{18d}$$

Here, (18d) represents the strong duality constraint, where μ, ω, η, and ψ are the dual variables corresponding to constraints (12b)–(12e), respectively. This method addresses the natural gas short-term operation problem by overcoming the limitations of both the steady-state model and the non-convex dynamic model through a novel convex relaxation and rank-minimization approach. A summary of the introduced models to solve the OGF problem is presented in Table 1.

5 Coordinated Operation of IEGSs

So far, we have employed methodologies to transform the initial non-convex OGF problem into a tractable convex form, which can be solved to optimality very quickly. To coordinate the operation of the IEGS, both the electricity system operator and the gas system operator must solve their respective optimal operation problems. However, due to the presence of gas-fired generators, their optimal operation schedules may not necessarily align. The natural gas fuel required for the optimal dispatch of gas-fired generators, as determined by the electricity operator, can impact the

Table 1 Summary of different OGF models

OGF model choice	Non-convex OGF	PSD-convex OGF	SOC-convex OGF	Standard rank-minimization	Bi-level rank-minimization	Single-level rank-minimization	Steady-state (Weymouth)
Objective	(12a)	(12a)	(12a)	Multi-objective with (12a) and (16)	(16)	(16)	(12a)
Constraints	(12b)–(12e), (8), (9)	(12b)–(12e), (8), (13), (14)	(12b)–(12e), (8), (13), (15)	(12b)–(12e), (8), (13), (14)	(14), $\gamma \in$ argmin $\{(8), (12), (13)\}$	(18b)–(18d)	(12b)–(12e), (10)

optimal operating conditions of the natural gas system, and vice versa. This section presents two methods for solving the coordinated optimization problem of IEGSs.

5.1 Centralized Optimization

If both systems are operated by the same utility, this coordination will not pose a serious challenge. Let $f(x)$ denote the objective function of the optimal operation of the electricity system (11a)–(11d), where x represents the decision variables. Similarly, let $g(z)$ denote the natural gas system objective function in (18a)–(18d). To solve the joint coordinated operation problem of these systems, (19) needs to be solved:

$$\min f(x) + g(z), \text{ subject to } (11b) - (11d), (18b) - (18d) \tag{19}$$

It is noteworthy that (19) contains the coupling constraint (12d), which relates the gas demand of gas-fired units to their electricity dispatch and can be viewed as $z = h(x)$. If the operator has access to all the information required for solving Eq. (19), this is the most straightforward way of coordinating the operation of IEGS. We use this method for our case studies, but we also present an alternative approach for situations where centralized operation is not possible.

5.2 Decentralized Optimization

Due to limitations regarding privacy, regulations, and communication infrastructure, the coordinated operation of IEGSs is not always feasible in practice. To address this challenge, we present an optimization method using the alternating direction method of multipliers (ADMM) approach to achieve shared decision-making. By employing the convex-relaxed form of the OGF problem, we ensure the convergence of the ADMM algorithm and significantly reduce the computational burden [18]. If the objectives of problems (11a)–(11d) (the electricity system) and (18a)–(18d) (the natural gas system) are denoted by $f(x)$ and $g(z)$ respectively, the augmented Lagrangian function for solving the coordinated operation problem can be defined as (20). Here, x and z represent the dispatch decisions for gas-fired units at each step, obtained by the electricity and natural gas systems, respectively. The penalty parameter and the Lagrangian multiplier are denoted by ρ and λ, respectively:

$$L_\rho(x, z, \lambda) = f(x) + g(z) + \langle \lambda, x - z \rangle + \frac{\rho}{2} \|x - z\|_2^2 \tag{20}$$

The consensus-based coordinated operation ADMM algorithm used to solve the coordination problem of IEGS is outlined in Algorithm 1. In each step, each system determines and communicates the dispatch of gas-fired units based on operational constraints and decisions shared by the other system. This message-passing continues until the decisions of both systems converge to within a specified threshold. Once the convergence criteria are met, a consensus is reached, and the solution process terminates. The input parameters for this algorithm are the maximum number of iterations (i_{max}), the penalty parameter (ρ), and the convergence threshold (ϵ).

Algorithm 1 The ADMM method.

Input: i_{max}, ρ, ϵ
Output: $p_{g,t}$
Initiate: $z^0, \lambda^0 \leftarrow \mathbf{0}$
1 **for** $i = 0 : i_{max}$ **do**
2 $\quad x^{i+1} \leftarrow \text{argmin}_x \, L_\rho(x, z^i, \lambda^i)$, subject to: (11b)-(11d)
3 $\quad z^{i+1} \leftarrow \text{argmin}_z \, L_\rho(x^{i+1}, z, \lambda^i)$, subject to:(18b)-(18d)
4 $\quad \lambda^{i+1} = \lambda^i + \rho(x^{i+1} - z^{i+1})$
5 $\quad$ **if** $\frac{|x^{i+1} - z^{i+1}|}{(x^{i+1} + z^{i+1})/2} \leq \epsilon$
$\quad$ **break**
6 **end**

The ADMM algorithm begins by initializing the natural gas system decision (z) and the Lagrangian multiplier (λ) to zero. At line 2, the electricity system's decision is determined by minimizing the Lagrangian function, which is then passed to the natural gas system. At line 3, the natural gas system's decision is obtained by minimizing the Lagrangian function. These steps are akin to solving Eqs. (11a)–(11d) and (18a)–(18d), but with modified objectives that take into account the other system's decision. At line 4, the Lagrangian multiplier term is updated based on the difference between the decisions of the two systems. At the end of each iteration, the relative error of the decisions from both systems is computed at line 5. Convergence is achieved when the relative error for each unit's decision at all times is less than ϵ, indicating that the decisions of both systems are equal and a consensus has been reached.

6 Unit Conversion and Parameters

A critical aspect of solving fluid dynamics is the meticulous attention to units and conversions. This section presents the calculations and conversions required to handle the units utilized in the gas flow equations. These details will help the reader better understand the units and facilitate the reproducibility of our results. First, we present the set of parameters required to model Eqs. (8) and (9) in Table 2 below.

Table 2 Value of parameters required for OGF model

Parameter	Symbol	value (unit)
Time segment granularity	Δt	300 (s)
Pipe segment granularity	Δx	5,000 (m)
Pipe diameter	D	24 (in)
Pipe area	A	1.167 (m^2)
Friction factor	λ	0.005
Speed of sound in fluid	c	410 (m/s)

Next, we discuss how pressure and mass of natural gas are represented in these equations. For pressure (π), the SI unit Pascal (Pa) is required for gas dynamic equations. In our experiments and simulations, we use bar units, which are equivalent to 10^5 Pa. For the mass of natural gas, the SI unit kilogram (Kg) is required. However, natural gas quantities are often reported in terms of volume. A widely used metric is the standard thousand cubic feet (Mscf) unit, which measures the volume of natural gas at 60 °F and 1 atm.

As an example, let us calculate the mass flow rate of natural gas required for the generation of 100 MW of power at a gas-fired power plant. If the amount of gas fuel required for the generation of 1 kWh of electricity is 7.4 scf, the gas mass rate in terms of Kg/s is calculated as follows:

1. Convert the power generation requirement to kWh:

$$100\ \text{MW} \times \frac{1000\ \text{kW}}{1\ \text{MW}} = 100{,}000\ \text{kWh}$$

2. Determine the volume of gas needed per hour:

$$100{,}000\ \text{k Wh} \times \frac{7.4\ \text{scf}}{1\ \text{kWh}} = 740{,}000 \frac{\text{scf}}{\text{h}}$$

3. Convert the volume to standard cubic meters (scm):

$$740,000 \frac{\text{scf}}{\text{h}} \times \frac{0.3048^3 \text{m}^3}{1\ \text{f}^3} = 20,953.39 \frac{\text{scm}}{\text{h}}$$

4. Convert the volume to cubic meters per second (m^3/s):

$$20,953.39 \frac{\text{scm}}{\text{h}} \times \frac{1\ \text{h}}{3600\ \text{s}} = 5.8204 \frac{\text{scm}}{\text{s}}$$

5. Convert the volume to mass using the density of natural gas ($\approx$0.8 kg/m^3 at standard conditions):

$$5.8204\frac{\text{scm}}{\text{s}} \times \frac{0.8\ \text{kg}}{1\ \text{scm}} = 4.656\frac{\text{Kg}}{\text{s}}$$

7 Results and Discussion

This section investigates the short-term operation problem of natural gas systems with dynamics induced by the volatile demand of gas-fired units. Three case studies are provided to demonstrate the effectiveness of the rank-minimization method and compare it to both steady-state and SOC-relaxation models. The findings emphasize the need for accurate dynamic models to ensure the reliable operation of coupled electricity and natural gas systems, particularly in systems with high renewable integration that are prone to considerable net demand fluctuations in the short term.

7.1 System Parameters

The diagram of the test IEGS is presented in Fig. 4. This system comprises a 6-bus electricity transmission grid linked to a 6-junction natural gas system. A one-day (24 h) operation period of the system is considered. The total daily demand of the electricity system is 14,873.2 MWh, with a peak value of 774.9 MW. Both generation units in the electricity system are gas-fired and fed via pipes connecting them to the natural gas system, with their detailed data provided in Table 3. There are two types of RESs in the electricity system: solar photovoltaic and wind turbine farms, with output powers peaking at 576 MW and 236 MW, respectively. The natural gas system has only one supplier, located at junction A in Gas Well 1, with a maximum gas supply value of 200 kg/s. The operational pressure bounds for the natural gas system are within 30 to 60 bars. In addition to the gas-fired generation units, the natural gas system serves other loads (gas load 1 to 4 in Fig. 4), which we refer to as heat demand, requiring a total of 70,516 Mcf of natural gas for the day.

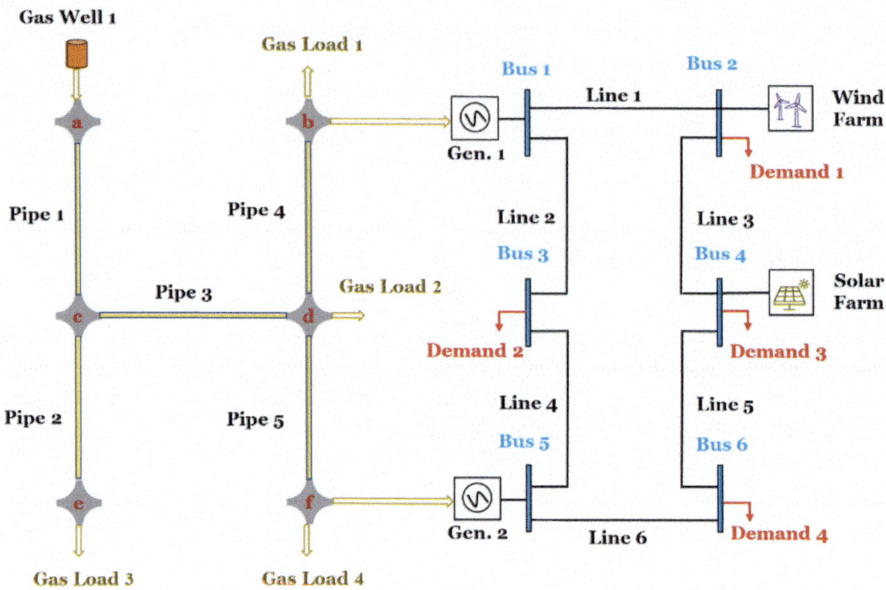

Fig. 4 Diagram of the 6-Bus 6-junction IEGS

Table 3 Generator unit data

Unit number	$\underline{P}_g (MW)$	$\overline{P}_g (MW)$	$c_{2g}(\frac{\$}{MWh^2})$	$c_{1g}(\frac{\$}{MWh})$	c_{0g} (\$)	$h_g\left(\frac{Mcf}{MWh}\right)$
1	60	300	0.013	32	0	13.51
2	80	480	0.0154	26	0	20.63

7.2 *Utilizing a Steady-State Model Versus Dynamic Model*

In the first case study, we compare the results of system operation based on the choice of the OGF model. Specifically, we focus on solving the coordinated operation problem using the Weymouth steady-state model and the bi-level rank-minimization approach, referred to as "rank" from now on. The comparative case study demonstrates that using the Weymouth model can lead to inaccurate results and potentially costly consequences for power system operations involving gas-fired units.

7.2.1 Results of the Dynamic Model

According to the results of the simulation using the rank OGF model displayed in Fig. 5, it is observed that during the peak period, the electricity system is unable to meet the demand due to gas system dynamics. As a result, 858.4 MWh of the electricity demand, equivalent to 5.77%, is shed.

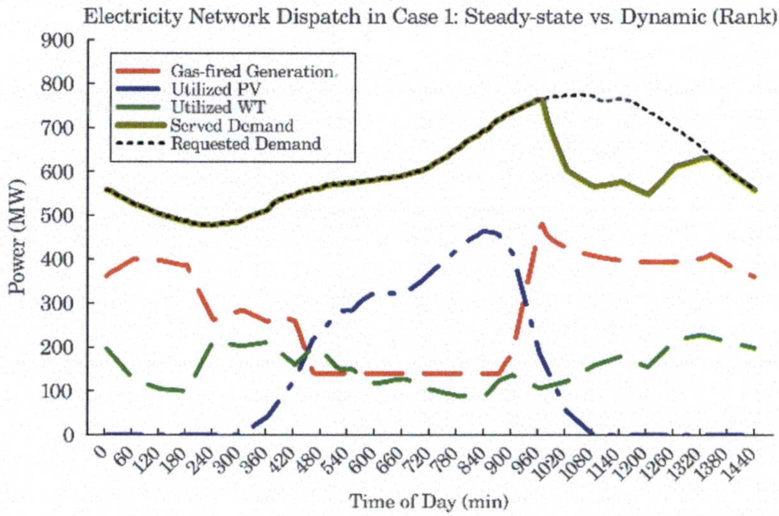

Fig. 5 Electricity system supply demand curve in Case 1 when utilizing rank model

A closer look at the output of the generation units in Fig. 6 reveals that the natural gas system cannot maintain the required fuel delivery for gas-fired units to produce sufficient electricity. Around hour 15 of the operation period, there is a sudden drop in the output of the photovoltaic system. Consequently, the gas-fired units need to ramp up to compensate for the loss of renewable generation. However, the natural gas system is unable to deliver enough fuel for both units to peak simultaneously.

The supply–demand curve of the natural gas system in this case is displayed in Fig. 7. Here, the sudden drop in the output generated by gas-fired units is observed around the peak time. The natural gas dynamics do not allow for the continued delivery of gas to these units. It should be noted that the values in this curve are displayed in Mcf units, which are dependent on system pressure. Hence, it is observed that the total gas supply is lower than gas delivery at some periods. Interestingly, it is observed that unlike electricity delivery, the natural gas supply and demand are not balanced at every moment, as they are governed by the dynamic Eqs. (8) and (9).

The pressure at the natural gas system junctions is also displayed in Fig. 8. The pressure of the gas at the gas well remains constant at 60 bar and drops as the fluid reaches the delivery junctions. The values reported in Mcf units in Fig. 7 can be adjusted based on the pressure at each junction to calculate Mscf or Kg/s units.

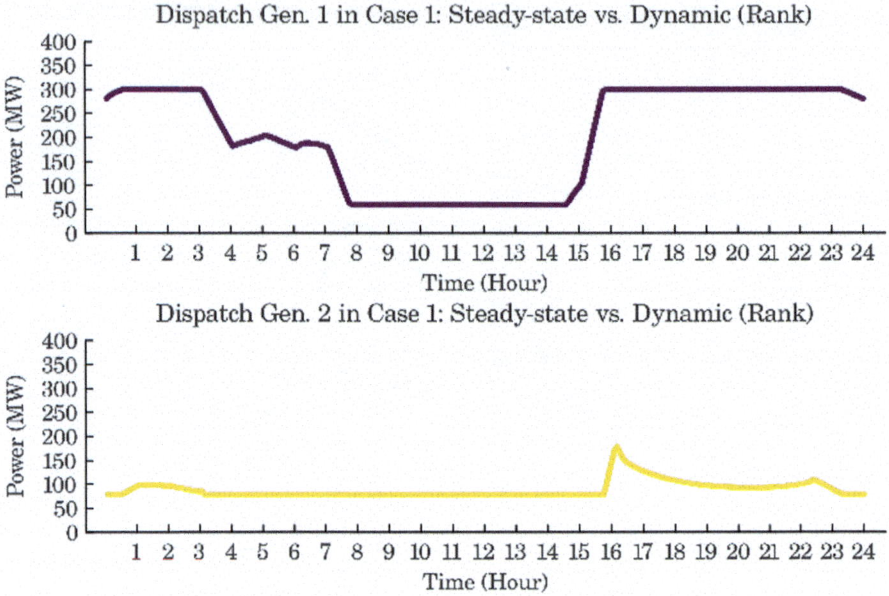

Fig. 6 Output of generation units in Case 1 when utilizing rank model

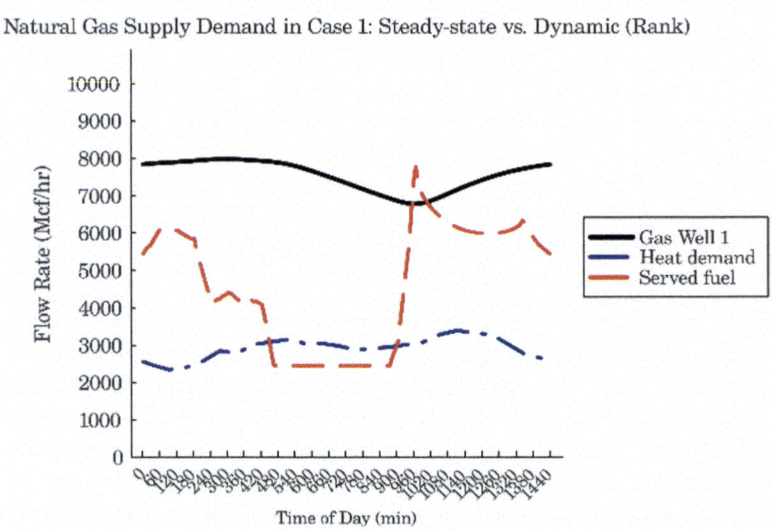

Fig. 7 Natural gas supply demand in Case 1 when utilizing rank model

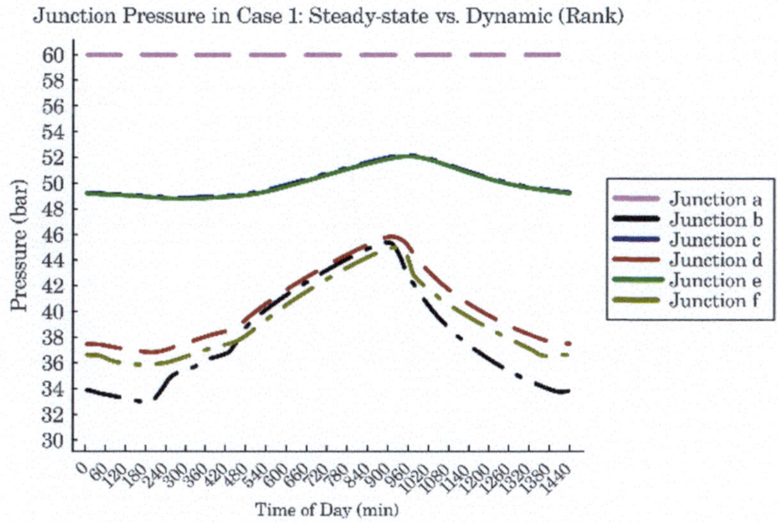

Fig. 8 Natural gas system pressure in Case 1 when utilizing rank model

7.2.2 Results of the Steady-State (Weymouth) Model

Unlike the dynamic model, when we utilize the Weymouth model for the operation problem, it is observed that the natural gas system is expected to deliver all the fuel demand required by the gas-fired units. As a result, all the electricity demand is served, which is inaccurate (Fig. 9).

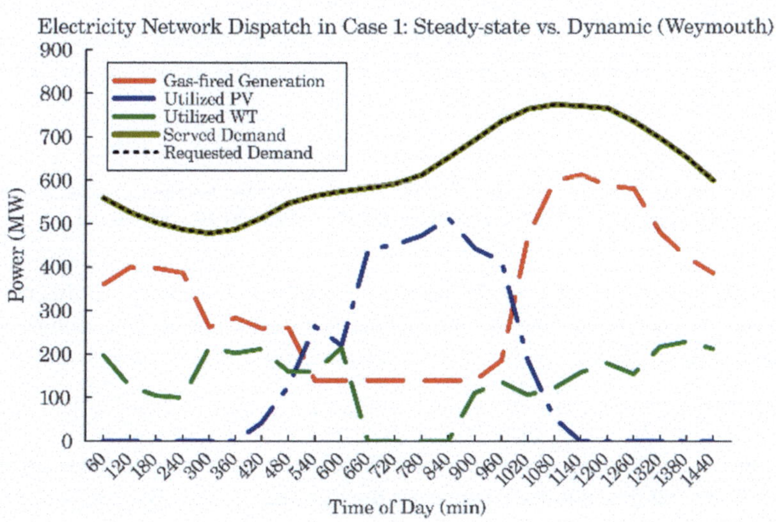

Fig. 9 Electricity system supply demand curve in Case 1 when utilizing Weymouth model

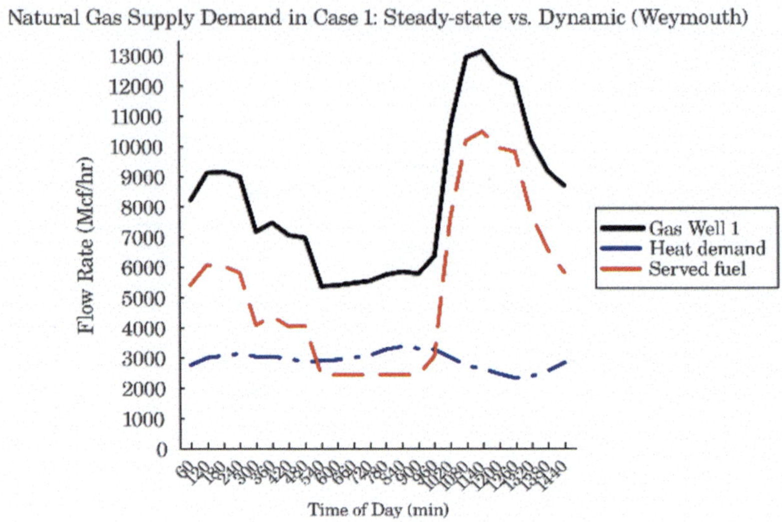

Fig. 10 Natural gas supply demand in Case 1 when utilizing Weymouth model

The supply–demand curve of the natural gas system is also displayed in Fig. 10. Unlike the dynamic approach, it is observed that supply and demand in the natural gas system are balanced at each moment, highlighting how steady-state models neglect temporal dynamics.

The Weymouth equation is an approximated model with assumptions that do not hold for short-term horizons. It cannot capture dynamic temporal relations within the natural gas system, such as the time it takes for gas to travel from the source to the demand point. The model also misses intra-hour variations, which are critical due to the high ramping rates of gas-fired units that cause significant changes in natural gas demand. Consequently, the results from the Weymouth model are unreliable for accurately determining natural gas fuel availability.

7.3 Advantage of Rank-Minimization Over SOC Method

This section investigates two of the relaxation methods introduced in this chapter. Specifically, we compare the SOC-relaxed OGF model with the rank-minimization relaxation method. To this end, we consider the 2-by-2 matrix present on the left-hand side of Eq. (14), i.e. $\begin{bmatrix} \pi^k_{(s,t)} & q^k_{(s,t)} \\ q^k_{(s,t)} & \gamma^k_{(s,t)} \end{bmatrix}$. If the rank of this matrix is 1 for all given pipe segments at all times, we can say that the relaxation is tight and the relationship between γ, π, and q is unique. To measure the tightness of the obtained solution, we use the logarithm of the ratio of the largest eigenvalue of this matrix to its second largest eigenvalue. For example, a tightness ratio of 4 means the error of recovering the lifted variables is equal to 0.0001 or 0.01%. Table 4 displays the tightness ratio for all five pipes of the natural gas system, with both average and minimum values reported for all spatio-temporal segments. For the whole system, the average tightness ratios of the SOC and rank methods are 5.83 and 7.81, respectively. It is noted that for the rank method, this value is 2 units higher than the SOC, indicating that the rank method results in a relaxation with 100 times more accuracy compared to SOC.

A 3D plot is presented in Fig. 11, displaying the tightness ratio values calculated for all spatio-temporal segments of Pipe 3 when applying the rank-minimization convex relaxation.

Table 4 Tightness ratio of SOC and rank-minimization relaxation methods

Method	SOC relaxation		Rank-minimization	
Tightness ratio	Average	Minimum	Average	Minimum
Pipe 1	6.61	5.69	8.2	7.91
Pipe 2	5.23	2.37	7.5	2.55
Pipe 3	6.2	5.27	8.1	7.0
Pipe 4	5.42	4.42	7.65	4.71
Pipe 5	5.71	3.32	7.61	2.85

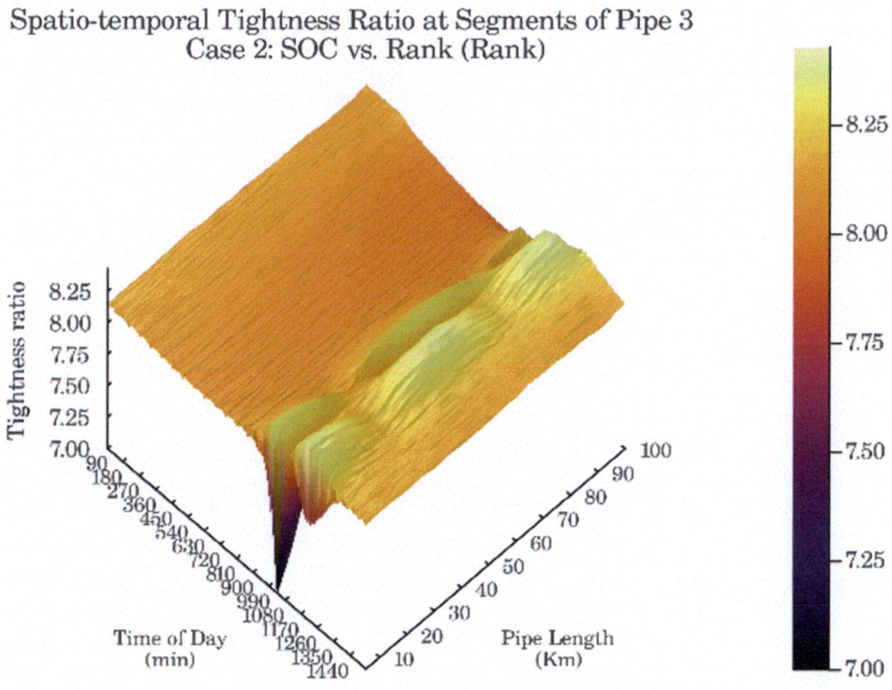

Fig. 11 Tightness ratio of Pipe 3 in Case 2 when utilizing rank model

7.4 Modifying the Natural Gas Pipes to Improve Delivery

The results of the coordinated operation of the presented IEGS configuration showed that the system is unable to provide the requested demand during the peak period. To address this issue, we propose and investigate a solution by modifying the IEGS system with pipelines that are 10% wider in diameter. The results displayed in Fig. 12 show that this modification resolves the problems encountered in the fuel delivery to the gas-fired units, allowing all of the electricity demand in the system to be fully met.

The output of gas-fired units is displayed in Fig. 13. The generation units collectively produce 7,896.5 MWh of electricity while consuming 128,231.4 Mcf of natural gas. Additionally, the PV and WT farms contribute by generating 3,258.0 MWh and 3,718.7 MWh of energy, respectively.

According to Fig. 14, in addition to supplying the gas-fired units, the natural gas system also delivers 70,516.0 Mcf of natural gas to heat loads.

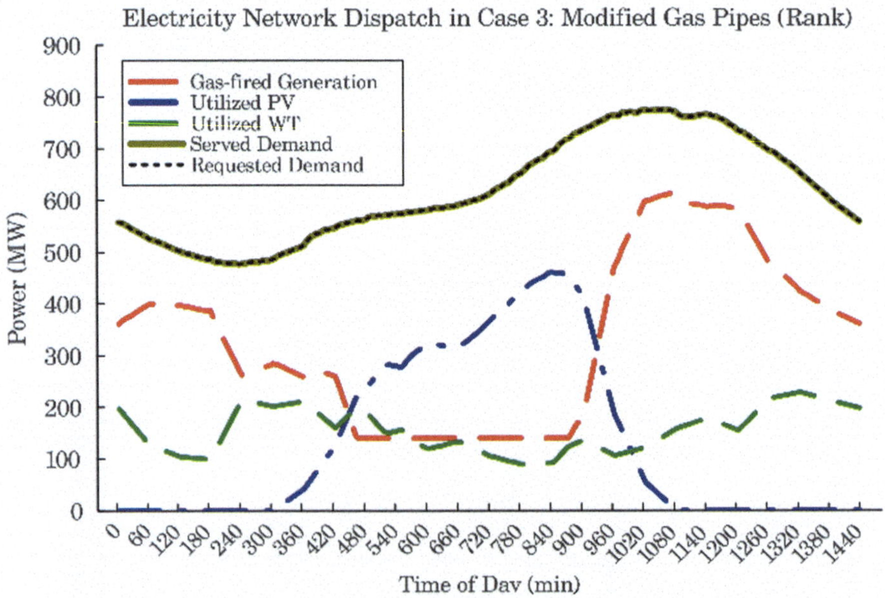

Fig. 12 Electricity system supply demand curve in Case 3 when utilizing rank model

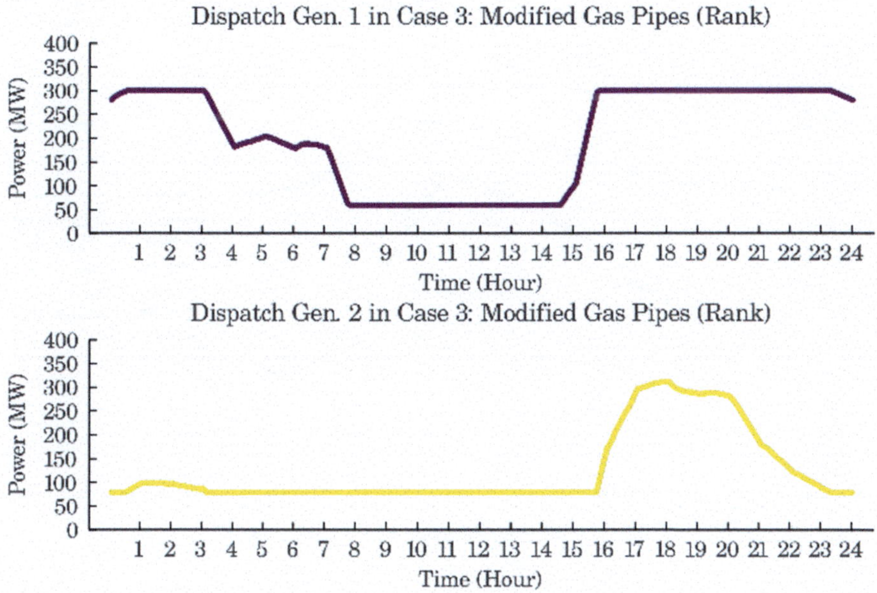

Fig. 13 Output of generation units in Case 3 when utilizing rank model

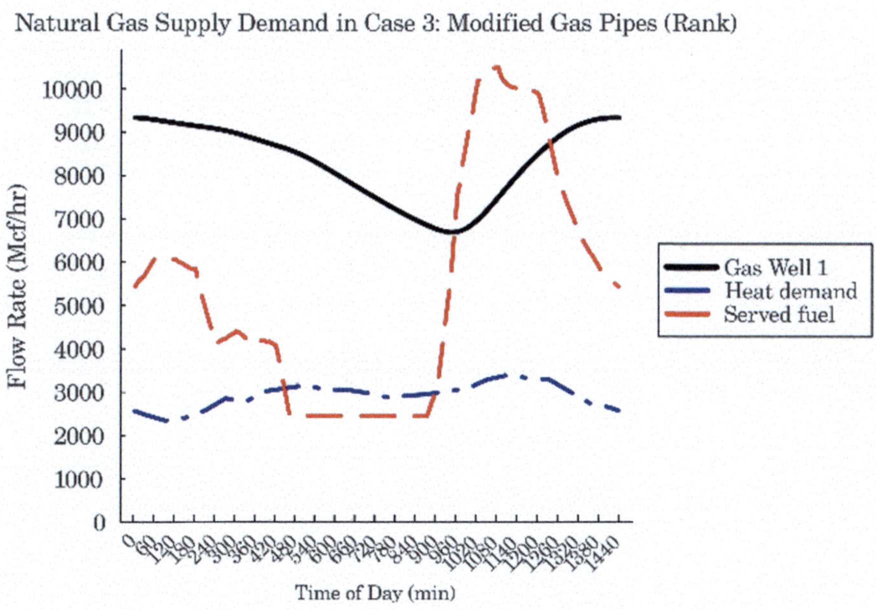

Fig. 14 Natural gas supply demand in Case 3 when utilizing rank model

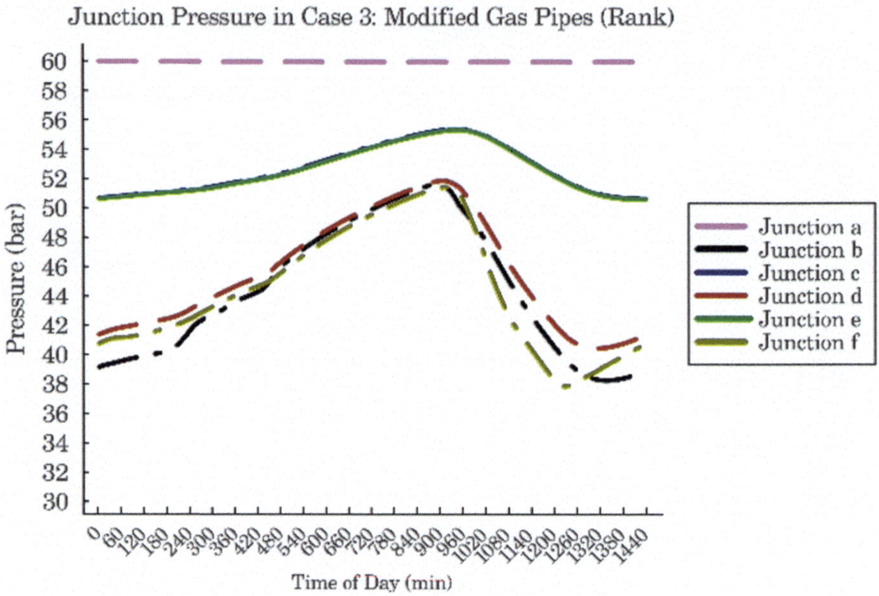

Fig. 15 Natural gas system pressure in Case 1 when utilizing rank model

To better understand the impact of the diameter modification, we can compare junction pressures in Fig. 15 against those in Fig. 8. It is observed that increasing the pipeline diameter relieves the pressure within the junctions. In Fig. 8, the difference between the highest and lowest pressure values in the system is approximately 27 bar, while in Fig. 15, this difference is reduced to around 22 bar, indicating a 5 bar decrease.

The spatio-temporal pressure of segments within Pipe 3 is displayed in Fig. 16.

Figure 17 displays the flow of natural gas for each pipeline in this case. As the pipes get closer to the consumption points, the flow curve becomes more similar in shape to the demand curve.

Additionally, the 3D flow of Pipe 3 is displayed in Fig. 18. It is observed that this pattern closely follows the output of the gas-fired units.

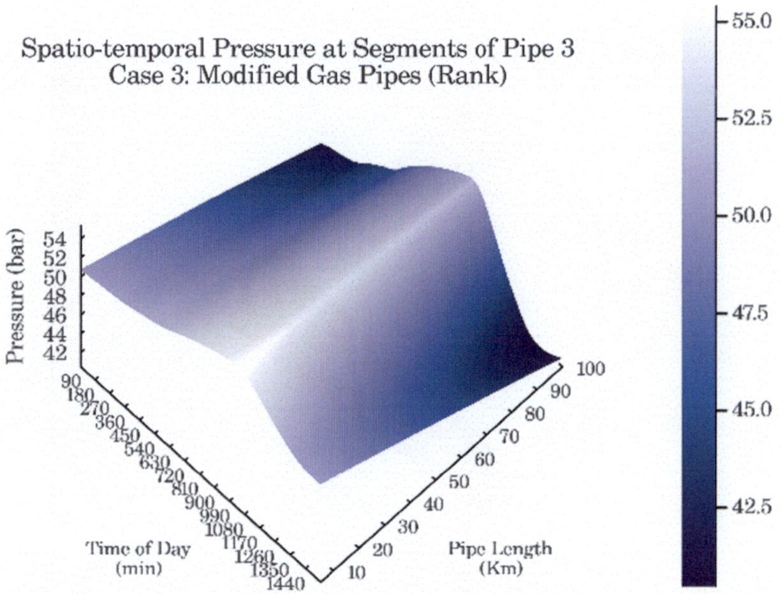

Fig. 16 Pressure of Pipe 3 in Case 3 when utilizing rank model

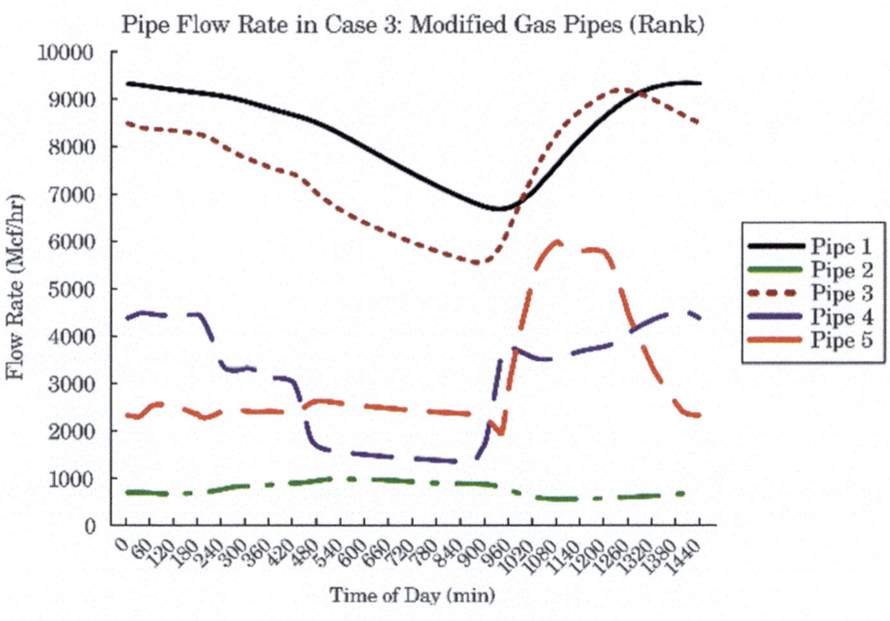

Fig. 17 Natural gas flow of pipes in Case 3 when utilizing rank model

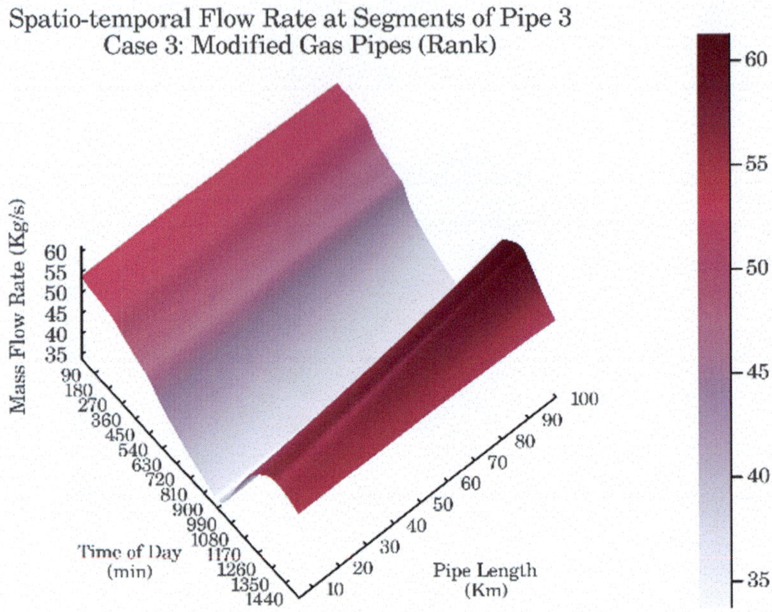

Fig. 18 Natural gas flow of Pipe 3 in Case 1 when utilizing rank model

References

1. U.S. Energy Information Administration, Annual energy outlook 2023 [Online] Available: https://www.eia.gov/outlooks/aeo/ (2024)
2. California Independent System Operator, CAISO supply outlook [Online] Available: https://www.caiso.com (2023)
3. X. Zhang, M. Shahidehpour, A. Alabdulwahab, A. Abusorrah, Hourly electricity demand response in the stochastic day-ahead scheduling of coordinated electricity and natural gas networks. IEEE Trans. Power Syst. **31**(1), 592–601 (2015)
4. A. Alabdulwahab, A. Abusorrah, X. Zhang, M. Shahidehpour, Coordination of interdependent natural gas and electricity infrastructures for firming the variability of wind energy in stochastic day-ahead scheduling. IEEE Trans. Sustain. Energy **6**(2), 606–615 (2015)
5. C. He, L. Wu, T. Liu, M. Shahidehpour, Robust co-optimization scheduling of electricity and natural gas systems via admm. IEEE Trans. Sustain. Energy **8**(2), 658–670 (2016)
6. T. Ding, Y. Hu, Z. Bie, Multi-stage stochastic programming with nonanticipativity constraints for expansion of combined power and natural gas systems. IEEE Trans. Power Syst. **33**(1), 317–328 (2017)
7. C.M. Correa-Posada, P. Sánchez-Martin, Integrated power and natural gas model for energy adequacy in short-term operation. IEEE Trans. Power Syst. **30**(6), 3347–3355 (2015)
8. A. Zlotnik, L. Roald, S. Backhaus, M. Chertkov, G. Andersson, Coordinated scheduling for interdependent electric power and natural gas infrastructures. IEEE Trans. Power Syst. **32**(1), 600–610 (2016)
9. B. Zhao, A.J. Conejo, R. Sioshansi, Unit commitment under gas-supply uncertainty and gas-price variability. IEEE Trans. Power Syst. **32**(3), 2394–2405 (2016)
10. M. Geidl, G. Andersson, Optimal power flow of multiple energy carriers. IEEE Trans. Power Syst. **22**(1), 145–155 (2007)

11. S.D. Manshadi, M.E. Khodayar, Resilient operation of multiple energy carrier microgrids. IEEE Transactions on Smart Grid **6**(5), 2283–2292 (2015)
12. Y. He, M. Yan, M. Shahidehpour, Z. Li, C. Guo, L. Wu, Y. Ding, Decentralized optimization of multi-area electricity-natural gas flows based on cone reformulation. IEEE Trans. Power Syst. **33**(4), 4531–4542 (2017)
13. A. Ojha, V. Kekatos, R. Baldick, Solving the natural gas flow problem using semidefinite program relaxation, in *2017 IEEE Power & Energy Society General Meeting. IEEE* (2017), pp. 1–5
14. R. Bayani, S.D. Manshadi, Natural gas short-term operation problem with dynamics: A rank minimization approach. IEEE Trans. Smart Grid **13**(4), 2761–2773 (2022)
15. A. Zlotnik, S. Dyachenko, S. Backhaus, M. Chertkov, Model reduction and optimization of natural gas pipeline dynamics, in *Dynamic Systems and Control Conference*, vol. 57267. American Society of Mechanical Engineers (2015), p. V003T39A002
16. W. Liu, P. Li, W. Yang, C. Chung, Optimal energy flow for integrated energy systems considering gas transients. IEEE Trans. Power Syst. **34**(6), 5076–5079 (2019)
17. J. Fang, Q. Zeng, X. Ai, Z. Chen, J. Wen, Dynamic optimal energy flow in the integrated natural gas and electrical power systems. IEEE Trans. Sustain. Energy **9**(1), 188–198 (2018)
18. R. Bayani, S.D. Manshadi, Coordinating flexible ramping products with dynamics of the natural gas network. Sustain. Energy, Grids Netw. **35**, 101079 (2023)

Short-Term Prediction of Photovoltaic Power Generation

Shahab Karamdel, Xiaodong Liang, and Sherif O. Faried

Abstract Solar photovoltaic (PV) is a dominant form of renewable energy source and has been widely deployed, but increasing PV power integrated into electric power grids creates challenges in the planning and operation of power grids due to PV power's intermittency and variability nature. Accurate solar PV power generation prediction techniques are essential to overcome these challenges. This chapter proposes a short-term PV power generation prediction framework by using nineteen regression models within five regression families. To further enhance the forecasting model's performance, hyperparameter optimization and tuning is discussed through the MATLAB Regression Learner toolbox. The historical datasets are utilized to develop the proposed forecasting models.

Keywords Hyperparameter tuning · Preprocessing · Regression models · Short-term · Solar PV power generation forecasting

1 Introduction

Renewable energy sources are increasing in electric power grids [1], and photovoltaic (PV) power has experienced the second-largest increase among all renewables in 2020. PV power generation is intermittent and weather dependent, high penetration of PV power in power grids will cause significant challenges in the system's planning and operation. To address these challenges, accurate PV power generation prediction techniques are urgently needed [2, 3].

In general, there are three categories of PV power generation forecasting: very short-term forecasting techniques, through which the PV power generation is forecasted in next a few seconds [1] to a few minutes [2–4]; short-term forecasting techniques, making forecasting for the next few hours to a few days [5–9]; and long-term forecasting techniques, making forecasting for longer than a few days [2, 10, 11].

S. Karamdel · X. Liang (✉) · S. O. Faried
Department of Electrical and Computer Engineering, University of Saskatchewan, Saskatoon, Canada
e-mail: xil659@mail.usask.ca

F. Shahnia and F. Wen (eds.), *Power and Energy Resources*, Power Systems,
https://doi.org/10.1007/978-981-96-2612-0_5

In the literature, PV power generation is forecasted using statistical methods, machine learning techniques, and hybrid methods. Statistical forecasting methods perform forecasting using historical data to obtain time series data, for example, the solar irradiance data. Machine learning forecasting models are trained using historical data to capture underlying relationships and patterns. A hybrid forecasting technique is obtained by combining the statistical and machine learning methods. PV power generation prediction models can predict the output power of PV modules [1, 2, 6–8, 10, 12, 13], or solar irradiance [5, 14–16].

Among various statistical forecasting techniques, autoregressive moving average (combining autoregressive with moving average sub-models), autoregressive integrated moving average, and seasonal autoregressive integrated moving average (SARIMA) are commonly used. Reference [15] proposes the autoregressive moving average method to predict solar irradiance, where solar irradiance's nonstationary data in time series are converted to stationary data in time series. Reference [17] compares PV power generation forecasting results of the models, "SARIMA", "SARIMAX" by combining SARIMA and an exogenous factor, and artificial neural networks (ANNs). It is found that artificial neural networks and SARIMAX outperform SARIMA [2].

PV power generation prediction can be carried out through machine learning, especially deep learning. Various machine learning techniques, such as decision tree, linear regression, support vector regression, multi-layer perception, random forest, and long short-term memory (LSTM), are utilized in [10] to forecast PV power in Alice Springs, Australia. A weighted Gaussian process regression (GPR) model is developed in [7] to predict short-term PV power, and the Gradient descend approach is used to optimize hyperparameters for the non-deterministic prediction model. LSTM, convolutional LSTM, convolutional neural networks (CNN), random forest, support vector regression, and extreme Gradient boosting models are trained in [14] to predict hourly solar irradiance. Reference [18] develops the support vector regression and linear least squares models using weather and solar irradiation data for PV power generation forecasting. Reference [1] proposes a very short-term PV power prediction technique in seconds ahead using feed-forward artificial neural networks with the cloud shading effect considered [2].

Ensemble models, which aggregate different machine learning predictors, can enhance the forecasting accuracy [5]. An ensemble of deep LSTM models aggregated via Choquet integral is presented in [5] to forecast solar irradiance on a long-term time scale. A probabilistic and non-parametric ensemble PV forecasting approach is also represented in [13], which uses quantile regression averaging to group several deterministic LSTM-based predictors. Reference [11] develops an ensemble of several meta-learners (autoregressive integrated moving average, vector autoregression, and LSTM) and artificial intelligence methods (multi-layer perceptron and CNN) for long-term PV power generation prediction. An ensemble model of NNs trained independently is proposed in [19] for day-ahead PV power forecasting; this ensemble model predicts several NN models in parallel and then assesses their performance [2].

A hybrid model is developed by combining several statistical and machine learning methods to achieve better forecasting [2]. A hybrid approach is used in [9] to predict the short-term PV power generation through the improved sparrow search algorithm, the gated recurrent unit, and the variational mode decomposition to decompose time series data to achieve an optimized model. To enhance the prediction accuracy considering unfamiliar weather conditions, a hybrid PV power prediction model using deep learning is presented in [11] through persistence models and autoencoder LSTM networks. Reference [14] proposes a hybrid model by combining LSTM and CNN for the output power forecasting of PV systems, non-linear patterns in historical weather data can be extracted by CNN, and the next time interval results can be forecasted by LSTM. CNN and LSTM are also used in [12] to forecast PV power through weather and PV inverter data. A hybrid approach is proposed in [22] by coordinating the gradient descend, the shuffled frog leaping algorithm, and feed-forward artificial neural networks to forecast short-term PV power generation. Reference [23] proposes a hybrid short-term PV power prediction through the genetic algorithm, fuzzy clustering, and radial basis function. A hybrid short-term PV power prediction method is reported in [24], and two LSTM NN models are utilized to predict the temperature and PV power [2].

Wavelet transform can be adopted in a hybrid approach for PV power prediction [4, 16]. A hybrid very short-term prediction method using discrete wavelet transform, recurrent neural networks, and autoregressive moving average for PV power prediction is proposed in [4]. The Wavelet transform and Elman neural networks are combined in [16] for solar irradiance prediction over the next hour, where the Wavelet transform is used to decompose time series data, and Elman neural networks are then used to predict PV power.

To explore similarities between PV output power in different locations, spatiotemporal forecasting methods are introduced [20]. Two novel deterministic spatiotemporal methods for PV power generation forecasting located in multiple sites are proposed in [20], where the two models based on the graph theory, the graph-convolutional long short-term memory model, and the graph-convolutional transformer model, are adopted for the PV power prediction.

In the literature, probabilistic methods are commonly used to model the uncertainty. Probabilistic forecasting predicts the probability density function (PDF) instead of a point when using deterministic forecasting [21]. Quantile regression has been used in probabilistic forecasting methods [21–23]. Ref. [21] proposes a probabilistic model with time series and quantile regression methods through a two-level hierarchy [2]. Reference [22] proposes two probabilistic models to predict PV power generation by bootstrap and quantile regression methods. Another probabilistic method to forecast PV power generation is GPR in [7].

Despite the extensive PV power generation forecasting research that has been conducted, a forecasting model developed for one specific location may not perform well for a different location [10]. To address this research gap, a comprehensive short-term PV power generation prediction framework is developed in this chapter to determine the best-performing prediction models from 19 regression models considered. Hyperparameter tuning is conducted in this chapter to enhance the best prediction model's accuracy. Three datasets representing summer, winter, and the combination of summer and winter are utilized to validate the model's performance.

2 The Proposed Framework for PV Power Generation Forecasting

The proposed framework for PV power generation prediction is based on regression models. Figure 1 shows the flow chart of this framework, involving three stages as explained below [2]:

- Stage 1—Data acquisition and data pre-processing.
- Stage 2—Training and validation of the forecasting model, and selection of the best-performing models.

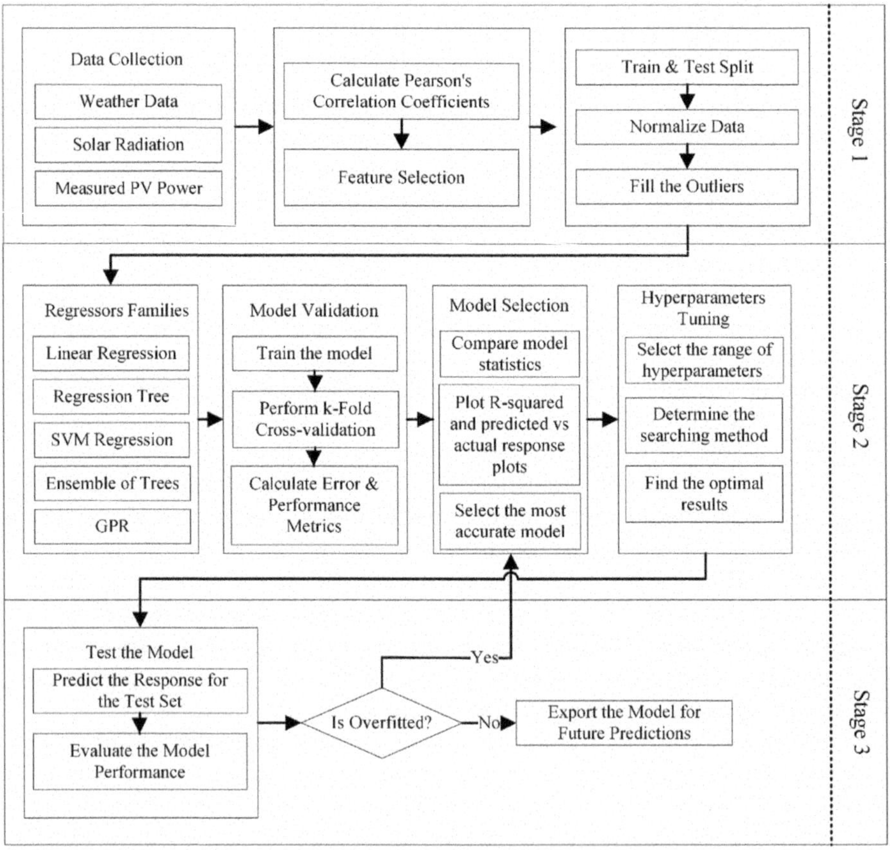

Fig. 1 The flowchart of the proposed framework for short-term pv power generation forecasting

- Stage 3—Testing of the selected model using unseen data, and exportation of the best model.

2.1 Stage 1

In this stage, raw measurement datasets are first obtained, the obtained datasets are then preprocessed to prepare for the data-driven forecasting. The five steps introduced below are involved in Stage 1 [2]:

Step 1—*Data acquisition*: The historical measurements are collected for weather data, solar radiation, and power generation of PV modules.
Step 2—*Feature selection*: Features with the most significant impacts on PV power generation forecasting are selected, a key step for achieving an accurate forecasting model. Training a model with less relevant features/variables may significantly reduce the model's accuracy. To assist with feature selection, Pearson's correlation coefficients are computed to determine the correlation between independent variables and the PV module's output power.
Step 3—*Dataset preparation*: 80% of the dataset for training, and 20% of the dataset for testing.
Step 4—*Dataset normalization*: The datasets prepared for training and testing in Step 3 are further normalized to be between 0 and 1 using the Min–max scaler [24].
Step 5—*Outlier replacement*: Outliers are determined and replaced by interpolated data using the linear interpolation method.

Through Stage 1, the datasets are prepared and ready for training, validation, and testing of the regression models.

2.2 Stage 2

In Stage 2, training, validation, and hyperparameters tuning of the regression models are conducted through four steps below [2]:

Step 1—*Model training:* The training dataset prepared in Stage 1 is utilized for training the regression models.
Step 2—*Model validation***:** Five-fold cross-validation is used to reduce overfitting.
Step 3—*Best performing models:* The best-performing regression models can be identified according to the error indices and the predicted response plots.
Step 4—*Hyperparameters tuning*: For the identified best performing models in Step 3, hyperparameters tuning is essential, as improper hyperparameters may result in overfitted models. In this chapter, hyperparameters tuning is done automatically within the MATLAB Regression Learner toolbox. The *mean squared*

error (MSE) can be minimized by tuning hyperparameters iteratively [25], and specific optimizable hyperparameters are associated with each regression model. As a default optimization method in MATLAB, the Bayesian method is used to optimize hyperparameters in this chapter.

2.3 Stage 3

In this stage, the trained models are tested through unseen data. The predicted response plots of the forecasting models are obtained, and the model accuracy is evaluated by the root mean squared error (RMSE), which can also serve as an indicator if a model is overfitting. When a regression model has low RMSE values at cross-validation, while has high RMSE values at Stage 3, that means the trained model is overfitting for the training dataset, and it may predict poorly using unseen data [2].

In this framework, if a forecasting model is overfitting, the next best-performing model will be evaluated. If the model is not overfitting, it is then considered as the most accurate forecasting model and exported for predicting PV power generation.

3 Regression Models

3.1 The Adopted Regression Models

To develop the best forecasting model, we consider 19 regressors representing five families of regression models, as explained below [2]:

Linear regression models: They are simplest and commonly utilized to determine the deviation of dependent variables vs. variations of independent variables for a continuous dataset [10].

Regression tree models: A training dataset, which can include both continuous and discrete data [26], is used to train a regression decision tree model, where the input variables are split iteratively into binary forms, and create a tree structure including nodes and branches [26, 27].

Support Vector Machine regression models: Different kernels (linear, Gaussian, polynomial, and the radial basis function kernels) [26] can be used in a SVM regressor. The kernel function (K) can be applied to non-linearly classifiable data by [10]

$$y = \sum_i \left(a_i - a_i^*\right).K(x_i, x) + b \tag{1}$$

where a and b are the model coefficients.

Gaussian process regression models: They are non-parametric, probabilistic, use the Gaussian distribution by utilizing kernel functions, and consider random variables through the Gaussian PDF [26].

Ensemble of regressors: Different regression models are combined by ensemble techniques to enhance the model's prediction accuracy [26, 28].

3.2 Error Indices

The following four error indices are used in this chapter to evaluate the developed regression models [2]:

RMSE—RMSE can be determined as follows, showing the error's standard deviations:

$$RMSE = \sqrt{\frac{\sum_t (f_t - \hat{f}_t)^2}{n}} \tag{2}$$

where n is the sample size, f_t and $\hat{f}_t$ are the predicted and actual responses at the time t, respectively. More accurate models have lower values of RMSE [26]. RMSE is positive, and it has the unit of the response of the model [29].

R-squared—R-squared values are obtained by comparing the trained model's predicted results with the actual response's statistical mean [29]. When R-squared values are closer to 1, the model has a better accuracy [26].

Mean absolute error (MAE)—They are less sensitive to outliers than RMSE [26, 29]. MAE can be calculated by

$$MAE = \frac{\sum_t |f_t - \hat{f}_t|}{n} \tag{3}$$

4 Results of the Proposed Models

4.1 Dataset Preparation

In this chapter, the historical data with a sampling rate of 5-min recorded in São Paulo, Brazil [30] during December 1–10, 2013 (Summer) and June 1–10, 2014 (Winter) are used to predict PV power. The historical data include the time of the day, the external temperature (°C), the PV module's temperature (°C), the wind speed (m/s), the solar radiation (W/m^2), and the PV module's output power (W). The PV module has a nominal installed capacity of 2.55 kilowatts (kW) [2].

Table 1 Correlation coefficients for three datasets

	Datasets		
Independent variables	December	June	Combination
Time of the day	0.1396	0.0425	0.0819
Solar radiation	0.9669	0.9867	0.9683
External temperature	0.8373	0.7031	0.7256
Module temperature	0.9394	0.8607	0.8868
Wind speed	0.5821	0.5578	0.5107

Three datasets are used in this chapter: (1) Summer/December dataset (recorded during December 1–10, 2013); (2) Winter/June dataset (recorded during June 1–10, 2014); and (3) Combined dataset (A combination of December and June data). The total number of rows for the Summer, Winter, and combined datasets are 2,721, 2,724, and 5,445, respectively. Each dataset is further divided for training and testing (80% data for training and 20% data for testing).

Pearson's correlation coefficients are computed representing the correlation between the PV module's output power and independent variables (Table 1) to determine the most essential features. These coefficients are between -1 and $+1$ (a value closer to 1 means a higher correlation). Based on Table 1, the solar radiation, the module temperature, and the external temperature have the most impacts for the PV power generation, and thus, are used as features. On the other hand, wind speed affects the model accuracy negatively, and the time of the day has a low correlation, so they are not considered as features. Figure 2 illustrates the scatter plots of the PV module's output power versus the selected three features/variables, confirming a linear correlation.

The data must be pre-processed into a proper format for the aggression models by following the procedure in Stage 1's Steps 3–4, which are implemented in MATLAB. The training datasets for December and June after replacing their outliers are shown in Fig. 3.

4.2 Case 1—December Dataset

December 2013 dataset represents the Summer time in Brazil. Through the fivefold cross-validation, 19 regression models from the five regression families as introduced in Sect. 3.1 are developed by training, and Table 2 shows error indices of these models. Because the R-squared values among 15 regressors are similar, RMSE can be utilized to assess the models; the rows of Table 2 are sorted according to their RMSE values computed during the model validation process. According to Table 2, the interactions linear model and the stepwise linear model appear to be the most accurate prediction models using this December dataset. The next best-performing

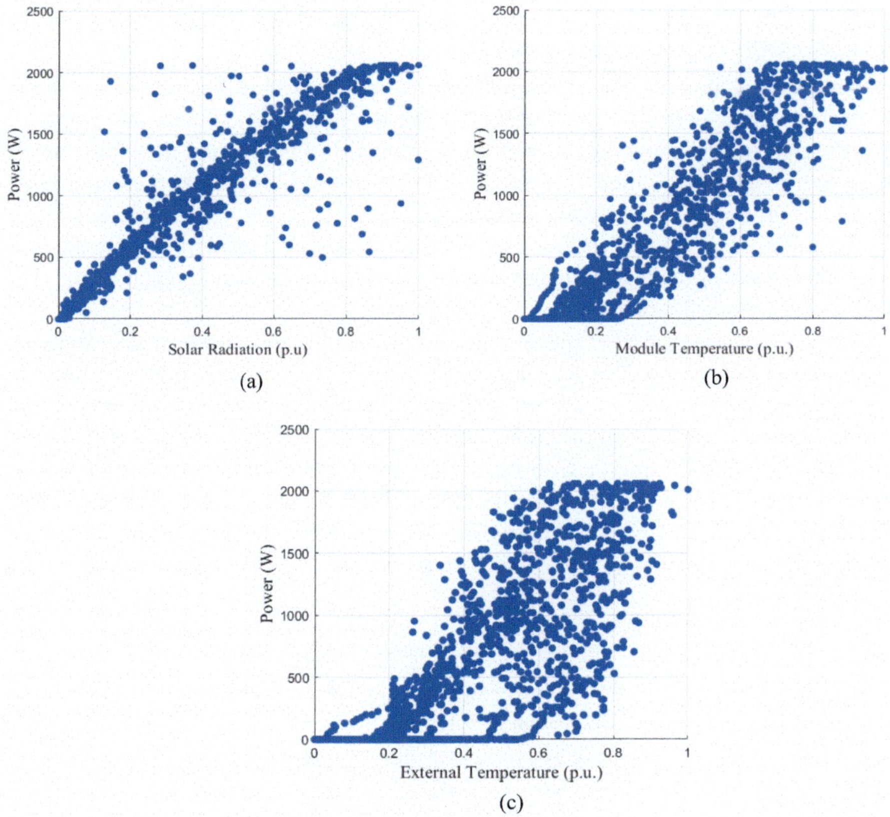

Fig. 2 The scatter plots of the response vs. **a** solar radiation, **b** module temperature, and **c** external temperature

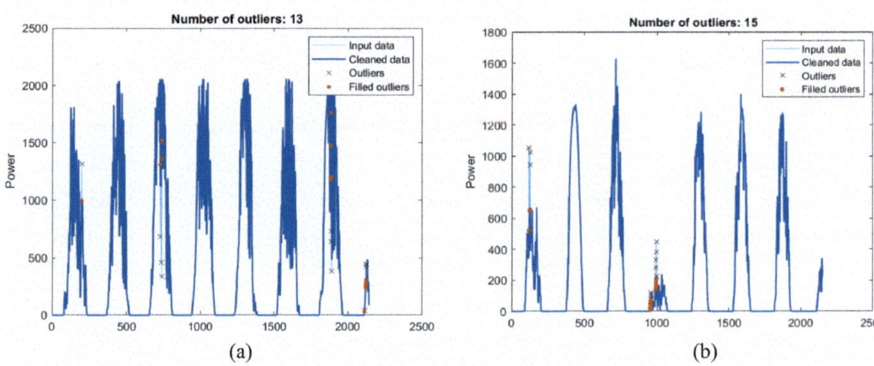

Fig. 3 Training datasets of **a** December and **b** June after filling the outliers

models are the bagged trees ensemble model, the medium Gaussian SVM model, and the boosted trees ensemble model.

Figures 4a–d illustrate the scatter plots of the predicted and actual responses of the five selected best-performing regressors. Since the interactions linear and stepwise linear regressors are identical, they are both plotted in Fig. 4a, and they have better prediction results than other three models. The scatter plot of the exponential GPR model in Fig. 4e shows the worst performance.

The trained regression models are tested through unseen data in the testing dataset. Figure 5 demonstrates the predicted responses of the regressors in each of the five families for a high-fluctuation day in December.

The interactions/stepwise linear regression models have the best performance in predicting PV power. The medium Gaussian SVM also has a high accuracy in following actual PV power patterns. Within the linear regression family, the robust linear regression performs poorly, which can be expected from its high RMSE value. Similarly, the SVM regressor with the linear kernel demonstrates the worst performance in the SVM family. Within the ensemble of trees family, the bagged trees ensemble outperforms the boosted trees model at peak power levels, although their accuracy is similar at low power levels. The accuracy of Gaussian process regression

Table 2 The model's validation results using the December training dataset

Regression model	RMSE		R^2	MSE	MAE
	Validation	Testing			
Interactions linear	134.74	111.79	0.96	18,154	64.539
Stepwise linear	134.74	111.79	0.96	18,154	64.539
Bagged trees	139.17	120.83	0.95	19,369	65.782
Medium Gaussian SVM	141.26	113.52	0.95	19,953	72.896
Boosted trees	143.26	115.33	0.95	20,524	64.368
Quadratic SVM	145.17	115.38	0.95	21,075	71.581
Coarse Gaussian SVM	145.57	114.16	0.95	21,191	75.563
Medium tree	145.67	132.86	0.95	21,220	62.318
Linear regression	145.72	129.01	0.95	21,233	78.791
Fine Gaussian SVM	146.35	158.72	0.95	21,418	74.846
Coarse tree	147.28	121.83	0.95	21,693	63.375
Exponential GPR	150.5	124.11	0.94	22,649	60.414
Linear SVM	152.88	134.31	0.94	23,373	83.698
Fine tree	153.31	138.22	0.94	23,504	63.927
Rational quadratic GPR	155.22	129.45	0.94	24,093	65.002
Matern 5/2 GPR	180.33	187.72	0.92	32,519	80.052
Robust linear	187.98	209.26	0.91	35,335	75.337
Squared exponential GPR	231.26	323.49	0.87	53,481	114.35
Cubic SVM	439.8	139.08	0.52	193,420	294.35

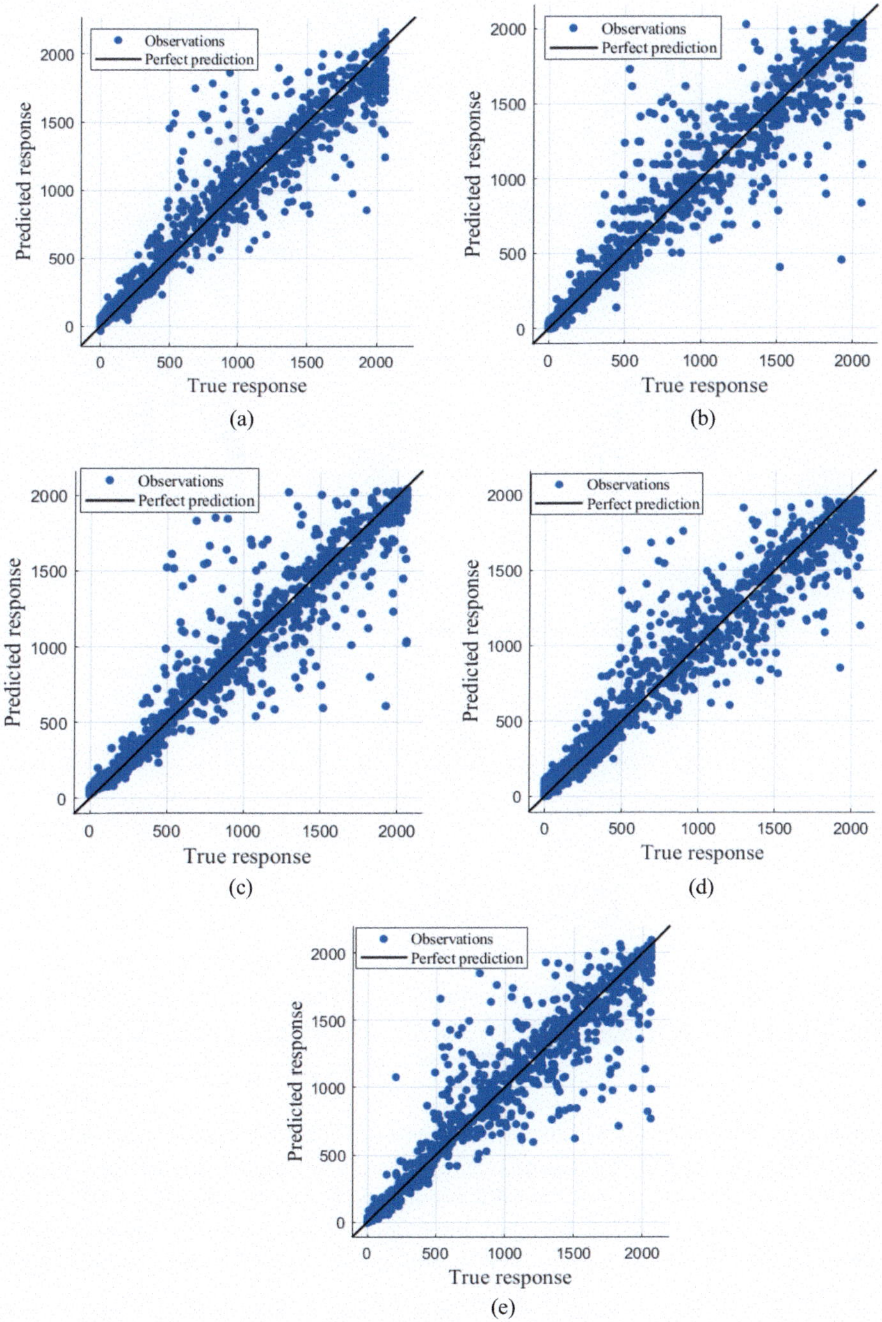

Fig. 4 The predicted results vs. actual data for the December dataset: **a** Interactions/stepwise linear models, **b** Medium tree model, **c** Medium Gaussian SVM model, **d** Bagged trees ensemble model, and **e** Exponential Gaussian process regression model

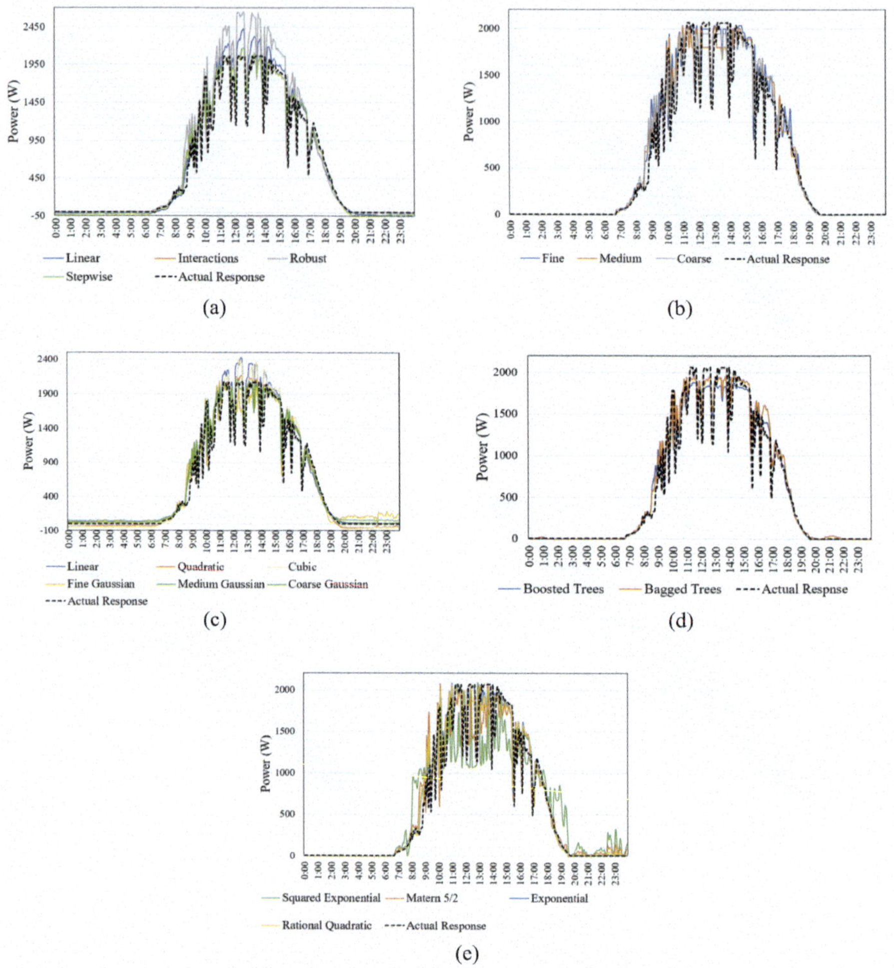

Fig. 5 Prediction results in a high-fluctuation test day in December for five regression families: **a** linear regression, **b** regression trees, **c** support vector regression, **d** ensemble of trees, and **e** Gaussian process regression

models significantly varies within their own group. For instance, squared exponential and Matern 5/2 show a poor performance, whereas exponential GPR has significantly better accuracy than the rest of the GPR predictors. Figure 6 shows these 19 regressors' performance at another test day. Interactions linear, stepwise linear, bagged trees ensemble, medium Gaussian SVM, and exponential GPR regression models show the best results.

Based on visual and statistical assessments, the best regression models include the interactions linear regression model, the ensemble of bagged trees model, and the medium Gaussian SVM model, which can more accurately predict the PV power

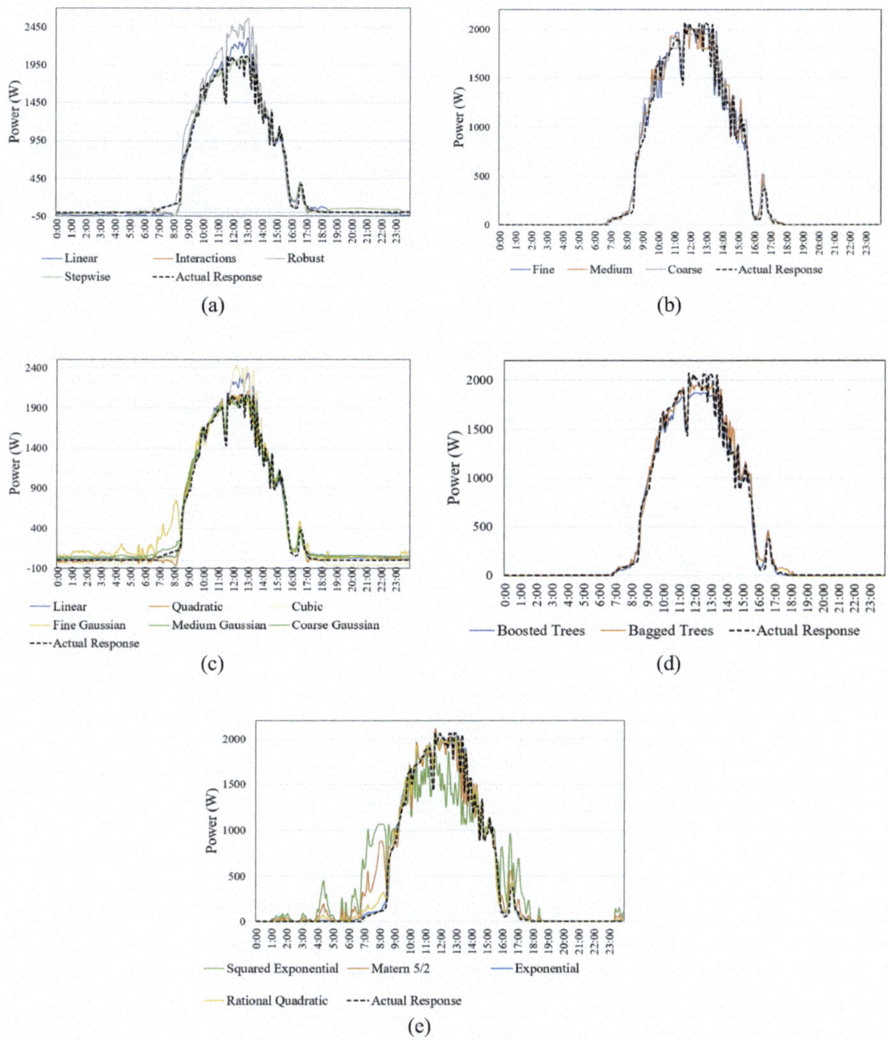

Fig. 6 Forecasting results in a low-fluctuation test day in December for five regression families: **a** linear regression, **b** regression trees, **c** support vector regression, **d** ensemble of trees, and **e** Gaussian process regression

compared to actual data. Hyperparameters for the ensemble of bagged trees model and the medium Gaussian SVM model are tuned to further enhance their performance (Table 3) (note: the interactions linear regressor cannot perform optimization for hyperparameters using MATLAB) [2, 25].

Figure 7 shows the comparison of the predicted responses of the optimized ensemble of bagged trees model, the optimized medium Gaussian SVM model, and the interactions linear regression model with actual PV module's output power.

Table 3 Optimization of hyperparameters for the best regression models (Case 1)

Optimized ensemble		Optimized SVM	
Ensemble method	Bag	Kernel function	Gaussian
Minimum leaf size	22	Kernel scale	0.92204
Number of learners	62	Box constraint	138.6182
Number of predictors	2	Epsilon	20.3897
		Standardize data	True
Minimum objective	18,800	Minimum objective	18,859
Minimum estimated objective	18,708	Minimum estimated objective	18,790
		Solver	Sequential minimal optimization

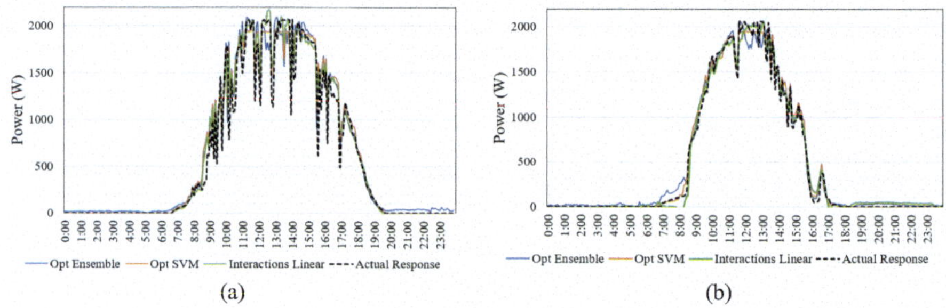

Fig. 7 The predicted results vs. actual data in December for the interactions linear, optimized SVM, optimized ensemble regression models: **a** a high-fluctuation day, **b** a low-fluctuation day

It is found all three models have high accuracy when tested with unseen data, the interactions linear regressor is a little better for the peak value prediction, while the optimized medium Gaussian SVM model is better at low levels of PV power prediction among the three [2].

4.3 Case 2—June Dataset

The dataset of June 2014 is used for representing winter in Brazil. The regression models are trained using this dataset, and the validation results reveal that the exponential GPR and medium Gaussian SVM outperform the other models as shown in Table 4. The interactions linear regressor, stepwise linear regressor, fine Gaussian SVM regressor, and coarse Gaussian SVM regressor show similar performances. However, since RMSE values for the exponential GPR regression model using the test dataset are relatively high, it is probably overfitted. The robust linear regression and cubic SVM models demonstrate the worst accuracy. The calculated R-squared and MAE values also confirm the results obtained from RMSE.

The best-performing five models are exponential GPR, medium Gaussian SVM, the interactions/stepwise linear regressors, and Fine Gaussian SVM as shown in Table 4. The best models in the five regressors families are the interactions linear regressor, medium Gaussian SVM regressor, medium tree regressor, bagged trees ensemble regressor, and exponential GPR regressor. The scatter plots of predicted vs. true responses for these models indicate that exponential GPR and medium Gaussian SVM are the two most accurate models for this dataset based on validation results, because they relatively better align with the diagonal line as shown in Fig. 8.

The exponential GPR and Gaussian SVM models are selected for hyperparameters optimization, and the obtained results are presented in Table 5. These models are then tested on unseen data. The testing results illustrate that optimized SVM replicates actual measurements more accurately for both high- and low-fluctuation test days (Fig. 9). This also confirms the overfitting occurrence for the GPR model.

Table 4 Validation results for june training dataset

Regression	RMSE		R^2	MSE	MAE
Model	Validate	Test			
Exponential GPR	54.709	74.412	0.98	2993.1	20.939
Medium Gaussian SVM	55.729	58.248	0.98	3105.7	24.599
Interactions linear	57.301	58.886	0.98	3283.5	29.419
Stepwise linear	57.373	58.886	0.98	3291.7	29.383
Fine Gaussian SVM	57.462	120.58	0.98	3301.8	26.518
Coarse Gaussian SVM	57.661	54.88	0.98	3324.8	26.975
Bagged trees	59.274	77.523	0.98	3513.4	26.797
Rational quadratic GPR	59.359	81.052	0.98	3523.5	23.417
Linear regression	59.628	61.191	0.98	3555.5	31.034
Quadratic SVM	60.627	60.522	0.97	3675.6	31.201
Boosted trees	60.816	62.68	0.97	3698.5	27.001
Linear SVM	61.123	61.934	0.97	3736	27.974
Medium tree	64.033	60.994	0.97	4100.2	26.185
Fine tree	64.837	72.884	0.97	4203.9	26.524
Coarse tree	65.239	59.25	0.97	4256.1	26.973
Matern 5/2 GPR	72.322	140.88	0.96	5230.4	28.494
Squared exponential GPR	97.664	224.97	0.93	9538.3	39.121
Cubic SVM	251.74	328.2	0.56	63,375	161.92
Robust linear	395.34	60.057	0.09	156,290	180.12

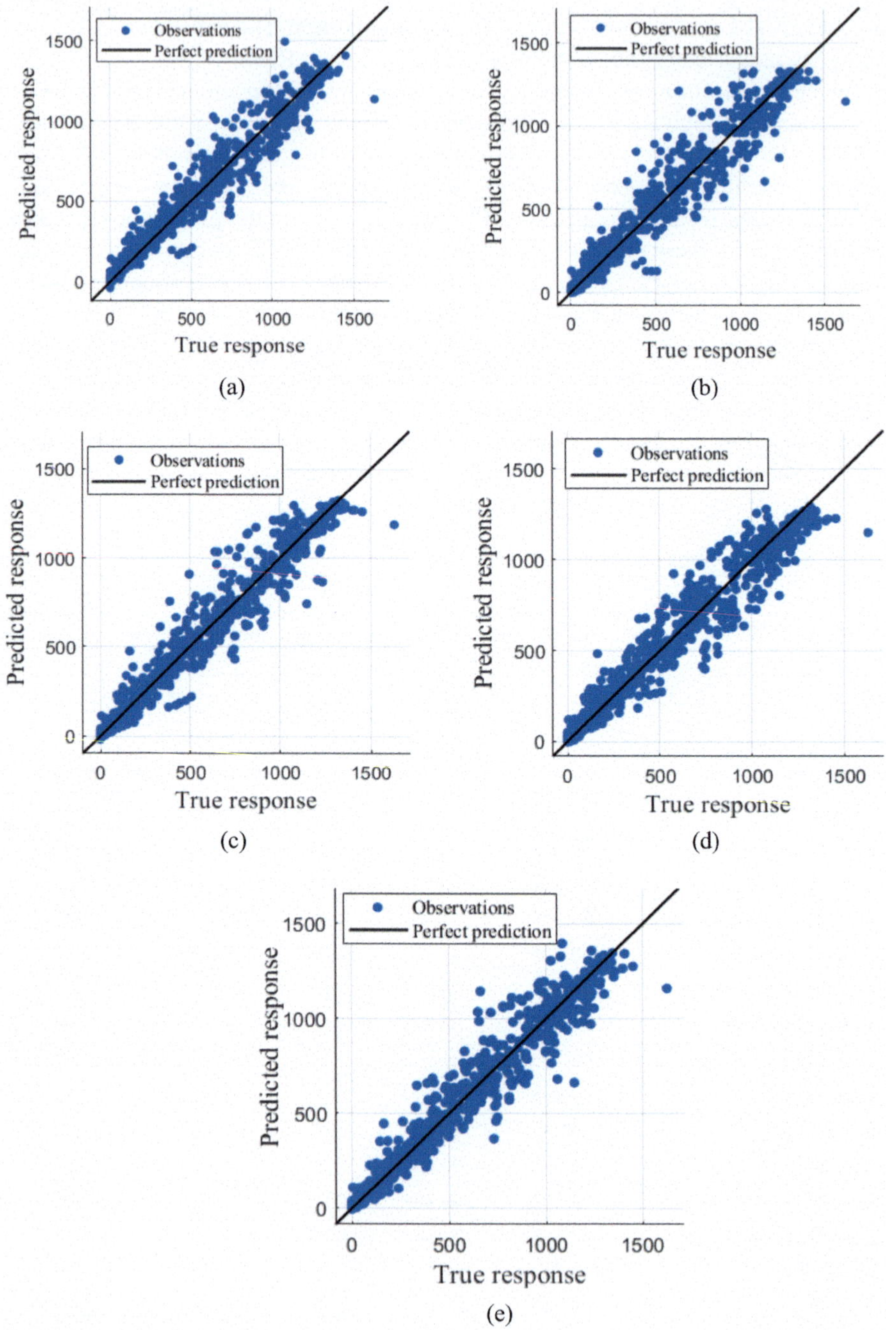

Fig. 8 The predicted results vs. actual data June: **a** the interactions linear regressor, **b** medium tree regressor, **c** medium Gaussian SVM regressor, **d** bagged trees ensemble regressor, **e** exponential GPR regressor

Table 5 Optimized hyperparameters for two best regressors (Case 2)

Optimized GPR		Optimized SVM	
Basis function	Constant	Kernel function	Gaussian
Kernel function	Isotropic exponential	Kernel scale	4.0074
		Box constraint	794.9968
Kernel scale	0.0032113	Epsilon	15.9889
Sigma	3143.4932	Standardized	True
Standardized	True	Optimization method	Sequential minimal optimization
Fit method	Standard deviation	Minimum objective	3376.5
Optimizer	Quasi-Newton	Minimum estimated objective	3389.4
Minimum objective	2993.1		
Minimum estimated objective	3012.9		

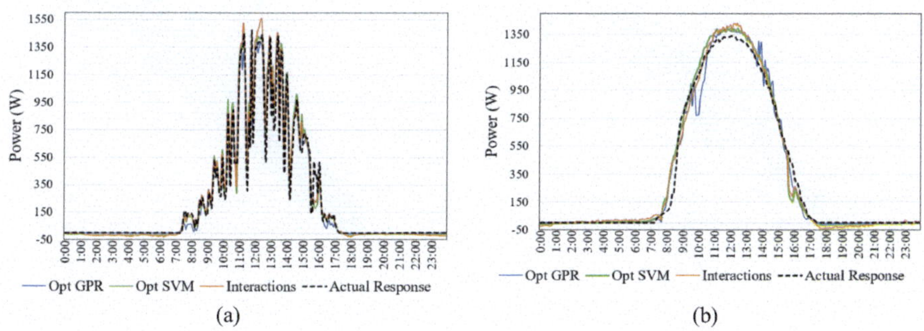

Fig. 9 The predicted results vs. actual data in June for the optimized GPR regressor, optimized SVM regressor, and interactions linear regressor: **a** a high-fluctuation day, **b** low-fluctuation day

4.4 *Case 3—Combined December and June Dataset*

The third dataset is the combination of December 2013 and June 2014 data. After training the models, the validation RMSE values indicate that the interactions linear regressor, the stepwise linear regressor, the medium Gaussian SVM regressor, the ensemble of bagged trees regressor, and the ensemble of boosted trees are ranked the five best-performing models for the combined dataset (Table 4). Since the difference in their RMSE is relatively small, these models can be used interchangeably. The testing RMSE values demonstrate that the medium Gaussian SVM does not show any overfitting over the data, as it has the lowest RMSE values on the test dataset. Similar to Cases I and II, the interactions linear regression, medium tree, medium Gaussian SVM, and exponential GPR are found to have the best accuracy within their own regressor family (Fig. 10) (Table 6).

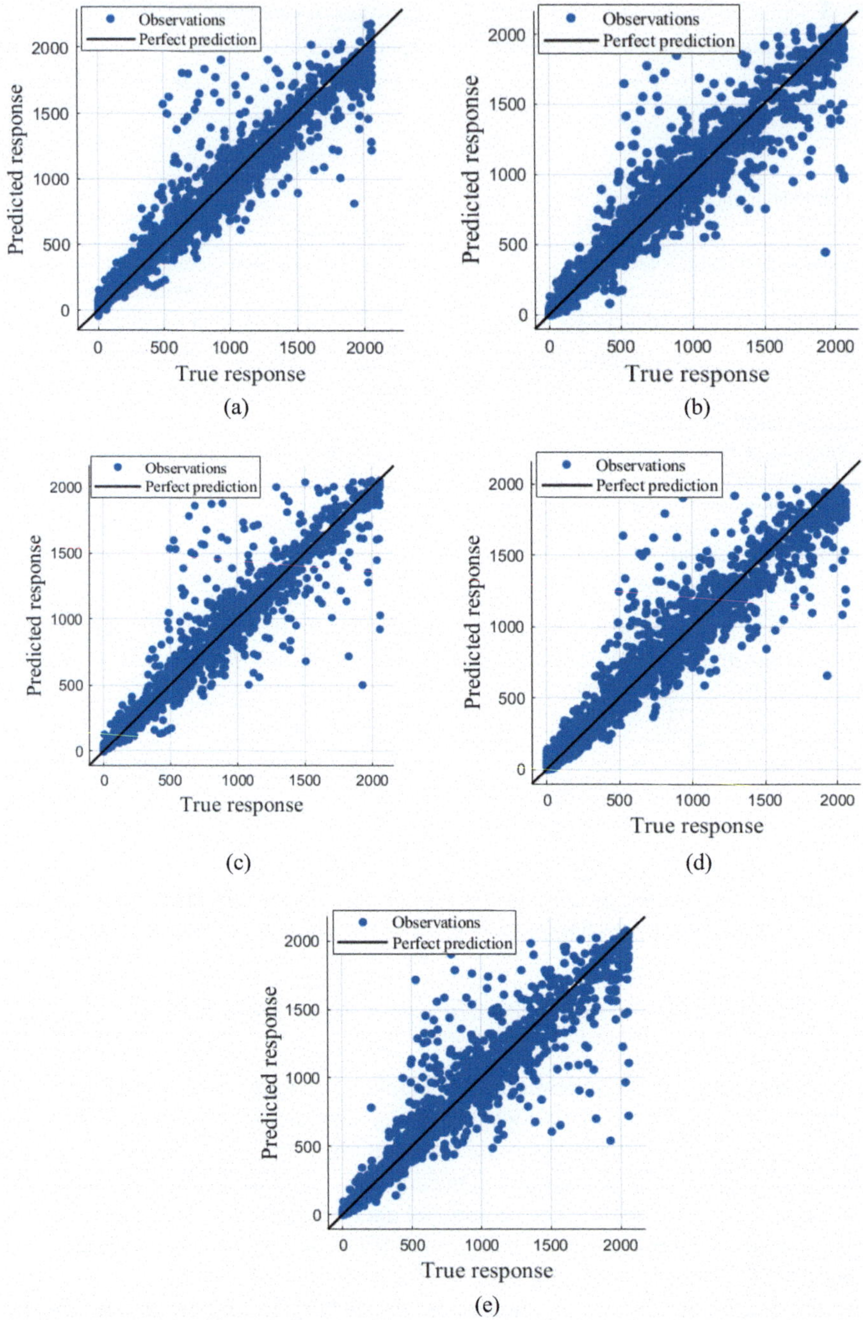

Fig. 10 The predicted results and actual data in the combined dataset: **a** the interactions linear and stepwise linear regressors, **b** the medium tree regressor, **c** the medium Gaussian SVM regressor, **d** the bagged trees ensemble regressor, **e** the exponential Gaussian process regressor

Table 6 Validation results for the combined training dataset

Regression	RMSE		R^2	MSE	MAE
Model	Validate	Test			
Bagged trees	107.63	105.47	0.96	11,584	45.711
Medium Gaussian SVM	108.3	92.384	0.96	11,730	41.846
Interactions linear	108.44	100.43	0.96	11,760	51.641
Stepwise linear	108.44	100.43	0.96	11,760	51.641
Boosted trees	110.37	94.355	0.96	12,181	45.708
Fine Gaussian SVM	111.01	128.46	0.96	12,323	47.899
Quadratic SVM	112.6	94.316	0.96	12,680	52.777
Medium tree	113.1	106.92	0.96	12,792	44.291
Coarse tree	113.92	99.17	0.95	12,978	44.688
Coarse Gaussian SVM	114.07	95.423	0.95	13,012	44.921
Exponential GPR	115.52	105.96	0.95	13,344	42.135
Fine tree	121.64	103.59	0.95	14,796	47.806
Linear regression	123.45	122.64	0.95	15,239	67.485
Rational quadratic GPR	126.42	115.57	0.94	15,983	49.075
Linear SVM	128.35	127.34	0.94	16,473	63.654
Matern 5/2 GPR	160.37	197.43	0.91	25,717	59.508
Squared exponential GPR	194.65	288.94	0.87	37,889	83.929
Robust linear	491.43	746.83	0.16	241,500	220.38
Cubic SVM	611.86	959.24	– 0.3	374,380	377.63

4.5 *Results Discussions*

To compare the regression models trained for Cases 1–3, their validation RMSE values are replaced by the normalized RMSE (nRMSE) using the average of actual measurements in the historical data. Table 7 represents these nRMSE values, in which the last column is the average of nRMSE obtained for the three cases. Figure 11 also visualizes these average values. It is found that for this dataset, the cubic Support Vector Machine and robust linear regression models result in poor performances and have the worst average nRMSE. On the other hand, the interactions linear regressor, the stepwise linear regressor, the medium Gaussian SVM regressor, and bagged trees regressor are the best regression models.

Table 7 nRMSE of 19 regression models for cases I-III

Regression		nRMSE		Avg
Model	Dec	Jun	Combined	nRMSE
Linear	0.321	0.274	0.368	0.321
Interactions linear	0.296	0.264	0.323	0.294
Stepwise linear	0.296	0.264	0.323	0.294
Robust linear	0.413	1.819	1.463	1.232
Fine tree	0.337	0.298	0.362	0.333
Medium tree	0.320	0.295	0.337	0.317
Coarse tree	0.324	0.300	0.339	0.321
Linear SVM	0.336	0.281	0.382	0.333
Quadratic SVM	0.319	0.279	0.335	0.311
Cubic SVM	0.967	1.159	1.822	1.316
Fine Gaussian SVM	0.322	0.264	0.330	0.306
Medium Gaussian SVM	0.311	0.256	0.322	0.297
Coarse Gaussian SVM	0.320	0.265	0.340	0.308
Boosted trees	0.315	0.280	0.329	0.308
Bagged trees	0.306	0.273	0.320	0.300
Squared exponential GPR	0.509	0.449	0.580	0.513
Matern 5/2 GPR	0.397	0.333	0.477	0.402
Exponential GPR	0.331	0.252	0.344	0.309
Rational quadratic GPR	0.341	0.273	0.376	0.330

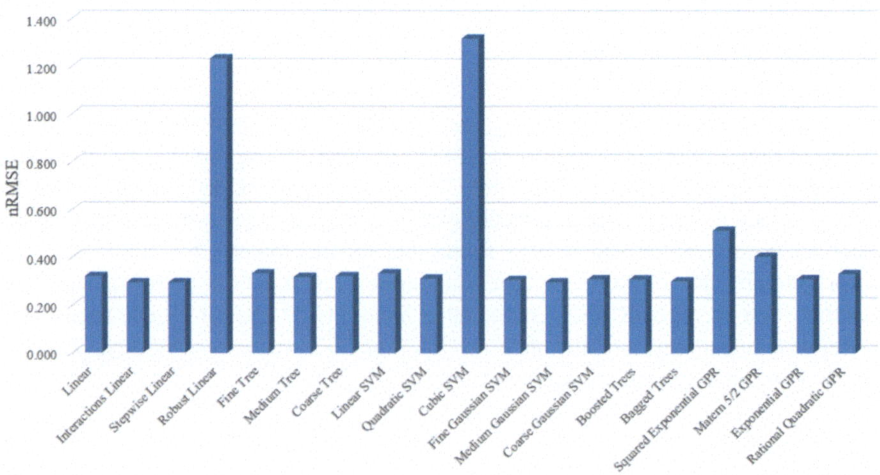

Fig. 11 Comparing the average normalized RMSE obtained from the three case studies

References

1. X. Meng, F. Gao, T. Xu, K. Zhou, W. Li, Q. Wu, Inverter-data-driven second-level power forecasting for photovoltaic power plant. IEEE Trans. Ind. Electron. **68**(8), 7034–7044 (2021). https://doi.org/10.1109/TIE.2020.3005098
2. S. Karamdel, X. Liang, S. O. Faried, M. N. S. K. Shabbir, A regression model-based short-term PV power generation forecasting, in *2022 IEEE Electrical Power and Energy Conference (EPEC)*, Victoria, BC, Canada (2022), pp. 261–266. https://doi.org/10.1109/EPEC56903.2022.10000086
3. M. Monfared, M. Fazeli, R. Lewis, J. Searle, Fuzzy predictor with additive learning for very short-term PV power generation. IEEE Access **7**, 91183–91192 (2019). https://doi.org/10.1109/ACCESS.2019.2927804
4. H. Nazaripouya, B. Wang, Y. Wang, P. Chu, H. R. Pota, and R. Gadh, Univariate time series prediction of solar power using a hybrid wavelet-ARMA-NARX prediction method, in *2016 IEEE/PES Transmission and Distribution Conference and Exposition (T D)* (2016), pp. 1–5. https://doi.org/10.1109/TDC.2016.7519959
5. M. Abdel-Nasser, K. Mahmoud, M. Lehtonen, Reliable solar irradiance forecasting approach based on choquet integral and deep LSTMs. IEEE Trans. Ind. Informatics **17**(3), 1873–1881 (2021). https://doi.org/10.1109/TII.2020.2996235
6. P. Jia, H. Zhang, X. Liu, X. Gong, Short-term photovoltaic power forecasting based on VMD and ISSA-GRU. IEEE Access **9**, 105939–105950 (2021). https://doi.org/10.1109/ACCESS.2021.3099169
7. H. Sheng, J. Xiao, Y. Cheng, Q. Ni, S. Wang, Short-term solar power forecasting based on weighted Gaussian process regression. IEEE Trans. Ind. Electron. **65**(1), 300–308 (2018). https://doi.org/10.1109/TIE.2017.2714127
8. Y. Zhang, C. Qin, A.K. Srivastava, C. Jin, R.K. Sharma, Data-driven day-ahead PV estimation using autoencoder-LSTM and persistence model. IEEE Trans. Ind. Appl. **56**(6), 7185–7192 (2020). https://doi.org/10.1109/TIA.2020.3025742
9. W. Lee, K. Kim, J. Park, J. Kim, Y. Kim, Forecasting solar power using long-short term memory and convolutional neural networks. IEEE Access **6**, 73068–73080 (2018). https://doi.org/10.1109/ACCESS.2018.2883330
10. K. Mahmud, S. Azam, A. Karim, S. Zobaed, B. Shanmugam, D. Mathur, Machine learning based PV power generation forecasting in alice springs. IEEE Access **9**, 46117–46128 (2021). https://doi.org/10.1109/ACCESS.2021.3066494
11. H. Eom, Y. Son, S. Choi, Feature-selective ensemble learning-based long-term regional PV generation forecasting. IEEE Access **8**, 54620–54630 (2020). https://doi.org/10.1109/ACCESS.2020.2981819
12. G. Li, S. Xie, B. Wang, J. Xin, Y. Li, S. Du, Photovoltaic power forecasting with a hybrid deep learning approach. IEEE Access **8**, 175871–175880 (2020). https://doi.org/10.1109/ACCESS.2020.3025860
13. F. Mei et al., Day-ahead nonparametric probabilistic forecasting of photovoltaic power generation based on the LSTM-QRA ensemble model. IEEE Access **8**, 166138–166149 (2020). https://doi.org/10.1109/ACCESS.2020.3021581
14. C.N. Obiora, A.N. Hasan, A. Ali, N. Alajarmeh, Forecasting hourly solar radiation using artificial intelligence techniques. IEEE Can. J. Electr. Comput. Eng. **44**(4), 497–508 (2021). https://doi.org/10.1109/ICJECE.2021.3093369
15. I. Sansa, Z. Boussaada, M. Mazigh, N.M. Bellaaj, Solar radiation prediction for a winter day using ARMA model, in *2020 6th IEEE International Energy Conference (ENERGYCon)* (2020), pp. 326–330. https://doi.org/10.1109/ENERGYCon48941.2020.9236541
16. X. Huang, J. Shi, B. Gao, Y. Tai, Z. Chen, J. Zhang, Forecasting hourly solar irradiance using hybrid wavelet transformation and Elman model in smart grid. IEEE Access **7**, 139909–139923 (2019). https://doi.org/10.1109/ACCESS.2019.2943886

17. S.I. Vagropoulos, G.I. Chouliaras, E.G. Kardakos, C.K. Simoglou, A.G. Bakirtzis, Comparison of SARIMAX, SARIMA, modified SARIMA and ANN-based models for short-term PV generation forecasting, in *2016 IEEE International Energy Conference (ENERGYCON)* (2016), pp. 1–6 https://doi.org/10.1109/ENERGYCON.2016.7514029
18. N. Sharma, P. Sharma, D. Irwin, P. Shenoy, Predicting solar generation from weather forecasts using machine learning, in *2011 IEEE International Conference on Smart Grid Communications (SmartGridComm)* (2011), pp. 528–533. https://doi.org/10.1109/SmartGridComm.2011.6102379
19. A. Nespoli, S. Leva, M. Mussetta, E.G.C. Ogliari, A selective ensemble approach for accuracy improvement and computational load reduction in ANN-based PV power forecasting. IEEE Access **10**, 32900–32911 (2022). https://doi.org/10.1109/ACCESS.2022.3158364
20. J. Simeunović, B. Schubnel, P.-J. Alet, R.E. Carrillo, Spatio-temporal graph neural networks for multi-site PV power forecasting. IEEE Trans. Sustain. Energy **13**(2), 1210–1220 (2022). https://doi.org/10.1109/TSTE.2021.3125200
21. H. Panamtash, Q. Zhou, Coherent probabilistic solar power forecasting, in *2018 IEEE International Conference on Probabilistic Methods Applied to Power Systems (PMAPS)* (2018), pp. 1–6. https://doi.org/10.1109/PMAPS.2018.8440483
22. Y. Wen et al., Performance evaluation of probabilistic methods based on bootstrap and quantile regression to quantify PV power point forecast uncertainty. IEEE Trans. Neural Networks Learn. Syst. **31**(4), 1134–1144 (2020). https://doi.org/10.1109/TNNLS.2019.2918795
23. W. Zhang, H. Quan, O. Gandhi, C.D. Rodríguez-Gallegos, A. Sharma, D. Srinivasan, An ensemble machine learning based approach for constructing probabilistic PV generation forecasting, in *2017 IEEE PES Asia-Pacific Power and Energy Engineering Conference (APPEEC)* (2017), pp. 1–6 https://doi.org/10.1109/APPEEC.2017.8308947
24. U. Zaman, H. Teimourzadeh, E.H. Sangani, X. Liang, C.Y. Chung, Wind speed forecasting using ARMA and neural network models, in *2021 IEEE Electrical Power and Energy Conference (EPEC)* (2021), pp. 243–248. https://doi.org/10.1109/EPEC52095.2021.9621650
25. MathWorks, Hyperparameter Optimization in Regression Learner App, https://www.mathworks.com/help/stats/hyperparameter-optimization-in-regression-learner-app.html
26. M. Madhukumar, A. Sebastian, X. Liang, M. Jamil, M.N.S.K. Shabbir, Regression model-based short-term load forecasting for university campus load. IEEE Access **10**, 8891–8905 (2022). https://doi.org/10.1109/ACCESS.2022.3144206
27. M.S. Acharya, A. Armaan, A.S. Antony, A comparison of regression models for prediction of graduate admissions, in *2019 International Conference on Computational Intelligence in Data Science (ICCIDS)* (2019), pp. 1–5. https://doi.org/10.1109/ICCIDS.2019.8862140
28. MathWorks, Choose Regression Model Options, https://www.mathworks.com/help/stats/choose-regression-model-options.html
29. MathWorks, Assess Model Performance in Regression Learner, https://www.mathworks.com/help/stats/assess-model-performance-in-regression-learner.html
30. Z. Vale, N. Kagan, F. Fernandes, H. Morais, V.V. Garcia, L. Gomes, Dynamic loads and microgeneration method for a house management system (2016). https://doi.org/10.1109/PSC.2016.7462829

Optimal Placement of Photovoltaic Systems in Multi-Energy Systems

Shahab Karamdel, Xiaodong Liang, and Sherif O. Faried

Abstract Solar photovoltaic (PV) systems can provide reactive power support through their interfacing smart inverters. This reactive power exchange can provide effective voltage regulation at the inverter's connection point and supplement the voltage regulation already offered by legacy voltage regulation devices, on-load tap changers and capacitor banks. Conservation voltage reduction (CVR), the intentional operation of the electrical system to provide customer voltages in the lower end of the acceptable range to reduce energy and demand, has recently gained significant research interest due to voltage regulation support by the increasing integration of PVs. This chapter discusses that CVR benefits can be significantly enhanced by considering the location, inverter capacity and number of PV systems, and presents an optimal PV placement method to enhance CVR benefits in integrated electricity and natural gas systems, where the electrical and natural gas networks are linked via gas-fired distributed generations and power-to-gas units. The chapter then introduces an optimal PV placement problem to minimize load consumptions and the network power losses while observing all operational constraints in the system. The location of PVs and capacity of PV inverters are decision variables. Through case studies using the modified IEEE 33-bus electricity and 7-node natural gas test system, the chapter demonstrates that optimally placed PVs lead to significant improvement in load consumption reductions compared to randomly placed PVs, and a higher number of PVs provides higher power loss reductions for CVR implementation.

Keywords Conservation voltage reduction · Multi-energy system · Optimal placement · Photovoltaics · Smart inverters · Volt/VAR optimization

S. Karamdel · X. Liang (✉) · S. O. Faried
Department of Electrical and Computer Engineering, University of Saskatchewan, Saskatoon, Canada
e-mail: xil659@mail.usask.ca

F. Shahnia and F. Wen (eds.), *Power and Energy Resources*, Power Systems,
https://doi.org/10.1007/978-981-96-2612-0_6

1 Introduction

Solar photovoltaics (PV) are increasingly integrated to electric distribution networks due to the advancement of PV technology and their relatively low cost [1], and the solar power generation capacity addition in the U.S. is estimated to reach 71% in 2025 [2]. PVs are connected to distribution grids through interfacing smart inverters, providing reactive power control potential for grid services [3], such as voltage regulation as PV smart inverters can either produce or absorb reactive power upon the request.

Natural gas is a dominant energy source in the U.S. [4] with considerably lower carbon emissions than other fossil fuels. Gaseous hydrogen, as a clean energy source, can be injected into natural gas distribution pipelines [5]–[6]. Through the energy conversion technologies, such as gas-fire distributed generations (DGs) and power-to-gas (P2G) units, electricity and natural gas networks can be linked together and become interdependent, requiring unique considerations in energy management programs, such as demand response and conservation voltage reduction (CVR).

CVR is a grid management technique that reduces load consumptions and the peak load demand by intentionally lowering voltage magnitudes through voltage-sensitive loads [7], decreasing 1% feeder voltage can lead to 0.3–1% of load reductions, so utilities can supply less energy to customers and financially benefit from it. CVR was traditionally implemented through legacy voltage regulation devices, on-load tap changers (OLTCs), and capacitor banks (CBs), which usually have switching limitations to prevent them from physical damages. PV smart inverters, on the other hand, do not have such limitations, and can provide rapid voltage regulating services [8]. Therefore, optimal PV placement can determine the PV location and the capacity of PV smart inverters, which is crucial for unlocking their potentials in CVR implementation.

In the literature, the CVR research can be categorized into [7]: (1) short-term CVR operation scheduling; (2) long-term CVR planning. Most of the CVR research done belong to the first category, including scheduling the legacy voltage regulation devices for CVR in [9–11], coordinating the legacy voltage regulation devices with PV smart inverters in [12–14] and with soft open points in [15]. Although an optimal allocation problem is not a new concept [16], few studies have focused on for CVR purposes. A meter placement method for CVR is proposed in [17]. Reference [18] develops an stochastic programming model for optimal placement of soft open points and battery energy storage systems with CVR and demand response enabled. A DG placement model for load consumption reductions without considering power losses is presented in [19], but the reactive power support of DGs is not considered in this model. Optimal placement of battery energy storage systems and DGs is addressed in [20] as a two-layer problem, and reactive power support of PV smart inverters is considered. Reference [21] focuses on long-term planning of distribution systems for CVR implementation. To the best of the authors' knowledge, CVR research has not been conducted for multi-energy system with integrated electricity and natural gas networks in the literature.

2 The Proposed Optimal PV Placement Method

The optimal PV placement method is shown in Fig. 1. The required input data include the integrated energy system topology and characteristics, parameters of voltage-sensitive loads, and the forecasted load and PV power generation profiles. To have more accurate results, several representative days should be considered, and clustering techniques, such as *k-means* clustering, can be applied to consider uncertainties.

In this planning method, it is assumed that each PV smart inverter consists of several small units, and the searching domain is discrete. Hence, the number of these small units and their capacity need to be determined at the beginning of the allocation problem. The maximum number of PVs installed in the whole network is another parameter required to be set. After having this information, an optimization problem is solved to minimize the total load consumptions and the network power losses. The

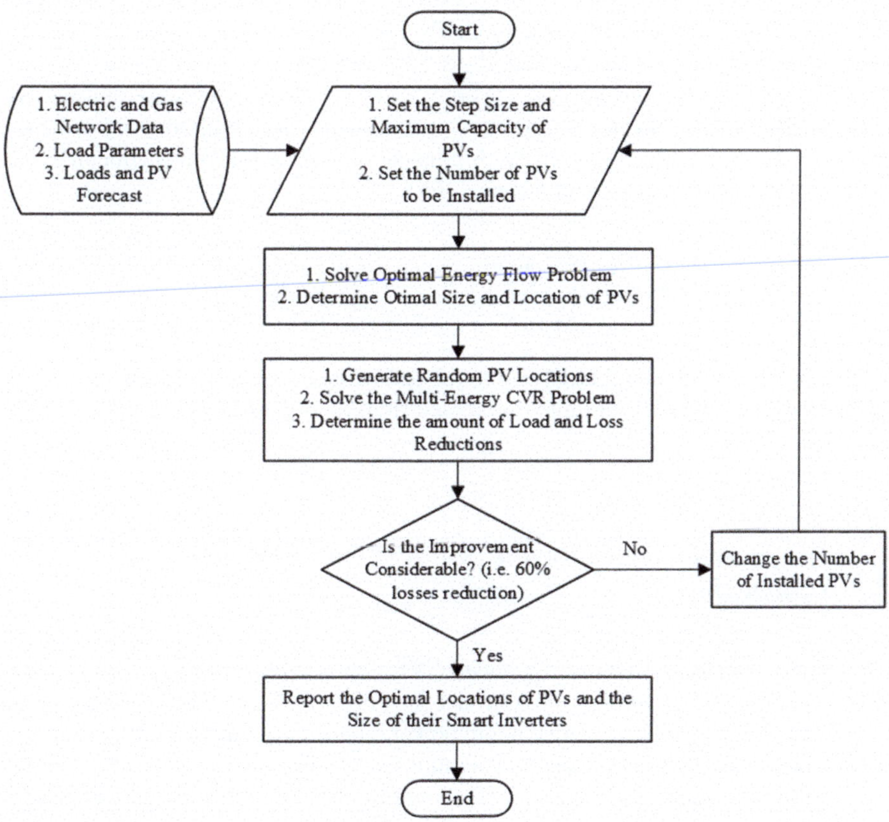

Fig. 1 Flowchart of the optimal placement of PVs

outcome of this program is the optimal PV locations and capacity/size of PV smart inverters.

A trade-off study is then conducted to evaluate CVR benefits to determine if the number of optimally placed PVs is enough or needs to be increased by comparing with randomly placed PVs for day-ahead CVR scheduling in multi-energy systems. The amount of reductions for load consumptions and power losses is then calculated between scenarios with optimally and randomly placed PVs. If the reductions are satisfactory (roughly 60% power losses reduction), the optimal PV location and the capacity of PV smart inverters are determined. Otherwise, the maximum number of PVs will be changed, most likely increases, and the whole process is repeated.

3 Mathematical Formulation

The mathematical formulation of the optimal PV placement method in multiple-energy systems is provided in this section. The objective function of this optimization model is to minimize load consumptions and power losses in (1). The decision variables are PV locations and the capacity of PV smart inverters. w_L and w_P are weight coefficients of this multi-objective optimization for minimizing power losses and load consumptions, respectively. A constant σ can be defined to calculate the objective function's yearly amount ($\sigma = 365$).

$$\min \sigma . \sum_t \left(\sum_{(i,j)\in \aleph^l} w_L . P_{ij,t}^{Loss} + \sum_j w_P . P_{j,t}^{Load} \right) \tag{1}$$

This optimization model is subject to the following constraints: (1) optimal CVR scheduling constraints in the integrated electricity and natural gas system; (2) PV allocation constraints.

3.1 Integrated Electricity and Natural Gas System Constraints

These constraints represent the day-ahead CVR operation scheduling. Second-order conic relaxation of the branch flow model for power flow calculations is provided by (2)–(6). The active and reactive nodal power balance is guaranteed in (2) and (3), respectively. Equation (4) corresponds to Ohm's law. The active and reactive power flow over power distribution lines is calculated by (5) [24]. The voltage and current magnitudes are bounded between their upper and lower limits in (6). Similar to the power flow constraints, Eqs. (7)–(13) model the natural gas flow in the natural gas network. The natural gas nodal energy balance is represented by (7). Equation (8) calculates the line-pack in gas pipelines, which is the capacity of the pipelines to store natural gas. Equation (9) relates line-pack values to the nodal gas flow [25]. The average gas flow over the pipelines is calculated in (10), and its value is used in the Weymouth equation in (11) to calculate nodal gas pressures. For tractability purposes, like in (5), second-order conic relaxation of the Weymouth equation is adopted in this chapter [26]. The upper and lower bounds of nodal gas pressures and gas production of suppliers are constrained by (12). The model of the gas compressor is provided in (13). The active power output and the input of gas-fired DGs and P2G units are limited by (14). The limits of reactive power output of gas-fired DGs are calculated with respect to their power factors in (15).

The mathematical formulation needs to model legacy voltage regulators, OLTCs and CBs. The voltage magnitude of the secondary of OLTCs is calculated by (16), where $TAP_{i,j,t}$ is the integer variable corresponding to the tap position. Equation (16) contains the square of this variable, which makes the formulation non-linear. To deal with this non-linearity, similar to the work done in [27], the binary expansion technique and auxiliary variables are adopted to reach a convex model for OLTCs. The reactive power injection by CBs is calculated using the integer variable representing their switch position in (17).

The CVR performance highly depends on the load and its voltage sensitivity modeling. As a popular ZIP (constant-impedance, constant-current, and constant-power) load model is used in this chapter. The voltage-dependent ZIP loads are calculated through (18) and (19), where α and β are parameters of active and reactive loads, respectively. Characteristics of PV units need to be formulated. The second-order conic relation between the active and reactive power output of PVs is calculated in (20) [28]. This equation shows the *P*-*Q* capability curve of PVs. The active and reactive power output of PVs are limited by (21), where the upper limit of $P^{PV}_{j,t}$ is calculated using the power factor. Note that $Q^{PV}_{j,t}$ can have either positive or negative values, and PV smart inverters tend to regulate voltages by injecting or absorbing reactive power. PV smart inverters may curtail their active power output to satisfy the nodal power balance.

$$\sum_{g\in\aleph^{GE}_{g,j}} P^{GF}_{g,t} + P^{PV}_{j,t} + \sum_{(i,j)\in\aleph^{l}} \left(P_{i,j,t} - l_{i,j,t}.r_{i,j}\right) = \sum_{(j,k)\in\aleph^{l}} P_{j,k,t} + P^{Load}_{j,t} + \sum_{u\in\aleph^{P2G}_{u,j}} P^{P2G}_{u,t}, \forall j,t \tag{2}$$

$$\sum_{g\in\aleph^{GE}_{g,j}} Q^{GF}_{g,t} + Q^{PV}_{j,t} + \sum_{(i,j)\in\aleph^{l}} \left(Q_{i,j,t} - l_{i,j,t}.x_{i,j}\right) + Q^{CB}_{j,t} = \sum_{(j,k)\in\aleph^{l}} Q_{j,k,t} + Q^{Load}_{j,t}, \forall j,t \tag{3}$$

$$v_{j,t} = v_{i,t} - 2\left(P_{i,j,t}.r_{i,j} + Q_{i,j,t}.x_{i,j}\right) + \left(r^2_{i,j} + x^2_{i,j}\right)l_{i,j,t}, \forall (i,j) \in \aleph^{l}, t \tag{4}$$

$$\left\| \begin{matrix} 2P_{i,j,t} \\ 2Q_{i,j,t} \\ l_{i,j,t} - v_{i,t} \end{matrix} \right\|_2 \leq l_{i,j,t} + v_{i,t}, , \forall(i,j) \in \aleph^l, t \tag{5}$$

$$\underline{V}_j^2 \leq v_{j,t} \leq \overline{V}_j^2, 0 \leq l_{i,j,t} \leq \overline{I}_{i,j}^2 \tag{6}$$

$$G_{n,t}^{SP} + \sum_{u \in \aleph_{u,n}^{P2G}} \eta_u.R.P_{u,t}^{P2G} + \sum_{(m,n) \in \aleph^{GB}} F_{m,n,t}^{out} - \sum_{(n,o) \in \aleph^{GB}} F_{n,o,t}^{in} + \sum_{(m,n) \in \aleph^C} F_{m,n,t}^C$$

$$= \sum_{g \in \aleph_{g,n}^{GG}} \frac{1}{\eta_g}.R.P_{g,t}^{GF} + G_{n,t}^L, \forall n, t \tag{7}$$

$$L_{m,n,t} = K_{m,n}^L.\left(\pi_{m,t} + \pi_{n,t}\right), \forall(m,n) \in \aleph^{GB}, t \tag{8}$$

$$F_{m,n,t}^{in} - F_{m,n,t}^{out} = L_{m,n,t} - L_{m,n,t-1}, \forall(m,n) \in \aleph^{GB}, t \tag{9}$$

$$F_{m,n,t} = 0.5\left(F_{m,n,t}^{in} + F_{m,n,t}^{out}\right), \forall(m,n) \in \aleph_n^{GB}, t \tag{10}$$

$$\left\| \begin{matrix} F_{m,n,t}/K_{m,n}^W \\ \pi_{n,t} \end{matrix} \right\|_2 \leq \pi_{m,t}, \forall(m,n) \in \aleph^{GB}, t \tag{11}$$

$$\underset{-n}{\pi} \leq \pi_{n,t} \leq \overline{\pi}_n, \underline{G}_n^{SP} \leq G_{n,t}^{SP} \leq \overline{G}_n^{SP} \tag{12}$$

$$\pi_{m,t} \leq \pi_{n,t} \leq K_{m,n}^C.\pi_{m,t}, 0 \leq F_{m,n,t}^C \leq F_{m,n}^{C,max}, \forall(m,n) \in \aleph^C, t \tag{13}$$

$$\underline{P}_g^{GF} \leq P_{g,t}^{GF} \leq \overline{P}_g^{GF}, \underline{P}_u^{P2G} \leq P_{u,t}^{P2G} \leq \overline{P}_u^{P2G} \tag{14}$$

$$-P_{g,t}^{GF}.\sqrt{1 - pf^2} \leq Q_{g,t}^{GF} \leq P_{g,t}^{GF}.\sqrt{1 - pf^2}, \forall g, t \tag{15}$$

$$v_{j,t} = (RATE_{i,j}^0 + TAP_{i,j,t}.\Delta TAP_{i,j})^2.v_{i,t}, \forall(i,j) \in \aleph_{i,j}^{OLTC}, t \tag{16}$$

$$Q_{j,t}^{CB} = N_{j,t}^{CB}.Q_j^{CB,unit}, \forall j \in \aleph_j^{CB}, t \tag{17}$$

$$P_{j,t}^{Load} = P_{j,t}^f \left\{ \left(\alpha_j^P + \frac{\alpha_j^I}{2} \right) + \left(\alpha_j^Z + \frac{\alpha_j^I}{2} \right).v_{j,t} \right\}, \forall j, t \tag{18}$$

$$Q_{j,t}^{Load} = Q_{j,t}^f \left\{ \left(\beta_j^P + \frac{\beta_j^I}{2} \right) + \left(\beta_j^Z + \frac{\beta_j^I}{2} \right).v_{j,t} \right\}, \forall j, t \tag{19}$$

$$P_{j,t}^{PV\,2} + Q_{j,t}^{PV\,2} \leq S_j^{PV\,2}, \forall j, t \tag{20}$$

$$-Q_{j,t}^{Max} \leq Q_{j,t}^{PV} \leq Q_{j,t}^{Max}, 0 \leq P_{j,t}^{PV} \leq pf_j.S_j^{PV}, \forall j, t \tag{21}$$

3.2 PV Placement Constraints

These constraints in (22)–(24) are for PV optimal placement [19]. Equation (22) calculates the capacity of PV smart inverters. It is assumed that the smart inverter's capacity at every bus consists of several small units, each having the capacity of PV_j^{Unit}; $PV_{j,p}^{Size}$ is a binary variable deciding whether a unit needs to be connected to that bus; PV_j^{Ins} is a binary variable to determine if the bus needs to have an inverter. The maximum number of installed PVs within the whole network is determined by (23). The multiplication of the two binary variables makes (22) non-linear. Adopted from [19], this bi-linear term is substituted by an auxiliary binary variable named $PV_{j,p}^{aux}$ (24). Equations (25)–(27) build $PV_{j,p}^{aux}$ from PV_j^{Ins} and $PV_{j,p}^{Size}$.

$$S_j^{PV} = \sum_p PV_j^{Ins}.PV_{j,p}^{Size}.PV_j^{Unit}, \forall j \tag{22}$$

$$\sum_j PV_j^{Ins} \leq PV^{Total} \tag{23}$$

$$PV_{j,p}^{aux} = PV_j^{Ins}.PV_{j,p}^{Size}, \forall j, p \tag{24}$$

$$PV_{j,p}^{aux} \leq PV_j^{Ins}, \forall j, p \tag{25}$$

$$PV_{j,p}^{aux} \leq PV_{j,p}^{Size}, \forall j, p \tag{26}$$

$$PV_{j,p}^{aux} \leq PV_j^{Ins} + PV_{j,p}^{Size} - 1, \forall j, p \tag{27}$$

4 Validation Results

4.1 Test System Description

IEEE 33-bus test system is widely used in the literature. To validate the optimal PV placement model and measure its effectiveness in securing the maximum CVR benefits by reducing load consumptions and power losses in integrated electricity and natural gas networks, IEEE 33-bus test system in [29] is coupled with a 7-node natural gas system in [30]. The integrated multi-energy system topology and its parameters can be found in [31] (Fig. 2). The PV locations, however, are obtained by solving the optimal PV placement model. The following assumptions are made in this study:

1. PVs are connected to the grid through PV smart inverters, capable of injecting or absorbing reactive power. These smart inverters can curtail active power outputs of PVs to ensure power balance.
2. The maximum capacity of each PV smart inverter is 0.5 MVA, and the estimated active power generation of each PV is half of its smart inverter capacity [32].
3. Each PV smart inverter consists of 10 units, and each unit has a capacity of 0.05 MVA.
4. The electric distribution system is a radial system.

The models in this chapter are solved by GAMS's CPLEX solver on a personal computer with Intel Core i7-10750H CPU @ 2.60GHz processor and 16 GB RAM.

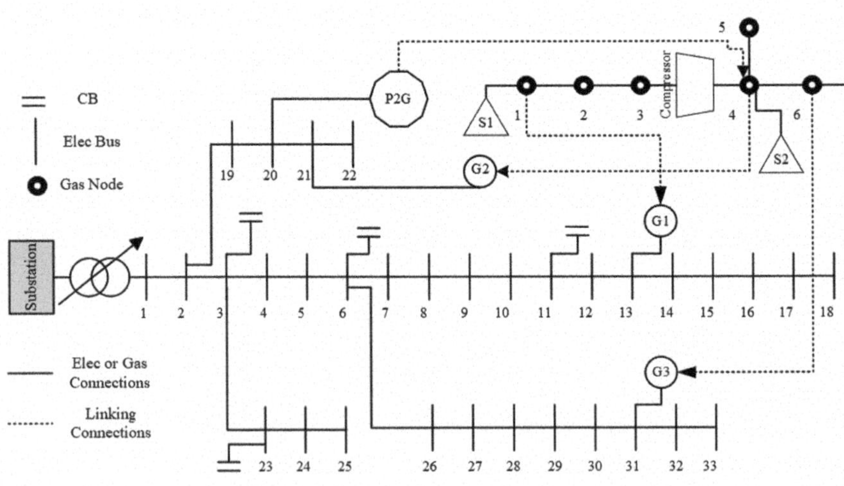

Fig. 2 The modified IEEE 33-bus test system integrated with a 7-node natural gas network

4.2 Optimal PV Allocation Results

In this section, optimal PV allocation problem is solved considering the number of PVs to be 5 and 10. The following three case studies, inspired from [19], are designed to measure the CVR benefits:

- *Case I*: The substation voltage is fixed at its upper limit to ensure that voltages at other buses are kept within the standard limit [0.95, 1.05] p.u., no CVR, no voltage regulation devices involved.
- *Case II*: OLTCs, CBs, gas-fired DGs, and PV smart inverters at optimal PV locations are all utilized to regulate voltages, no CVR, only minimizing power losses.
- *Case III (this method):* The same as Case II, except CVR is enabled.

The values of decision variables for optimal PV placement are shown in Table 1. The penetration levels of PVs are 33.65% and 63.26% of the peak load with 5 and 10 PVs, respectively. For Case I, the daily load consumption and power losses are 70.935 MWh and 1.688 MWh, respectively. Case I is used as the baseline for comparison. In Case II, voltage regulation devices, including OLTCs, CBs, gas-fired DGs, and PV smart inverters at optimal PV locations, are used only for the power loss minimization, the daily load consumption and network power losses are 69.55 MWh and 0.638 MWh, and decrease by 1.952% and 62.204%, respectively, when compared to Case I. Case II has the best power loss reduction that could be achieved by 5 PVs. When CVR is enabled in Case III, the daily load consumption and network power losses are 67.707 MWh and 1.24 MWh, and reduce by 4.55% and 26.54%, respectively, when compared to Case I. This result is as expected as reductions of load consumptions and power losses are conflicting objectives [33]. The comparison is summarized in 2 (Table 2).

Table 1 Optimal location of PVs and PV smart inverter sizes for different numbers of installed PVs

Number of PVs	Optimal bus location [Inverter capacity (MVA)]
5	7 [0.5], 11 [0.5], 14 [0.5], 28 [0.5], 30 [0.5]
10	7 [0.45], 8 [0.5], 9 [0.5], 12 [0.45], 14 [0.45], 18 [0.5], 24 [0.5], 25 [0.35], 29 [0.5], 30 [0.5]

Table 2 Comparing the CVR scheduling results in the three cases with 5 PVs

	Daily load consumption (MWh)	Daily load consumption reduction		Daily power losses (MWh)	Daily power losses reduction	
		(MWh)	(%)		(MWh)	(%)
Case I	70.935	–	–	1.688	–	–
Case II	69.550	1.385	1.952	0.638	1.05	62.204
Case III	67.707	3.228	4.550	1.240	0.448	26.540

Since the power losses reduction in Case III is far below 60%, the number of installed PVs is increased to 10, and a similar analysis is carried out as shown in Table 3. It is found that by increasing the number of PVs, Case II leads to 1.586% and 75.237% reduction in daily load consumption and power losses, respectively, compared to Case I, and a higher power loss reduction is achieved compared to 5 PVs. In Case III with CVR enabled, the daily load consumption reduces by 4.552%, which is very close to the case with 5 PVs, but power losses reduce from 26.54% (5 PVs) to 68.542% (10 PVs). Increasing the number of PVs from 5 to 10 does not lead to significant load consumption reductions for CVR implementation, but integrating more PVs can reduce power losses.

Figure 3 shows a comparison of voltage profiles at a peak load time for three cases, without voltage regulations, with CVR and 5 optimally placed PVs, and with CVR and 10 optimally placed PVs. It is found that CVR implementation makes the voltage profile at a lower voltage level close to the 0.95 p.u. limit and becomes flatter, but the number of PVs does not change the voltage profile significantly.

Table 3 Comparing the CVR scheduling results in the three cases with 10 PVs

	Daily load consumption (MWh)	Daily load consumption reduction		Daily power losses (MWh)	Daily power losses reduction	
		(MWh)	(%)		(MWh)	(%)
Case I	70.935	–	–	1.688	–	–
Case II	69.810	1.125	1.586	0.418	1.27	75.237
Case III	67.706	3.229	4.552	0.531	1.157	68.542

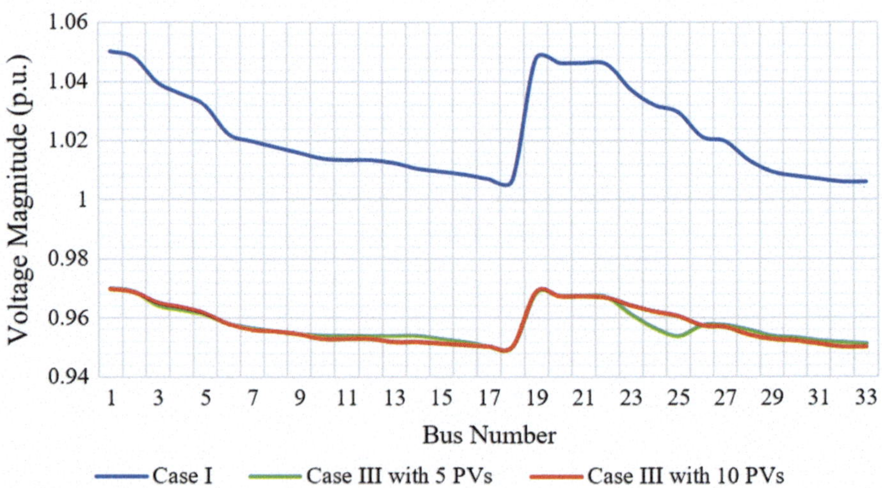

Fig. 3 Comparing voltage profiles at the peak load time (t = *14 h*) in Case 1 and Case 3 with 5 and 10 PVs optimally places

In this chapter, the optimization problem is solved for a day. To evaluate the CVR yearly benefits obtained from optimally placed PVs, the amounts of load consumption and network power losses can be further calculated for a whole year. As shown in Fig. 4a, CVR with optimal PV placement leads to 1.178 GWh/year reduction in the yearly load consumptions. Although the yearly load consumptions for 5 and 10 PVs installed at their optimal locations are almost the same, integrating 10 PVs instead of 5 PVs reduces the yearly power losses by 259 MWh/year (Fig. 4b). Overall, having 10 PVs installed contributes to 422 MWh/year reduction in network power losses.

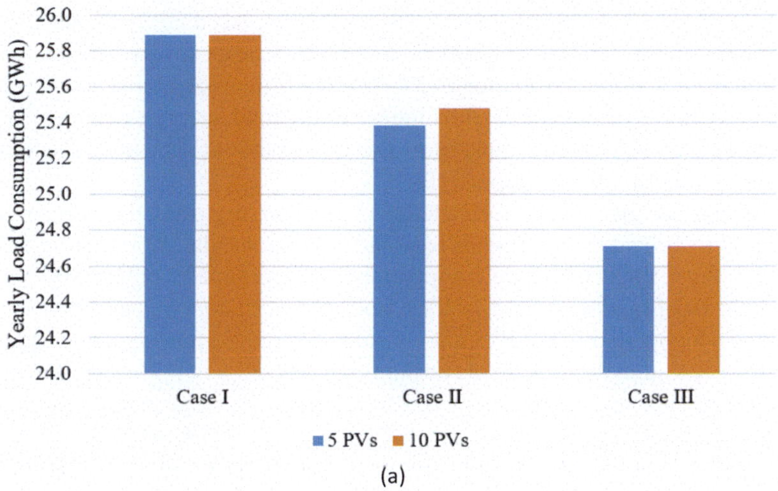

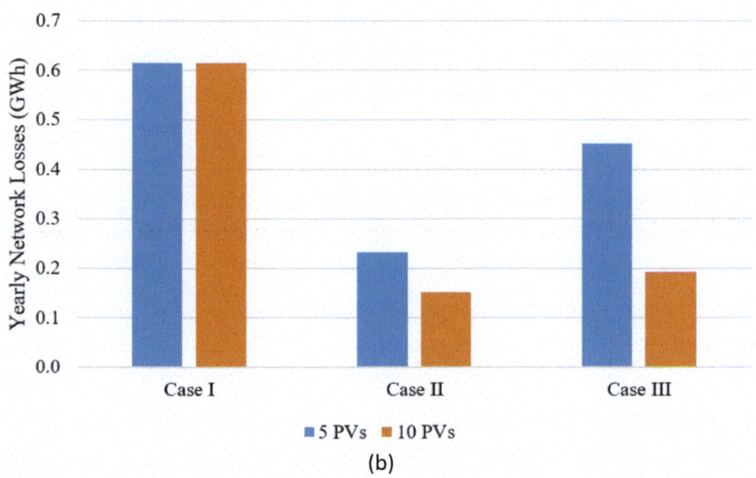

Fig. 4 Yearly values in the three case studies for 5 and 10 PVs optimally installed: **a** load consumption, **b** power losses

4.3 Day-Ahead CVR Scheduling Results: Randomly vs. Optimally Placed PVs

In this section, to evaluate the effectiveness of this method, the day-ahead CVR scheduling results in the modified IEEE 33-bus electricity and 7-node natural gas test system are obtained using optimally placed PVs vs. random placed PVs, as shown in Table 4. First, it is assumed that 5 PVs are optimally placed in the test system. In Table 4, the optimally placed 5 PVs lead to the daily load consumption reduction of 3.228 MWh in the test system. Compared to randomly placed 5 PVs in three scenarios, Radom 1.1, Radom 1.2, and Radom 1.3, the optimally placed 5 PVs reduce the daily load consumptions by 16.073%, 11.004%, and 15.08%, respectively. The total energy savings due to CVR implementation of 3.676 MWh is then calculated, and it is found that the optimal PV placement method improves the total energy saving by 7.737%, 1.407%, and 2.71%, when compared to the three random scenarios, Radom 1.1, Radom 1.2, and Radom 1.3, respectively.

To investigate if the CVR performance can be further improved, the number of PVs is increased to 10 and the analysis is repeated, and the results are shown in Table 5. In Table 5, the CVR benefits are compared between optimally placed 10 PVs with three scenarios of random placed 10 PVs, Radom 2.1, Radom 2.2, and Random 2.3. Integrating a higher number of PVs can considerably reduce power losses. Optimally placed 10 PVs lead to the daily load consumption reduction of 3.229 MWh in the test system, and reduce it by 5.385%, 18.105%, and 6.956%, respectively, compared to the three random PV scenarios, Radom 2.1, Radom 2.2, and Random 2.3. Optimally placed 10 PVs also lead to the daily power losses reduction of 1.157 MWh in the test system, and reduce it by 7.791%, 22.411%, and 12.983%, respectively, when compared to Radom 2.1, Radom 2.2, and Random 2.3. Therefore, having 10 PVs in this network seems more desirable. CVR benefits using the optimal PV placement

Table 4 Comparing the CVR scheduling results using the modified IEEE 33-bus electricity and 7-node natural gas system with 5 PVs

Case	PVs bus locations	Daily load consumptions reduction		Daily power losses reduction (MWh)	Total load consumptions and power losses reduction	
		(MWh)	(%)		(MWh)	(%)
Proposed optimal5 PVs	7, 11, 14, 28, 30	3.228	–	0.448	3.676	–
Random 1.1	3, 11, 16, 23, 28	2.781	16.073	0.631	3.412	7.737
Random 1.2	4, 10, 15, 21, 29	2.908	11.004	0.717	3.625	1.407
Random 1.3	5, 9, 12, 18, 31	2.805	15.080	0.774	3.579	2.710

Table 5 Comparing the CVR scheduling using the modified IEEE 33-bus electricity and 7-node natural gas test system with 10 PVs

Case	DGs bus locations	Daily load consumptions reduction		Daily power losses reduction (MWh)	Total load consumptions and power losses reduction	
		(MWh)	(%)		(MWh)	(%)
Proposed optimal 10 PVs	7, 8, 9, 12, 14, 18, 24, 25, 29, 30	3.229	–	1.157	4.386	–
Locations in [31]	4, 7, 8, 10, 14, 16, 20, 23, 29, 30	3.038	6.287	0.575	3.613	21.395
Random 2.1	3, 6, 7, 11, 12, 18, 21, 25, 29, 33	3.064	5.385	1.005	4.069	7.791
Random 2.2	4, 8, 9, 14, 15, 19, 22, 25, 27, 31	2.734	18.105	0.849	3.583	22.411
Random 2.3	2, 5, 7, 11, 14, 17, 20, 23, 26, 30	3.019	6.956	0.863	3.882	12.983

method are also compared with that in [31], and this method increases reductions of daily load consumptions and power losses by 6.287% and 21.395%, respectively.

Figure 5 shows the integrated energy system, where the green and red circles demonstrate the optimal PV locations and the ones adopted from [31], respectively. Five of the PV locations in [31] are the same as the optimal locations. However, the optimal allocation suggests that it is not beneficial to have a PV installed at bus 20, for instance. We could expect this because the loads supplied by that PV are small $(0.18 + j0.12)$ p.u., and there is a small room for load consumption reduction using that PV. The optimal PV placement model suggests that more PVs should be placed at the feeder with buses 23, 24, and 25, as that feeder's load is relatively large $(0.84 + j0.4)$ p.u., and can contribute to higher load consumption reductions.

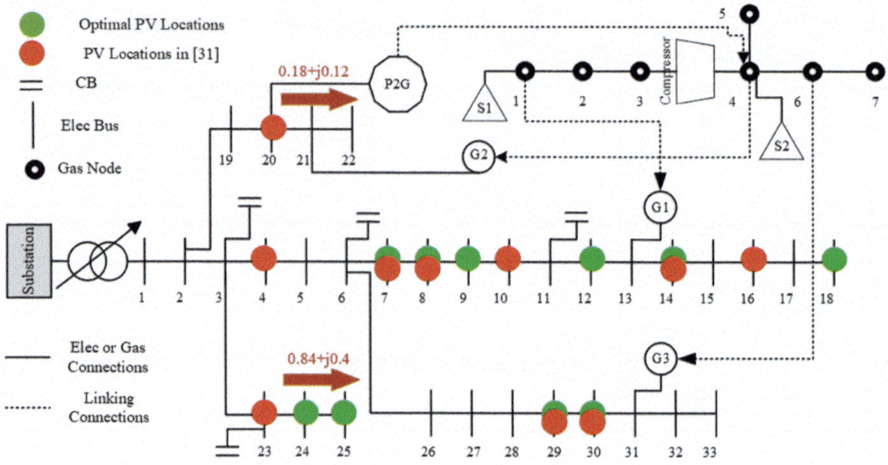

Fig. 5 Comparing optimal DG locations with the ones used in [31]

References

1. IEA, Solar PV, more efforts needed (2021), https://www.iea.org/reports/solar-pv
2. DOE, Quarterly solar industry update (2023), https://www.energy.gov/eere/solar/quarterly-solar-industry-update
3. DOE, Solar integration: Inverters and grid services basics, https://www.energy.gov/eere/solar/solar-integration-inverters-and-grid-services-basics
4. EIA, What is U.S. electricity generation by energy source? (2023), https://www.eia.gov/tools/faqs/faq.php?id=427&t=3
5. I. Eames, M. Austin, A. Wojcik, Injection of gaseous hydrogen into a natural gas pipeline. Int. J. Hydrogen Energy **47**(61), 25745–25754 (2022). https://doi.org/10.1016/j.ijhydene.2022.05.300
6. A.J. Khabbazi, M. Zabihi, R. Li, M. Hill, V. Chou, J. Quinn, Mixing hydrogen into natural gas distribution pipeline system through Tee junctions. Int. J. Hydrogen Energy **49**, 1332–1344 (2024). https://doi.org/10.1016/j.ijhydene.2023.11.038
7. S. Karamdel, X. Liang, S.O. Faried, Conservation voltage reduction in renewable-rich distribution networks: a review (2024)
8. A. Dutta, S. Ganguly, C. Kumar, MPC-based coordinated voltage control in active distribution networks incorporating CVR and DR. IEEE Trans. Ind. Appl. **58**(4), 1–1 (2022). https://doi.org/10.1109/TIA.2022.3163108
9. H. Ahmadi, J.R. Martí, H.W. Dommel, A framework for volt-VAR optimization in distribution systems. IEEE Trans. Smart Grid **6**(3), 1473–1483 (2015). https://doi.org/10.1109/TSG.2014.2374613
10. H. Gharavi, L.F. Ochoa, X. Liu, G. Paterson, B. Ingham, S. McLoone, CVR and loss optimization through active voltage management: a trade-off analysis. IEEE Trans. Power Deliv. **36**(6), 3466–3476 (2021). https://doi.org/10.1109/TPWRD.2020.3043245
11. L. Gutierrez-Lagos, L.F. Ochoa, OPF-based CVR operation in PV-rich MV–LV distribution networks. IEEE Trans. Power Syst. **34**(4), 2778–2789 (2019). https://doi.org/10.1109/TPWRS.2019.2894795
12. Q. Zhang, Y. Guo, Z. Wang, F. Bu, Distributed optimal conservation voltage reduction in integrated primary-secondary distribution systems. IEEE Trans. Smart Grid **12**(5), 3889–3900 (2021). https://doi.org/10.1109/TSG.2021.3088010

13. F. Ding, Y. Zhang, J. Simpson, A. Bernstein, S. Vadari, Optimal energy dispatch of distributed PVs for the next generation of distribution management systems. IEEE Open Access J. Power Energy **7**, 287–295 (2020). https://doi.org/10.1109/OAJPE.2020.3009684
14. S. Paul, A. Sharma, N.P. Padhy, Risk constrained energy efficient optimal operation of a converter governed AC/DC hybrid distribution network with distributed energy resources and Volt-VAR controlling devices. IEEE Trans. Ind. Appl. **57**(4), 4263–4277 (2021). https://doi.org/10.1109/TIA.2021.3081526
15. V.B. Pamshetti et al., Cooperative operational planning model for distributed energy resources with soft open point in active distribution network. IEEE Trans. Ind. Appl. **59**(2), 2140–2151 (2023). https://doi.org/10.1109/TIA.2022.3223339
16. N. Acharya, P. Mahat, N. Mithulananthan, An analytical approach for dg allocation in primary distribution network. Electr. Power Energy Syst. **28**(10), 669–678 (1981)
17. V. Zamani, M.E. Baran, Meter placement for conservation voltage reduction in distribution systems. IEEE Trans. Power Syst. **33**(2), 2109–2116 (2018). https://doi.org/10.1109/TPWRS.2017.2737402
18. V.B. Pamshetti, S.P. Singh, Coordinated allocation of BESS and SOP in high PV penetrated distribution network incorporating DR and CVR schemes. IEEE Syst. J. **16**(1), 420–430 (2022). https://doi.org/10.1109/JSYST.2020.3041013
19. Z. Wang, B. Chen, J. Wang, M.M. Begovic, Stochastic DG placement for conservation voltage reduction based on multiple replications procedure. IEEE Trans. Power Deliv. **30**(3), 1039–1047 (2015). https://doi.org/10.1109/TPWRD.2014.2331275
20. Y. Zhang, Y. Xu, H. Yang, Z.Y. Dong, Voltage regulation-oriented co-planning of distributed generation and battery storage in active distribution networks. Int. J. Electr. Power Energy Syst. **105**, 79–88 (2019). https://doi.org/10.1016/j.ijepes.2018.07.036
21. B. Le, C.A. Cañizares, K. Bhattacharya, Incentive design for voltage optimization programs for industrial loads. IEEE Trans. Smart Grid **6**(4), 1865–1873 (2015). https://doi.org/10.1109/TSG.2015.2394362
22. S.M.R.H. Shawon, X. Liang, M. Janbakhsh, Optimal placement of distributed generation units for microgrid planning in distribution networks. IEEE Trans. Ind. Appl. **59**(3), 2785–2795 (2023). https://doi.org/10.1109/TIA.2023.3236363
23. R.I. Sudjoko, A. Soeprijanto, M. Syai'in, D.F. Uman Putra, M. Ashari, F. Faizah, PV placement for reducing losses based on load bus capacity, in *2021 International Conference on Advanced Mechatronics, Intelligent Manufacture and Industrial Automation (ICAMIMIA)* (2021), pp. 190–194. https://doi.org/10.1109/ICAMIMIA54022.2021.9807817
24. M. Farivar, S.H. Low, Branch flow model: relaxations and convexification—part I. IEEE Trans. Power Syst. **28**(3), 2554–2564 (2013). https://doi.org/10.1109/TPWRS.2013.2255317
25. S. Chen, A.J. Conejo, Z. Wei, Gas-power coordination: from day-ahead scheduling to actual operation. IEEE Trans. Power Syst. **37**(2), 1532–1542 (2022). https://doi.org/10.1109/TPWRS.2021.3098768
26. S. Chen, A.J. Conejo, R. Sioshansi, Z. Wei, Unit commitment with an enhanced natural gas-flow model. IEEE Trans. Power Syst. **34**(5), 3729–3738 (2019). https://doi.org/10.1109/TPWRS.2019.2908895
27. P. Li et al., Coordinated control method of voltage and reactive power for active distribution networks based on soft open point. IEEE Trans. Sustain. Energy **8**(4), 1430–1442 (2017). https://doi.org/10.1109/TSTE.2017.2686009
28. V.B. Pamshetti, S. Singh, A.K. Thakur, S.P. Singh, Multistage coordination Volt/VAR control With CVR in active distribution network in presence of inverter-based DG units and soft open points. IEEE Trans. Ind. Appl. **57**(3), 2035–2047 (2021). https://doi.org/10.1109/TIA.2021.3063667
29. M.E. Baran, F.F. Wu, Network reconfiguration in distribution systems for loss reduction and load balancing. IEEE Trans. Power Deliv. **4**(2), 1401–1407 (1989). https://doi.org/10.1109/61.25627
30. IIT, Department of electrical and computer engineering, 33-Bus 7-Node (2018), motor.ece.iit.edu/data/

31. S. Karamdel, X. Liang, S.O. Faried, A two-stage conservation voltage reduction technique in multi-energy systems with PV smart inverters (2024)
32. C. Zhang, Y. Xu, Hierarchically-coordinated voltage/VAR control of distribution networks using PV inverters. IEEE Trans. Smart Grid **11**(4), 2942–2953 (2020). https://doi.org/10.1109/TSG.2020.2968394
33. X. Sun, J. Qiu, Y. Tao, Y. Ma, J. Zhao, A multi-mode data-driven volt/var control strategy with conservation voltage reduction in active distribution networks. IEEE Trans. Sustain. Energy **13**(2), 1073–1085 (2022). https://doi.org/10.1109/TSTE.2022.3149267

Modeling of Reactive Power-Voltage Response Characteristics of Renewable Energy Resources

Chenge Gao and Ye Guo

Abstract This chapter has focused on and discusses the challenge of modeling reactive power-voltage response characteristics of renewable energy sources. A physics-informed deep learning approach is employed in this chapter to incorporate the influence of the grid model into the training process. Moreover, to improve the adaptivity of the approach to time-varying operational and weather conditions and reduce the reliance on sample volume, a knowledge graph-based approach is used to store, search, and utilize pre-trained deep learning models. The chapter presents the simulation studies on a modified IEEE 39-bus system and a 2486-bus real system and illustrates the efficacy of the introduced method.

Keywords Renewable energy sources · Reactive power-voltage response characteristics · Deep learning · Knowledge graph

1 Introduction

Increasing countries and regions have committed to achieving carbon neutrality [1], and there has been a significant push toward renewable energy sources (RESs). Despite the advantages of reducing carbon emission, integrating RESs into existing power systems introduces several challenges. One of them is the influence of weather conditions, which may lead to significant uncertainties in power outputs from wind farms and solar stations. Additionally, RES stations are usually collections of multiple generator units that operate under different operating conditions, making it extremely difficult to model all units and aggregate them accurately. This complexity increases as the share of renewables in power generation rises. Therefore, developing precise models for RESs is essential for the reliability and efficiency of the grid operation.

Numerous approaches have been developed to model the voltage-response characteristics (VRCs) of large-scale RESs. Take wind farm modeling as an example, single-machine aggregated models assume that in one wind farm, the parameters

C. Gao · Y. Guo (✉)
Tsinghua-Berkeley Shenzhen Institute, Tsinghua University, Shenzhen, China
e-mail: guo-ye@sz.tsinghua.edu.cn

F. Shahnia and F. Wen (eds.), *Power and Energy Resources*, Power Systems,
https://doi.org/10.1007/978-981-96-2612-0_7

and operating conditions of all wind turbines are the same, which may not hold in practice [2]. To address this limitation, multi-machine aggregated models have been introduced. These models group wind turbines according to selected indicators and take weighted averaging to determine an equivalent aggregated model for the wind farm [3–5]. Different indicators for clustering wind turbines within a wind farm have been employed, which include wind speed [6], the topology in the wind farm [7], rotor speed [8], reactive power [9], pitch angle [10], and equivalent power angle [11]. For clustering techniques, K-means [12], support vector clustering [13], and hierarchical clustering [14] have been used in the literature.

Despite these advancements, traditional model-driven methods also face several significant challenges. First, the modeling process can be highly complex and time-consuming, particularly in large RES stations with a lot of generating units and multiple clustering criteria. Second, these models often involve high-dimensional differential equations that are difficult for system operators to collect, interpret, and manage. Third, considering various weather scenarios and machine parameters can lead to inaccuracies in the modeling outcomes. Lastly, in some cases, system operators may be unable to obtain the necessary information or parameters due to privacy concerns, further complicating the modeling process.

Recent advancements in deep learning (DL) have opened new avenues for modeling the VRCs of RES [15–17]. Also, DL techniques have been utilized in probabilistic forecasting of RESs power outputs [18–21]. Despite these advantages, applying deep learning in power systems presents several challenges, including issues with accuracy, transferability, and interpretability [22].

To improve the accuracy and interpretability of DL models, physics-informed neural networks (PINNs) have been designed and applied in the area of power and energy systems. PINNs help neural networks generate interpretable solutions that adhere to grid operation constraints. Recent studies have focused on two primary approaches to integrating physical knowledge into neural networks. The first approach involves creating a physics-informed loss function. This approach can be accomplished by involving governing equations of physical variables in the loss function using regularization techniques, which ensures that the solutions comply with the underlying physical laws [23]. The second approach is the development of a physics-informed architecture. This method assigns specific physical meanings to certain weights within the neural network and keeps these weights fixed during the training process. By doing so, explicit system operation laws are encoded directly into the network's structure [24]. These advancements in PINNs address key challenges in DL applications for power systems, paving the way for more reliable and interpretable modeling of RES response characteristics.

To tackle the transferability issue inherent in DL-based approach, transfer learning has been extensively utilized in a variety of power system applications, such as transient stability assessment [25, 26], photovoltaic (PV) power prediction [27], and event identification [28]. However, unsatisfactory modeling results may occur if the similarity of data distributions is not considered. In addition to transfer learning, exploring spatial and temporal correlations within datasets may enhance the capability of neural networks to transfer among learning tasks. While research specifically focused on spatial and temporal correlations in the modeling of VRC of RESs is limited, several techniques have been developed for RES power prediction. For example, temporal convolutional neural networks have been used to analyze temporal correlations [29], and graph neural networks [30] have been utilized to explore spatial correlations among different RESs. Graph neural networks are particularly effective in processing graph-structured data, but converting input data into graph form for voltage-response characteristic studies can be complex. Additionally, although these methods learn spatial and temporal correlations within a single task, the transferability between tasks with different samples remains underexplored. As a result, these methods may be not applicable to real power systems where weather and operational conditions are continuously changing. Therefore, more advanced approaches are needed to effectively capture and leverage temporal and spatial correlations for improving the transferability of deep learning models.

Knowledge graphs (KGs), widely used by companies like Google to store and manage unstructured data, have recently found applications in power systems. The majority of studies in this field have concentrated on employing semantic analysis to derive information from unstructured data sources [31, 32]. Additionally, the knowledge inference method has been used in constructed KGs to assist in generating decisions for intelligent operation and control [32]. But KGs have not been extensively studied to enhance the transferability of DL models. Integrating KGs with DL could offer significant improvements in model adaptability and robustness across varying operational scenarios in power systems, presenting a promising area for future research.

This chapter introduces a method fusing DL and KG to model the VRCs of RES stations. Initially, a DL-based method is presented to model VRCs of RESs while accounting for the influence of the VRC of the power grid. Moreover, to learn the temporal and spatial correlations in different tasks, the KG is introduced to facilitate the training, retrieval, and utilization of typical models in independent learning tasks. By leveraging KGs, the accuracy and interpretability are significantly enhanced. Some technical content and descriptions of this chapter are based on [33].

2 The Fusion Framework of DL and KG

Typically, deep learning models perform well with test samples that have distributions similar to the training samples. However, their accuracy may deteriorate when faced with test samples that are different in distribution from the training data. When modeling VRCs of RES stations in power systems, numerous factors influence the prediction results, making it impractical to account for all of them as inputs to neural networks. Therefore, to enhance the neural networks' adaptability to varying weather and system scenarios, a KG is employed to support the deep learning modeling

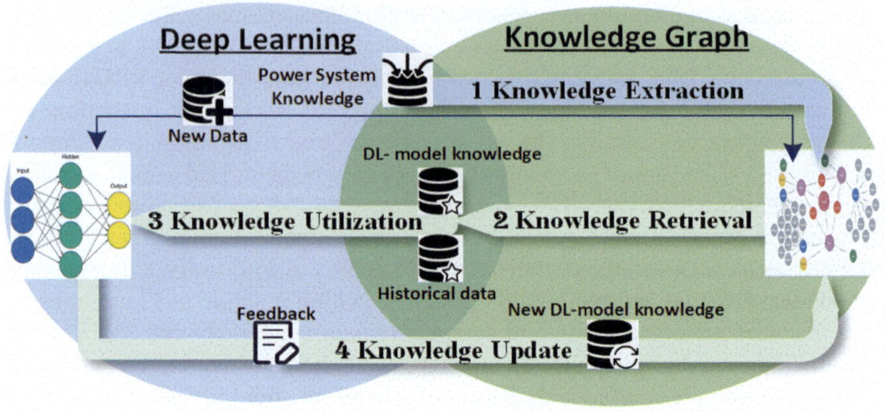

Fig. 1 The fusion framework of DL and KG

process. The designed framework based on the fusion of DL and KG is presented in Fig. 1.

As shown in Fig. 1, four steps are involved in the interactive process of a DL-based learning part and a KG-based knowledge part. First, during the knowledge extraction phase, the KG is initialized with trios comprising entities, relationships, and attributes, collectively referred to as knowledge records. The entities encompass electrical equipment and DL models. For instance, in the case of power equipment entities like busbars, attributes involve voltage, etc. The relations among power equipment entities are defined according to topology connections. Regarding the deep learning model entities, their attributes consist of model hyperparameters (such as activation function, learning rate, and optimizer), network parameters (including structure, the values of neurons, etc.), and typical scenarios relevant to DL models. The relations between various entities facilitate the evaluation of these correlations, which include the topological connections of different RESs, the correlation of various meteorological and operating scenarios that DL models adapt to, and the suitability of DL models to specific RES stations and meteorological scenarios.

During the knowledge retrieval phase, the KG will search for DL models appropriate for the current task by considering temporal and spatial correlations. The selected model knowledge record is then used to support future tasks.

During the knowledge utilization phase, various transfer learning techniques are employed to decrease training samples and time needed for online modeling, making the approach adaptive to various weather conditions. This is achieved by estimating the similarity between the datasets of models recorded in the KG and the dataset of the current task.

During the knowledge update phase, three methods are designed to update the KG records according to the modeling performance obtained from the knowledge utilization phase.

This method of fusing DL and KG enhances the transferability of DL models, enabling them to be applied to various scenarios accurately and efficiently. Next, we will elaborate on the DL and KG parts, respectively.

3 DL-Based Learning Part

In this section, we introduce a PINN-based method specifically designed to track the VRCs of the RESs while incorporating the grid's impact.

3.1 The Design of PINN

One important face to be noticed when designing the deep learning structure is: Measured voltages in the dataset reflect the joint impact from the RES's VRC and the grid's VRC. To account for the effect of the VRC of the grid, a PINN is developed. As depicted in Fig. 2, the training process involves iterative interactions between a neural network module, which models the RES's VRC, and a PI module—power flow calculation module, which describes the grid's VRC.

As depicted in Fig. 2, the output of the neural network module is the input of the PI module. In the power flow calculation, the bus connected to the RES is set as a PQ bus. The PI module then updates the voltage magnitude of the RES node, which serves as the input of the neural network module. This iterative process continues until the reactive powers Q converge.

In the neural network module, a convolutional neural network (CNN) is employed to track the VRCs of RESs. The structure of the CNN is illustrated in Fig. 3. The SAME padding method [34] is used in the convolutional layers, ensuring that the sizes of feature maps remain unchanged when processing the convolution. To address the overfitting issue, the dropout technique is applied during CNN training, with a dropout probability of 0.3 for each neuron. The inputs to the CNN are voltages and wind speeds, while the outputs are reactive powers. The entire dataset is divided into a training set, comprising 70% of the samples, and a testing set (30%).

The loss function is

$$l = \frac{1}{n_b}\sum\nolimits_{k=1}^{n_b}\left(\left(\widehat{Q}_k - Q_k\right)^2 + w_V * \left(\widehat{V}_k - V_k\right)^2\right), \tag{1}$$

where V_k and Q_k are the true values of the k th sample; $\widehat{V}$ and $\widehat{Q}$ are values after convergence in one iteration; n_b denotes the sample number in the dataset; w_V is the weight of the voltage loss part, which is set as 100 here since the variance of reactive power measurement is usually about 100 times that of voltage measurement under the per-unit system.

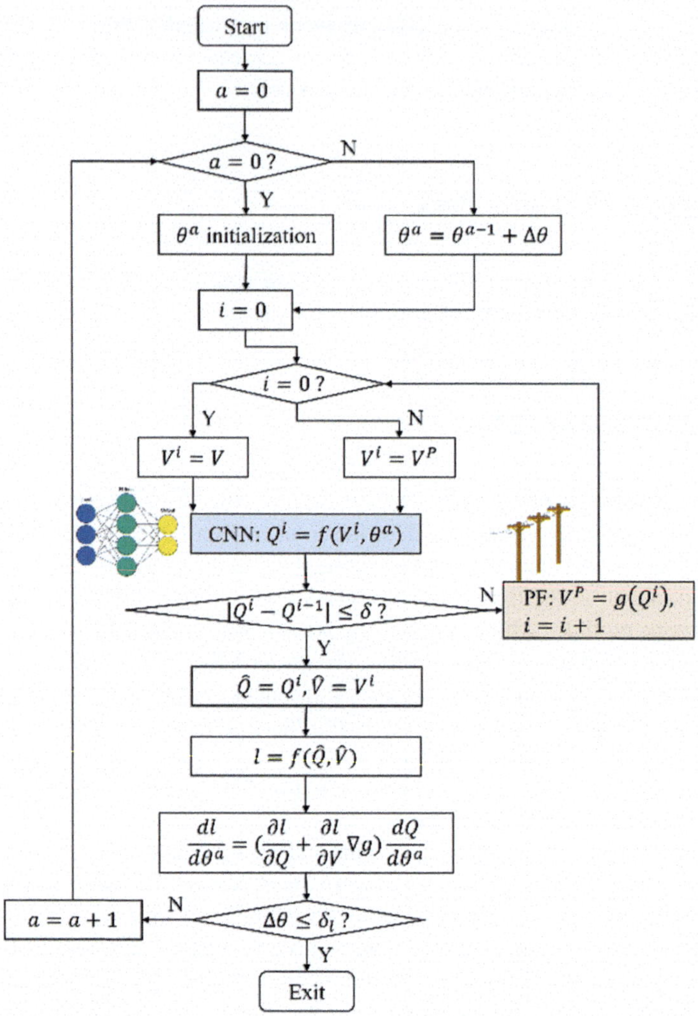

Fig. 2 The PINN algorithm

We use the gradient descent method [35] to train the PINN. The gradient of the loss is

$$\frac{\mathrm{d}l}{\mathrm{d}\theta} = \frac{\mathrm{d}l}{\mathrm{d}Q}\frac{\mathrm{d}Q}{\mathrm{d}\theta} = \left(\frac{\partial l}{\partial Q} + \frac{\partial l}{\partial V}\nabla\mathcal{G}\right)\frac{\mathrm{d}Q}{\mathrm{d}\theta}, \tag{2}$$

where $\frac{\partial l}{\partial Q}$ and $\frac{\partial l}{\partial V}$ are directly obtained by calculating the gradient of the loss l (1) with respect to Q and V; $\frac{\mathrm{d}Q}{\mathrm{d}\theta}$ is the gradient of the CNN $Q = f(V, w, \theta)$ with respect to θ, where θ includes all parameters of the neural network. Additionally, V can

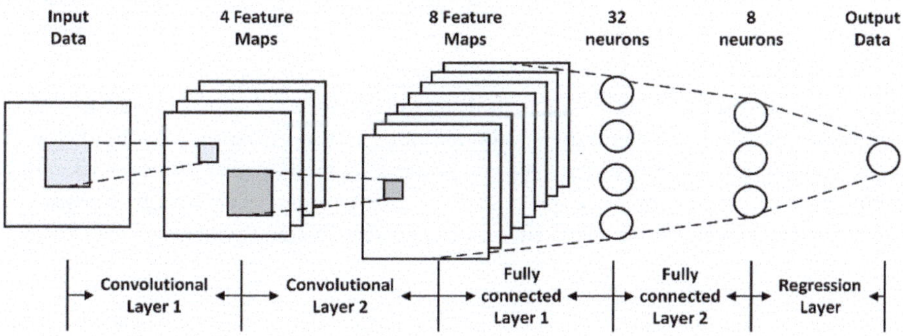

Fig. 3 The structure of the CNN for modeling voltage-response characteristics

be expressed as a function of Q ($V = \mathcal{G}(Q)$), which represents the power flow calculation in the power system. The derivative of $\mathcal{G}$ is obtained from a component of the inverse Jacobian matrix used in the power flow calculation, which is structured as follows:

$$J = \begin{bmatrix} \frac{\partial \Delta P}{\partial \delta^T} & \frac{\partial \Delta P}{\partial V^T} \\ \frac{\partial \Delta Q}{\partial \delta^T} & \frac{\partial \Delta Q}{\partial V^T} \end{bmatrix}, \tag{3}$$

where V and δ are the voltage magnitude and phase angle, and P and Q represent active and reactive power injection.

By integrating power flow into the design of the PINN, the neural network can precisely model the RES's VRC, and the results of PINN will align with the VRC of the grid.

3.2 The Interpretability of the PINN

Despite deep learning methods possessing powerful capabilities for learning from data, their results may lack interpretability [36]. System operators may be more interested in the voltage sensitivity of RES stations rather than their internal VRCs when performing online dispatch. To address this issue, we can compute the gradient of the CNN function to represent the voltage sensitivity, which acts as an attribute of the current VRC model:

$$S_U(U) = \frac{\partial Q}{\partial U}, \tag{4}$$

where $S_U(U)$ represents the voltage sensitivity. With results of voltage sensitivities of different RES stations, voltage regulation decisions are made with interpretability.

4 KG-based Knowledge Part

To enhance the transferability of neural networks across various scenarios, a KG is implemented to investigate the spatial and temporal correlations among different tasks. This section introduces the data-pre-processing method, and provides a detailed explanation of the interactive process between DL and KG which consists of knowledge retrieval, knowledge utilization, and knowledge update.

4.1 Data Pre-Processing

The DL-based method models the VRCs of RES stations using historical data. However, due to the time-varying nature of VRCs of RES stations, adaptive online modeling that leverages stream-type data is necessary. Unlike traditional online learning [37], the adaptive online modeling approach employs recent samples for new learning tasks, with historical models (i.e., models trained on past data) serving as knowledge records to support these new tasks. This method explores spatial–temporal correlations among different tasks. Based on these correlations, historical models are selected and utilized in the new tasks. Prior to examining these correlations, dataset division techniques will be introduced.

4.1.1 Dataset Division

The entire historical dataset is segmented into sub-datasets as illustrated in Fig. 4, where N_h represents the sample number in the entire dataset, Δt denotes the sampling interval. Each sub-dataset encompasses the samples in N_s continuous periods. To avoid losing trend variations, there are repeat samples in adjacent sub-datasets, with the number of overlapping samples denoted as $N_s - s$. Here, s is the non-repeat part calculated by $\frac{T}{\Delta t}$, where T is the online dispatch interval. We use N_d to represent the sub-dataset's number:

$$N_d = \left| \frac{N_h - N_s}{s} \right| + 1. \tag{5}$$

4.1.2 Scenario Clustering of Sub-Datasets

Temporal correlations among sub-datasets reflect varying meteorological and grid scenarios, as well as spatial correlations between RES stations. As depicted in Fig. 5, each historical sub-dataset comprises multiple types of data, involving weather scenarios, voltages, and reactive powers, with the total number of time-series types

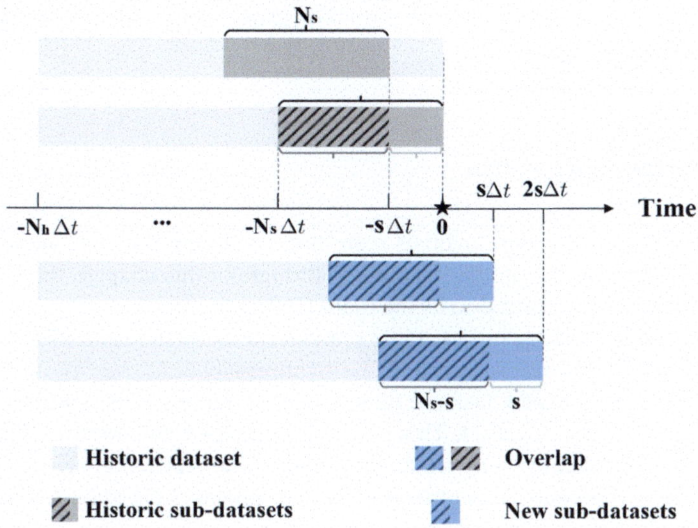

Fig. 4 The division method of the datasets

denoted as N_f. To discern the variation trends, data normalization is employed, scaling the values to the range [0,1]. To extract typical variation patterns for each type of data, the K-means algorithm is applied for clustering the sub-datasets.

To expedite the online modeling process, additional CNN models are developed to classify sub-datasets. These classification CNNs are trained using the offline clustering results obtained through the K-means algorithm. The input of each classification CNN is time-series data, and the output is scenario labels. The structure of these CNNs includes two convolutional layers (the valid padding method [34]), one pooling layer, one dropout layer, and three fully connected layers. To mitigate overfitting, the dropout technique is also incorporated.

Additionally, prior to normalization, for each sub-dataset, the average values of voltages, reactive powers, and weather condition samples are recorded. These

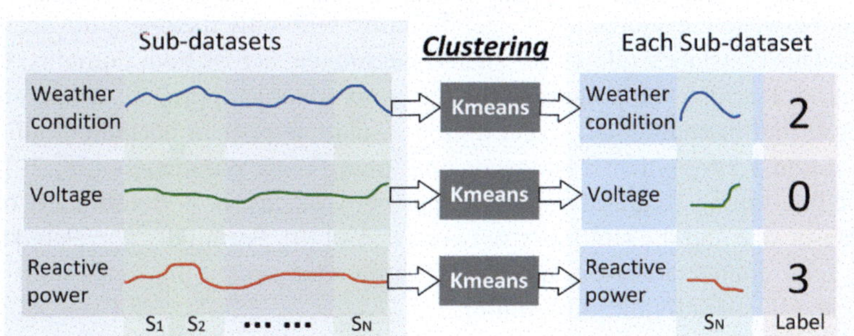

Fig. 5 Typical scenario clustering

average values are then analyzed to identify typical averages, which are subsequently categorized into different levels.

4.2 Knowledge Retrieval

The knowledge retrieval step involves searching for appropriate DL models stored in the KG to expedite the online learning. (Throughout this chapter, the CNN used for modeling VRCs is referred to as the main CNN.)

For temporal knowledge retrieval, the search criteria include classification labels and average levels of each scenario. Classification labels $C_i, i \leq N_f$ and average levels $A_i, i \leq N_f$ for the new task are then compiled into a scenario attribute matrix $SA = \left[C_1, \ldots C_{N_f}, A_1, \ldots, A_{N_f}\right]$. This matrix is compared with existing DL models in the KG. If a model entity with a matching SA is found, the parameters of the main CNN are used in the knowledge utilization step. If no matching main CNN model is found, spatial knowledge retrieval is performed.

Spatial correlations are also essential for modeling the VRCs of RES stations. A new RES station may not collect sufficient historical data to train DL models. The knowledge retrieval step also involves searching for knowledge records from different RES stations according to spatial correlations identified in the KG.

4.3 Knowledge Utilization

Through knowledge retrieval, detailed parameters such as the values of neurons, along with the configurations for the model training, are obtained. This process deviates from conventional methods that initialize the parameters of the neural network at random. Instead, the parameters are initialized by using the weights and biases knowledge records. Furthermore, the training configuration of the retrieved model is directly applied in the training of the new model.

In the context of VRCs modeling, the neural network's input and output remain consistent between the source and target domains, with only the operating scenarios differing. This constancy allows for shared parameters across both domains. Given this scenario, learning the parameters is the most suitable approach. It operates on the premise that in source and target domains, parameters can be shared, allowing for model transfer by either freezing or fine-tuning these parameters [38, 39]. The extent of fine-tuning required for the target network layers is determined by the target dataset size and the similarity between tasks [40]. For highly correlated tasks, only the fully connected layers need fine-tuning. Conversely, for tasks with lower correlation, all layers require fine-tuning.

The degree of correlation between the current and retrieved learning tasks is quantified using the Pearson coefficient (P), which is computed as follows:

Table 1 Knowledge utilization methods

Similarity between tasks	Criteria	Methods
Strong similarity	$P \geq a_1$	M1: Freeze convolutional layers and fine tune fully connected layers
Weak similarity	$a_2 \leq P < a_1$	M2: Fine tune convolutional layers and fully connected layers
Dissimilarity or no records	$P < a_2$	M3: Train without using retrieved parameters

$$P = \frac{\sum_{i=1}^{n}\left(X_i - \overline{X}\right)\left(Y_i - \overline{Y}\right)}{\sqrt{\sum_{i=1}^{n}\left(X_i - \overline{X}\right)^2}\sqrt{\sum_{i=1}^{n}\left(Y_i - \overline{Y}\right)^2}}, \tag{6}$$

where X and Y are random variables in the historic and current tasks respectively, such as voltages, weathers, and reactive powers. n is the total sample number. The Pearson coefficient provides an evaluation of the linear correlation between two variables. This coefficient is crucial for evaluating the similarity between historical and current learning tasks, guiding the degree of fine-tuning necessary for neural network layers to optimize transfer learning.

Three distinct methods for utilizing knowledge records are devised based on varying degrees of similarity between learning tasks. To implement these strategies, we introduce parameters a_1 and a_2. These strategies are outlined in Table 1. When the Pearson correlation coefficient indicates a strong similarity between learning tasks, the convolutional layers are frozen, and only the fully connected layers undergo fine-tuning. For tasks with moderate similarity, all layers of the CNN, including convolutional and fully connected layers, are fine-tuned. If the similarity is low or if no appropriate models are found in the KG, a new CNN is trained from scratch. This involves manual parameter tuning, with all layer weights and biases initialized using the Kaiming-normal initialization method [41]. The values for a_1 and a_2 are determined empirically. Here, P is obtained between each pair of sub-datasets. The average of the highest 10% of all P values is designated as a_1, while a_2 is set as the mean of the highest 40% of all P values.

4.4 Knowledge Update

Given that three methods of knowledge utilization are employed based on similarity between tasks, three knowledge update strategies are outlined in Table 2.

If datasets are highly similar and only the fully connected layers are fine-tuned, the main CNN model entity remains unchanged. For tasks with weak similarity, model losses before and after knowledge utilization step should be compared and then choose a more accurate one to be recorded in the KG. When tasks are dissimilar

Table 2 Knowledge update methods

Similarity between tasks	Criteria	Methods
Strong similarity	$P \geq a_1$	M1: No update
Weak similarity	$a_2 \leq P < a_1$	M2: Compare loss functions and store a more accurate one
Dissimilarity or no records	$P < a_2$	M3: Add a new main CNN entity

or no suitable model exists, a new main CNN model entity is created in the KG with the new CNN parameters. This updating process enhances the main CNN models' adaptability to different scenarios. The flowchart of the entire knowledge graph part is depicted in Fig. 6.

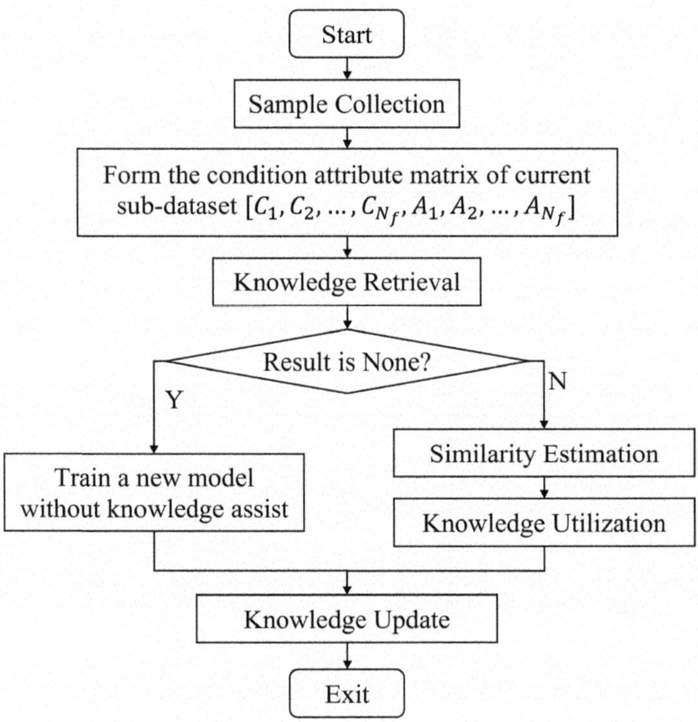

Fig. 6 The process in the knowledge graph part

5 Performance Evaluation

5.1 *Case Study a*

A modified IEEE 39-bus system was constructed by Simulink, incorporating three distinct wind farms equipped with different wind generator units. Wind speed data, collected from the Sotavento wind farm operated in Spain, was gathered between April and June 2021, with samples taken every 10 min [42]. Measurement data for the IEEE 39-bus system, including voltages, active powers, and reactive powers of all buses, as well as wind speeds from the wind farms, were sampled every minute using Simulink. Each wind farm dataset comprised 100,000 samples. To segment the entire dataset into sub-datasets, N_h and N_s were 100,000 and 2,000, respectively. Despite the neural network parameters being relatively modest in size, Gaussian noise was added to 2,000 samples to mitigate overfitting, and the model was trained using 40,000 samples after data augmentation to improve robustness. s was set to 400, resulting in N_d being 246. Voltages, wind speeds, and reactive powers were used to represent the operational and meteorological scenarios of learning tasks. The number of clusters for these features is specified in Table 3, while the settings for average levels are detailed in Table 4. $a_1 = 0.93$ and $a_2 = 0.75$.

5.1.1 Results of Modeling the VRCs of RES Stations

The introduced method was applied to model the VRCs of the wind farms. The results for VRCs under typical scenarios are illustrated in Fig. 7. The weather scenarios corresponding to the selected sub-datasets are detailed in Table 5.

Table 3 Settings of clustering in case study A

Data	U	Wind speed	Q
The number of clusters	4	4	2

Table 4 Average levels in case study A

Data	Original average value	Average level
Wind speed (m/s)	[0,6]	3
	(6,10]	8
	(10,16]	13
Q(MVar)	[0,4]	2
	(4,8]	6
	(8,12]	10
V(p.u.)	(0.98,1.02]	1.00
	(1.02,1.06]	1.04

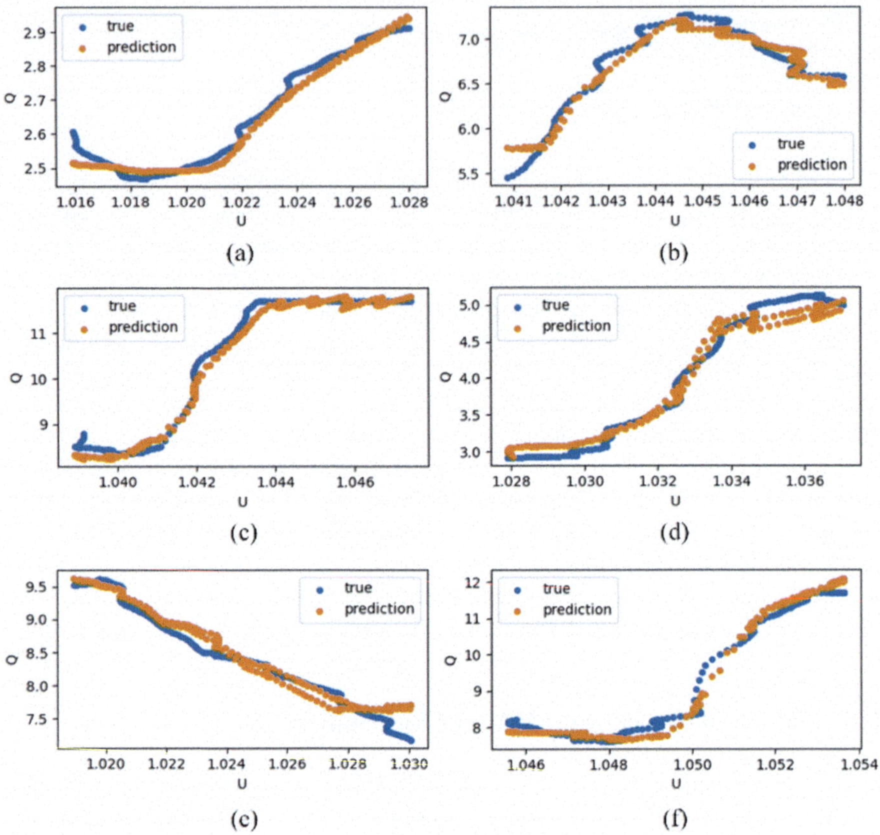

Fig. 7 Voltage-response characteristics in different scenarios

Table 5 Weather scenarios of the tasks in Fig. 8

Task	a	b	c	d	e	f
Wind speed average values (m/s)	3	8	13	3	8	13
Wind speed clustering labels	2	1	0	0	3	0

From the analysis of Fig. 7b, c, and f, it is evident that when voltages are within the range of 1.040–1.054, the wind farm exhibits different VRC curves under varying weather scenarios (as indicated by the average wind speeds and clustering labels in Table 6). The introduced method successfully adapts to these changes. Additionally, Fig. 7a, d, and e demonstrate that within the voltage range of 1.016–1.036, the method models the VRC curves effectively. This indicates that this method enhances the transferability of deep learning models across different weather scenarios. This capability is crucial for accurately modeling the VRCs of RESs.

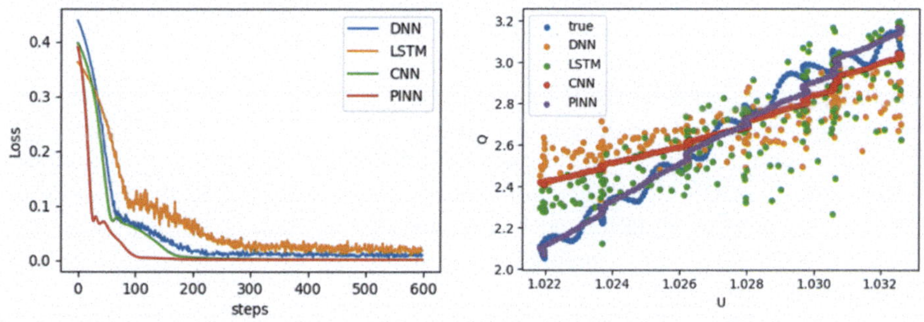

Fig. 8 The performance comparison **a** Loss functions, **b** The modeling results of VRCs

Table 6 The comparison of the accuracy (training after 150 epochs)

Method	The average of accuracy (%)
DNN	91.92
LSTM	89.83
CNN	95.65
PINN	99.34

The performances of deep neural networks (DNN), LSTM, CNN, and PINN models were shown in Fig. 8. Loss functions in the training are compared in Fig. 8a. The loss function of the PINN reduces significantly within the first 100 training steps (epochs), whereas the loss functions for DNN, CNN, and LSTM exhibit slower declines, only stabilizing after 200 steps. This demonstrates that the PINN converges more rapidly than the other approaches. Furthermore, the voltage-response characteristic curves generated by the models after 150 training steps were compared to the true curve. As presented in Fig. 8b, results of the DNN, CNN, and LSTM show deviation from the true value, but the PINN aligns closely with the true curve, indicating superior accuracy.

The accuracy was evaluated for the testing dataset of each sub-dataset after training the models for 150 epochs. The average level of the accuracy across all sub-datasets is presented in Table 6. The results indicate that the accuracy of the introduced PINN is the highest. This demonstrates that the PINN approach more effectively captures the VRCs of RESs. The advantage of the PINN-based method lies in its integration of the power flow module into the DL framework, which accounts for the influence of grid's VRC. This integration ensures that modeling results are compliant with the physical rules of the grid, leading to better overall performance.

5.1.2 Effectiveness of the KG Part in the Proposed Method

A part of the KG for the IEEE 39-bus system is depicted in Fig. 9. The KG includes both equipment and DL model entities. Notably, DL models can be linked through electrical equipment entities within the KG. This interconnected structure allows for the retrieval of DL models according to both temporal and spatial relationships. The time required to initialize the KG is provided in Table 7. This initialization time reflects the efficiency of integrating and organizing the diverse data and models within the KG framework.

In this study, Wind Farm 1 was assumed to have been operated earlier than Wind Farms 2 and 3, meaning that the learning tasks for Wind Farms 2 and 3 were undertaken after those for Wind Farm 1. As illustrated in Fig. 10, temporal-correlated model records could be searched for 74.8% of the learning tasks for Wind Farm 1. Furthermore, in 62.2% of these tasks, the searched knowledge records could be used in the current task. This highlights the effectiveness of the knowledge retrieval process in leveraging historical data to enhance the training efficiency and accuracy for new learning tasks.

For Wind Farm 2, Fig. 11a shows that the knowledge record was retrieved based on majorly temporal correlations (74.4%). As illustrated in Fig. 11b, the retrieved knowledge record was used for training the current learning tasks, representing 70.3%

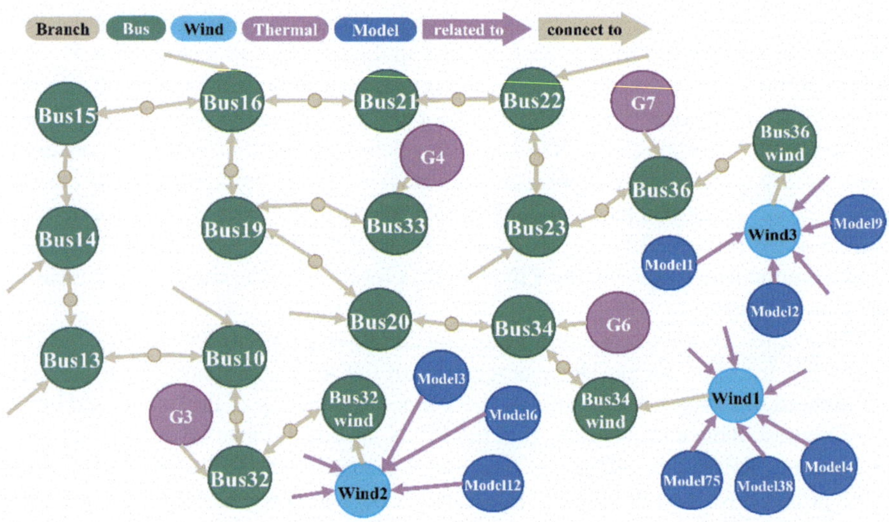

Fig. 9 A part of the knowledge graph of the IEEE39-bus system

Table 7 The time to initial a KG

Case	Case study A	Case study B
Time	10.780s	29.312s

Fig. 10 Wind farm 1 results of knowledge retrieval

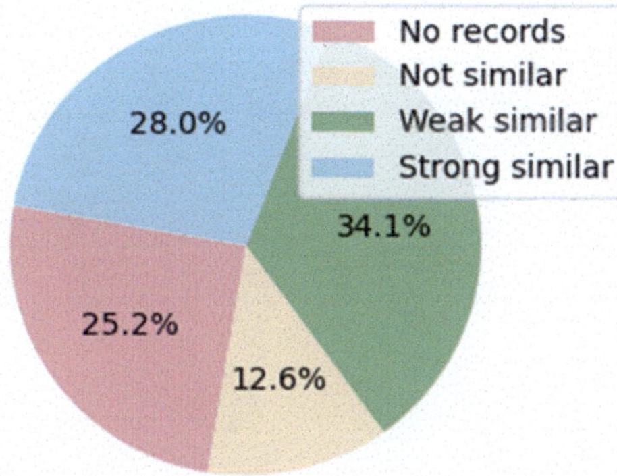

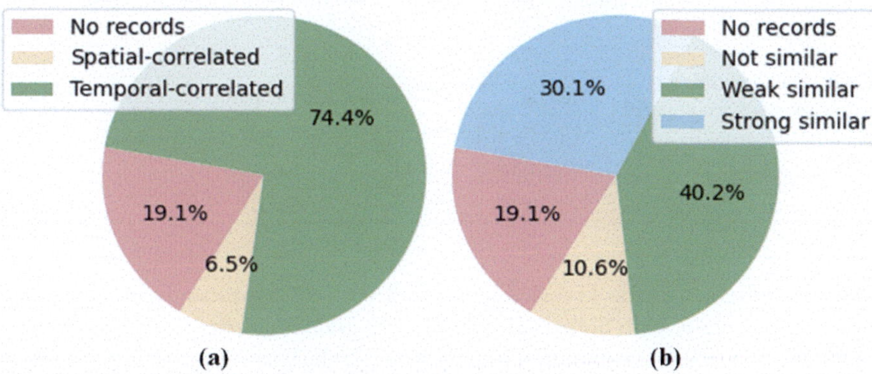

Fig. 11 Wind farm 2 results **a** Knowledge retrieval results, **b** Similarity estimation results

of the total tasks. This indicates the significant reliance on both spatial and temporal correlations in enhancing the effectiveness of model training for Wind Farm 2.

For Wind Farm 3, Fig. 12a illustrates that 13.8% knowledge records were retrieved by spatial knowledge retrieval and 71.1% knowledge records were searched by temporal knowledge retrieval. Notably, compared to Wind Farm 1, the number of tasks training without knowledge reduces since DL models from other wind farms are utilized through spatial correlations. As displayed in Fig. 12b, the retrieved parameters were employed for fine-tuning the current learning tasks, which constitute 73.6% of the tasks. These results demonstrate that the model parameters stored in the knowledge graph are frequently used in various learning tasks across different weather scenarios and RES stations, highlighting the effectiveness and efficiency of the knowledge retrieval process.

Moreover, the efficiency of a purely data-driven approach against the method fusing DL and KG was compared. As indicated in Table 8, this method resulted in a 51% reduction in total training time for Wind Farm 1 compared to the data-driven approach. Furthermore, as the increase of spatial knowledge retrieval, training time for Wind Farms 2 and 3 decreased even more than that for Wind Farm 1. These

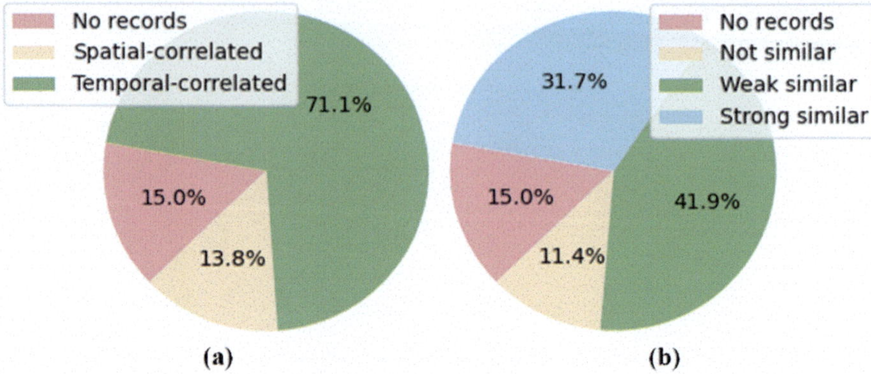

Fig. 12 Wind farm 3 results **a** Knowledge retrieval results, **b** Similarity estimation results

Table 8 The total training time in case study a compared in the aspect of wind farms

Method	This method	The purely data-driven method
Wind farm-1	40.18s	81.82s
Wind farm-2	35.41s	82.49s
Wind farm-3	33.49s	80.03s

findings highlight that the knowledge-assisted method significantly enhances the efficiency, especially with higher utilization of spatial knowledge retrieval.

Efficiency in online computation is crucial, particularly when operating and meteorological conditions change rapidly, necessitating quick adjustments to the model. Training a model from scratch requires accumulating a significant number of samples, which is time-consuming. By the time enough samples are collected and the model training is complete, the model may no longer be valid due to changing conditions. We compared the sample requirements of the two methods in the same training task. Figure 13 illustrates the sample sizes needed by both methods to achieve the same accuracy level. The horizontal axis denotes different tasks, and the vertical axis indicates the sample requirement. As shown in Fig. 13, it is evident that to reach the same error level, only about 25% of the samples needed by the purely data-driven method is required in this method. This efficiency aligns with the principles of transfer learning, demonstrating the capability of the introduced method in reducing the sample size requirements while maintaining accuracy.

As presented in Table 9, the introduced method demonstrated superior robustness compared to the purely data-driven method when handling samples with previously unseen distributions. This indicates that the introduced method maintains higher accuracy levels across diverse testing scenarios, underscoring its reliability and adaptability.

To better understand how model accuracy is influenced by uncertainty in weather and operational scenarios, we compared four purely data-driven methods (CNN, DNN, LSTM, graph neural networks) against their knowledge-assisted data-driven

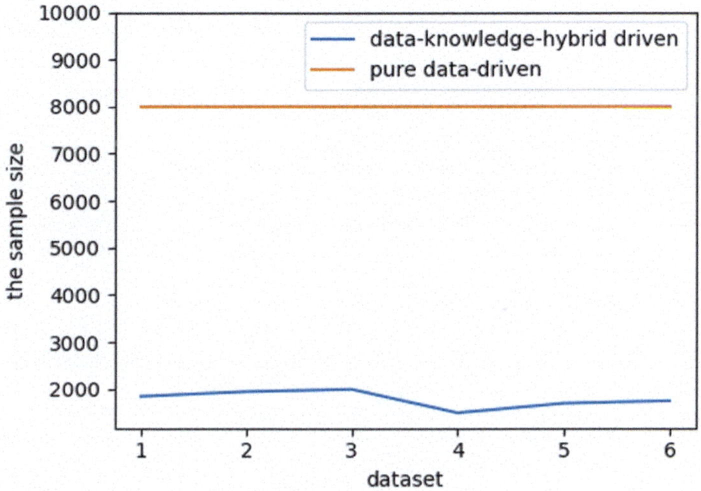

Fig. 13 Comparison of the training sample number requirements

Table 9 The total training time in case study a compared in the aspect of methods

Method	Pure data-driven method	This method
Average level of accuracy	91.731%	99.260%

methods. Uncertainty in the test dataset was simulated by adding Gaussian noise, with higher variances representing greater uncertainty. Figure 14 depicts the relation between mean squared error and the variance of Gaussian noise for these eight methods. The purely data-driven models exhibited higher accuracy when data uncertainty was low. However, as uncertainty increased, their accuracy diminished significantly. Conversely, the knowledge-augmented models demonstrated better performance under increasing uncertainty. The hybrid approaches showed less decline in fitting accuracy as uncertainty grew, indicating superior robustness. This improvement underscores the advantage of integrating knowledge records into the models, enhancing their resilience to data variability and uncertainty.

5.2 *Case Study B*

Measurement data was collected from a provincial power system in China from June to November 2021, with samples taken every minute. This data includes voltages, active powers, reactive powers at all buses, and wind speeds in wind farms. The system comprises 2486 buses and three wind farms, resulting in a dataset of 216,000 samples for each wind farm. To segment the entire dataset into sub-datasets, N_h was 216,000, N_s was 2,000, s was 400, and N_d was 536. Voltages, wind speeds, and

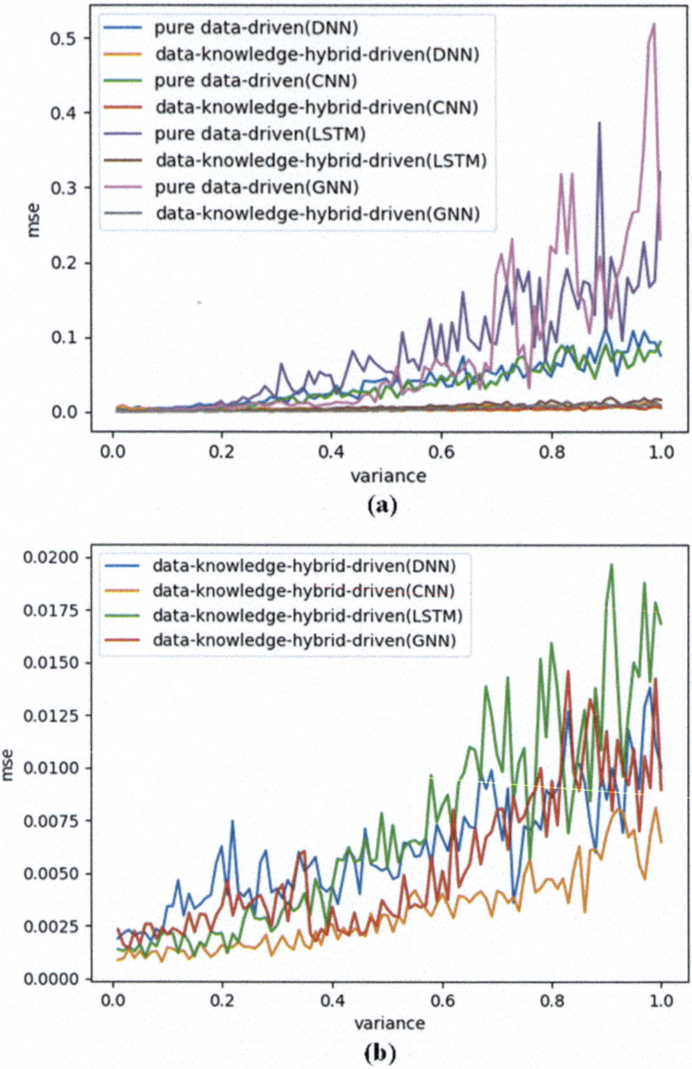

Fig. 14 The Relation between mean squared error (MSE) and sample variance **a** Comparison of all methods, **b** Comparison of four hybrid-driven methods

reactive powers were used to represent scenarios of the learning tasks. The number of clusters for these features is detailed in Table 10, while the settings for average levels are presented in Table 11. a_1 was calculated as 0.96 and a_2 was calculated as 0.64. Wind Farm 1 was assumed to have been operated earlier than Wind Farms 2 and 3.

The performance of this method and the purely data-driven method was compared. As presented in Table 12, the total training time for Wind Farm 1 using this method

Table 10 Settings of clustering in case study B

Data	U	Wind speed	Q
The number of clusters	4	4	4

Table 11 Average levels in case study B

Data	Original average value	Average level
Wind speed (m/s)	[0,4]	2
	(4,8]	6
	(8,12]	10
Q(MVar)	[−40, − 25]	− 30
	(−25, − 15]	− 10
	(−15,0]	− 7
V(p.u.)	(0.98,1.00]	0.99
	(1.00,1.02]	1.01

Table 12 The total training time in case study B compared in the aspect of wind farms

Method	This method	The purely data-driven method
Wind farm-1	64.96s	116.76s
Wind farm-2	58.59s	120.33s
Wind farm-3	55.04s	120.71s

was 44.4% less than that of the purely data-driven method. Furthermore, with the increase of spatial knowledge retrieval, the total training time for Wind Farms 2 and 3 decreased even more significantly compared to Wind Farm 1. These results demonstrate that the knowledge-assisted approach substantially improves the efficiency, particularly with the enhanced utilization of spatial knowledge retrieval.

The simulation results demonstrate that this method significantly enhances the transferability of DL models, which improves the online modeling efficiency. Additionally, when compared to Case Study A, it is clear that the overall time does not significantly increase with the expansion of the system size. This finding underscores the generalizability of the introduced method across different power systems, indicating its broad applicability and effectiveness in various scenarios.

References

1. The 14th five-year plan for the modern energy system, http://zfxxgk.nea.gov.cn/2022-01/29/c_1310524241.htm. Accessed 8 Apr 2022
2. A.P. Gupta, A. Mitra, A. Mohapatra, S.N. Singh, A multi-machine equivalent model of a wind farm considering LVRT characteristic and wake effect. IEEE Trans. Sustain. Energy **13**(3), 1396–1407 (2022)

3. L.P. Kunjumuhammed, B.C. Pal, C. Oates, K.J. Dyke, The adequacy of the present practice in dynamic aggregated modeling of wind farm systems. IEEE Trans. Sustain. Energy **8**(1), 23–32 (2017)
4. Y. Zhou, L. Zhao, I.B.M. Matsuo, W. Lee, A dynamic weighted aggregation equivalent modeling approach for the DFIG wind farm considering the weibull distribution for fault analysis. IEEE Trans. Ind. Appl. **55**(6), 5514–5523 (2019)
5. N. Shabanikia, A.A. Nia, A. Tabesh, S.A. Khajehoddin, Weighted dynamic aggregation modeling of induction machine-based wind farms. IEEE Trans. Sustain. Energy **12**(3), 1604–1614 (2021)
6. Y. Jin, D. Wu, P. Ju, C. Rehtanz, F. Wu, X. Pan, Modeling of wind speeds inside a wind farm with application to wind farm aggregate modeling considering LVRT characteristic. IEEE Trans. Energy Convers. **35**(1), 508–519 (2020)
7. M. Ali, I.-S. Ilie, J.V. Milanovic, G. Chicco, Wind farm model aggregation using probabilistic clustering. IEEE Trans. Power Syst. **28**(1), 309–316 (2013)
8. J. Zou, C. Peng, H. Xu, Y. Yan, A fuzzy clustering algorithm-based dynamic equivalent modeling method for wind farm with DFIG. IEEE Trans. Energy Convers. **30**(4), 1329–1337 (2015)
9. M. Liu, W. Pan, Y. Zhang, K. Zhao, S. Zhang, T. Liu, A dynamic equivalent model for DFIG-based wind farms. IEEE Access **7**, 74931–74940 (2019)
10. Y. Jin, P. Ju, C. Rehtanz, F. Wu, X. Pan, Equivalent modeling of wind energy conversion considering overall effect of pitch angle controllers in wind farm. Appl. Energy **222**, 485496 (2018)
11. L. Zhu et al., A study of dynamic equivalence using the similarity degree of the equivalent power angle in doubly fed induction generator wind farms. IEEE Access **8**, 88584–88593 (2020)
12. Z. Mi et al., Study on dynamic equivalence model of wind farms with wind turbine driven doubly fed induction generator. Autom. Elect. Power Syst. **34**(17), 72–77 (2010)
13. W. Teng, X. Wang, Y. Meng, W. Shi, An improved support vector clustering approach to dynamic aggregation of large wind farms. CSEE J. Power Energy Syst. **5**(2), 215223 (2019)
14. P. Wang, Z. Zhang, Q. Huang, N. Wang, X. Zhang, W.-J. Lee, Improved wind farm aggregated modeling method for large-scale power system stability studies. IEEE Trans. Power Syst. **33**(6), 6332–6342 (2018)
15. M.S. Hossain, H. Mahmood, Short-term photovoltaic power forecasting using an LSTM neural network and synthetic weather forecast. IEEE Access **8**, 172524–172533 (2020)
16. M.-S. Ko, K. Lee, J.-K. Kim, C.W. Hong, Z.Y. Dong, K. Hur, Deep concatenated residual network with bidirectional LSTM for one-hour-ahead wind power forecasting. IEEE Trans. Sustain. Energy **12**(2), 1321–1335 (2021)
17. H. Li, Z. Ren, Y. Xu, W. Li, B. Hu, A multi-data driven hybrid learning method for weekly photovoltaic power scenario forecast. IEEE Trans. Sustain. Energy **13**(1), 91–100 (2022)
18. M. Cui, J. Zhang, Q. Wang, V. Krishnan, B.-M. Hodge, A data-driven methodology for probabilistic wind power ramp forecasting. IEEE Trans. Smart Grid **10**(2), 1326–1338 (2019)
19. Y. Wen et al., Performance evaluation of probabilistic methods based on bootstrap and quantile regression to quantify PV power point forecast uncertainty. IEEE Trans. Neural Netw. Learn. Syst. **31**(4), 1134–1144 (2020)
20. L. Cheng, H. Zang, Z. Wei, T. Ding, R. Xu, G. Sun, Short-term solar power prediction learning directly from satellite images with regions of interest. IEEE Trans. Sustain. Energy **13**(1), 629–639 (2022)
21. H. Zhang, Y. Liu, J. Yan, S. Han, L. Li, Q. Long, Improved deep mixture density network for regional wind power probabilistic forecasting. IEEE Trans. Power Syst. **35**(4), 2549–2560 (2020)
22. B. Huang, J. Wang, Applications of physics-informed neural networks in power systems–a review. IEEE Trans. Power Syst.
23. G.S. Misyris, A. Venzke, S. Chatzivasileiadis, Physics-informed neural networks for power systems. IEEE Power Energy Soc. Gen. Meet. (PESGM) **2020**, 1–5 (2020)

24. A.S. Zamzam, N.D. Sidiropoulos, Physics-aware neural networks for distribution system state estimation. IEEE Trans. Power Syst. **35**(6), 4347–4356 (2020)
25. J. Ren, J. Chen, B. Li, M. Zhao, H. Shi, H. You, A method for power system transient stability assessment based on transfer learning. IEEE Power Energy Soc. Gen. Meet. (PESGM) **2020**, 1–5 (2020)
26. C. Ren, Y. Xu, B. Dai, R. Zhang, An integrated transfer learning method for power system dynamic security assessment of unlearned faults with missing data. IEEE Trans. Power Syst. **36**(5), 4856–4859 (2021)
27. J. Zhao, H. Yu, G. Geng, TransOS-ELM: a short-term photovoltaic power forecasting method based on transferred knowledge from similar days, in *2021 IEEE 5th Conference on Energy Internet and Energy System Integration (EI2)* (2021), pp. 89–94
28. H. Li, Z. Ma, Y. Weng, A transfer learning framework for power system event identification. IEEE Trans. Power Syst. https://doi.org/10.1109/TPWRS.2022.3153445
29. T. Hu, W. Wu, Q. Guo, H. Sun, L. Shi, X. Shen, Very short-term spatial and temporal wind power forecasting: a deep learning approach. CSEE J. Power Energy Syst. **6**(2), 434–443 (2020)
30. X. Dong, Y. Sun, Y. Li, X. Wang, T. Pu, Spatio-temporal convolutional network based power forecasting of multiple wind farms. J. Mod. Power Syst. Clean Energy
31. S. Zhang, G. Zheng, L. Liu, L. Li, J. Li, X. Wang, A machine learning based knowledge graph framework for heterogeneous power grid systems, in *2021 IEEE International Conference on Consumer Electronics and Computer Engineering (ICCECE)* (2021), pp. 119–123
32. H. Gao, L. Miao, J. Liu, K. Dong, X. Lin, Construction and application of knowledge graph for power system dispatching, in *2020 7th International Forum on Electrical Engineering and Automation (IFEEA)* (2020), pp. 690–695
33. C. Gao, Y. Guo et al., A data-knowledge-hybrid-driven method for modeling reactive power-voltage response characteristics of renewable energy sources. IEEE Trans. Power Syst. **39**(2), 3361–3373 (2024)
34. A. Wiranata, S.A. Wibowo, R. Patmasari, R. Rahmania, R. Mayasari, Investigation of padding schemes for faster R-CNN on vehicle detection, in *2018 International Conference on Control, Electronics, Renewable Energy and Communications (ICCEREC)* (2018), pp. 208–212
35. D. Kingma, J. Ba, Adam, A method for stochastic optimization (2014)
36. F.-L. Fan, J. Xiong, M. Li, G. Wang, On interpretability of artificial neural networks: a survey. IEEE Trans. Radiat. Plasma Med. Sci. **5**(6), 741–760 (2021)
37. V. Álvarez, S. Mazuelas, J.A. Lozano, Probabilistic load forecasting based on adaptive online learning. IEEE Trans. Power Syst. **36**(4), 3668–3680 (2021)
38. S. Niu, Y. Liu, J. Wang, H. Song, A decade survey of transfer learning (2010–2020). IEEE Trans. Artif. Intell. **1**(2), 151–166 (2020)
39. B. Azimian, R.S. Biswas, S. Moshtagh, A. Pal, L. Tong, G. Dasarathy, State and topology estimation for unobservable distribution systems using deep neural networks. IEEE Trans. Instrum. Meas. **71**, 1–14 (2022). Art no. 9003514
40. J. Yosinski, J. Clune, Y. Bengio, H. Lipson, How transferable are features in deep neural networks? *Advances in Neural Information Processing Systems*, pp. 3320–3328 (2014)
41. K. He, X. Zhang, S. Ren, J. Sun, Delving deep into rectifiers: surpassing human-level performance on ImageNet classification, in *2015 IEEE International Conference on Computer Vision (ICCV),* Santiago, Chile (2015), pp. 1026–1034
42. Wind speed, https://www.sotaventogalicia.com/en/technical-area/real-time-data/historical/. Accessed 1 Nov 2021

Carbon-Aware Distribution Network Operation and Optimization

Linwei Sang and Yinliang Xu

Abstract With the extensive integration of distributed resources into the distribution network, the operational management of the distribution system becomes pivotal to ensure its reliability, safety, and efficient operation. The interplay between the distribution network and distributed resources introduces complexity, manifesting in the intricate response of distributed resources to system incentive, thereby challenging their unified management. Leveraging machine learning and optimization theories, this chapter firstly introduces a constraint learning approach tailored to clusters of distributed resources. This involves establishing a data-driven response constraint model based on neural networks. Subsequently, the chapter discusses the fusing learning and optimization for distribution network operation framework through the lens of constraint learning. Based on the constraint model, carbon emission limitations, distribution network operational model, and bilinear relaxation strategies, an amalgamated machine learning and optimization-based distribution network operational model is constructed. This model not only can ensure the secure and economical operation of the distribution system but also addresses carbon emission management. Finally, through comprehensive case studies, the precision of constraint learning for distributed resource assimilation and the effectiveness of the integrated machine learning-optimization distribution network operational model are demonstrated in this chapter.

Keywords Machine learning · Neural networks · Distribution network · Carbon-aware operation

1 Introduction

To achieve the transition to low-carbon energy sources in the power system, a substantial integration of renewable distributed resources, represented by distributed photovoltaics, wind power, and energy storage, into the grid poses significant challenges

L. Sang · Y. Xu (✉)
Tsinghua Shenzhen International Graduate School, Tsinghua University, Beijing, China
e-mail: xu.yinliang@sz.tsinghua.edu.cn

F. Shahnia and F. Wen (eds.), *Power and Energy Resources*, Power Systems,
https://doi.org/10.1007/978-981-96-2612-0_8

to the operation of distribution networks due to the diversity and complexity of their operational characteristics. The traditional centralized control is difficult to adapt to the grid integration of a large number of distributed resources, presenting issues related to system reliability, massive communication, and information privacy.

A large number of distributed resources often form clusters through aggregation, participating in the management of distribution networks in a unified manner as clusters. However, the response characteristics of distributed resource clusters are complex, and how to effectively motivate and manage distributed resources is a key issue in the operation and management of power distribution systems. Moreover, the internal control strategies of distributed resources are complex and involve user privacy, making it difficult for distribution network operators to model and control uniformly. Price incentives are an effective strategy for managing distributed resources in distribution networks. Literature [2] reviews the current state and future of distributed resource scheduling in new distribution networks, achieving collaborative optimization of distribution networks with distributed resources based on hierarchical zoning scheduling principles. Literature [3] proposes a virtual power plant control strategy that considers the differences in adjustment characteristics of distributed resources, enhancing their response characteristics. Literature [5] designs an active distribution network collaborative strategy considering multi-tier electricity price response, achieving collaborative operation of source-network-load in distribution networks. The above research focuses on the collaborative operation of distributed resources, and carbon emission management is a fundamental approach to the low-carbon energy transition, with gas units in distribution networks contributing to carbon emissions. Effectively managing the carbon emissions of distribution networks containing distributed power sources is a key issue that needs to be addressed [4, 6, 7]. Literature [6] builds a real-time—day-ahead two-stage low-carbon economic dispatch model considering the carbon emission constraints of virtual power plants, achieving its low-carbon economic operation. Literature [7] discusses the peer-to-peer trading model of electricity-green certificates to extend the analysis of distributed resources. Therefore, achieving low-carbon operation of distribution networks with complex clusters of distributed resources is an important direction in the field of distribution networks.

With the continuous development in the field of machine learning-optimization, machine learning-optimization theory is widely applied in the operation and control of power systems, with each application area having its focus [8, 9, 13]. Machine learning has strong data mining capabilities, used to solve prediction problems in power systems [9, 10], such as load forecasting, renewable energy output forecasting, and electricity price forecasting. Optimization theory has reliable theoretical guarantees, applied to solve optimal decision-making problems in power systems [12], such as a series of operation scheduling problems based on optimal power flow. Inference models based on machine learning provide boundary conditions for operation decision models based on optimization, and the optimization of operation models yields optimal decisions. Both types of models are interconnected for operational decision-making in power systems. Literature [14, 15] further discusses data and knowledge-driven operation and control strategies, which are the evolving direction of future novel power system operation and control methods [16, 17].

Based on the above theories and the development direction of actual system operation, this chapter uses machine learning models to mine and represent unknown power system knowledge from data, use optimization theory to model based on the existing knowledge of the power system, and integrate both to leverage the respective advantages of machine learning and optimization theory to support the operational management of distribution networks. Specifically, the chapter considers using the different characteristics of machine learning optimization for the control issues of distributed resource clusters in distribution networks, introducing an integrated machine learning optimization support for the low-carbon operation of distribution networks, including the basic framework of integrating machine learning optimization, constraint learning algorithms for distributed resource clusters, and a low-carbon operation model for distribution networks integrating machine learning-optimization.

2 Preliminaries

The chapter first analyzes the basic framework of distribution network operation analysis, as shown in Fig. 1. Its main components include data, model, decision, and prediction, with its theoretical keys primarily being machine learning and optimization theory.

Machine learning, with its powerful data mining and learning capabilities, supports the predictive inference models of the distribution network, mining historical detection data and wide-area measurement data features to predict the quantities of uncertain factors in the distribution network, including forecasts of variables such as renewable energy output, load, and electricity prices. The predictive inference model within the model bridges the data module and the prediction module.

Optimization theory, based on mathematical analysis and approximate modeling of the actual physical system of the distribution network, simplifies related models

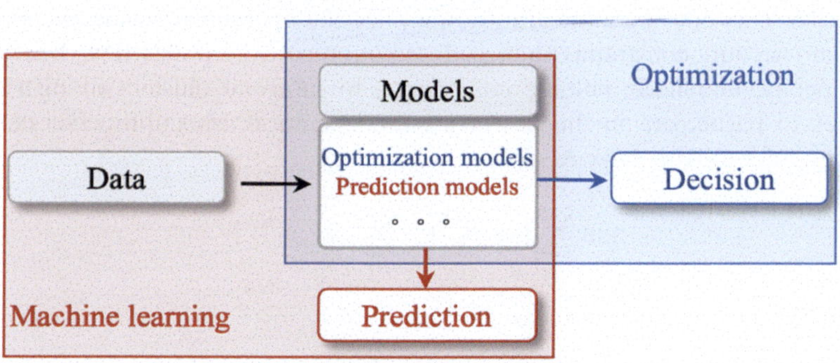

Fig. 1 Distribution network analytics framework

and designs algorithms to solve for the optimal decision-making of the distribution network under given input boundary conditions, such as the output of control devices in the distribution network for day-ahead and intra-day. Herein, the optimization scheduling model in the model optimizes and solves for the best decision based on input boundary conditions.

Through the above analysis, machine learning and optimization theory each have characteristics that are applicable to different parts of the operation analysis of the distribution network, corresponding to the red and blue areas in Fig. 1, respectively. However, the remaining white area is an emerging research field. Based on this, this chapter discusses an integrated learning and optimization framework for decision analysis in the distribution network to support decision analysis under complex scenarios in future distribution networks.

3 Problem Formulation

The operation model of the distribution network includes an upper-level distribution network operation model and a lower-level distributed resources response model. Initially, in Sect. 3.1, the distribution network operation model is constructed based on the linear power flow model. Then, in Sect. 3.2, the optimal operation model for distributed resource clusters is developed. Lastly, in Sect. 3.3, the carbon emissions of the distribution network are analyzed. Through price signal λ incentives, distributed resource clusters participate in the operation and scheduling of the distribution network to achieve economical and secure operation of the distribution network, and through carbon quota C, the overall carbon emissions of the distribution network are limited to manage the carbon emissions of the distribution network.

3.1 Distribution Network Operation Model

Distribution network operators aim to minimize the operating costs of the system, taking into account constraints such as distribution network power flow, line power flow capacity, and node voltage constraints, to motivate clusters of distributed resources to participate in the power system operation scheduling. For detailed modeling, refer to Appendix A. (1)

$$\begin{aligned} &\min \sum_{t \in \mathcal{T}} \{\lambda_{S,t} p_t^S + \sum_{i \in \mathcal{DER}} \lambda_{i,t} p_{i,t}^{DER}\} \\ &\text{s.t.} p_t^S, p_{i,t}^{DER} \in G \end{aligned} \tag{1}$$

In the model's objective function, $\lambda_{S,t}$ and $\lambda_{i,t}$ represent the main grid electricity price and the price for cluster i, respectively; $P^S{}_T$, $P^{DER}{}_{i,t}$ denote the electricity purchased from the main grid and from cluster i; $\lambda^S{}_t$, $P^S{}_T$, $\lambda_{i,t}$, $P^{DER}{}_{i,t}$ represent the electricity purchasing cost from the main grid and the cluster; G denotes the constraints on the operation of the distribution network.

3.2 *Distributed Energy Resource Response Model*

Based on the operation model of the distribution network, the response model for distributed resources is constructed, wherein distributed resource clusters respond to the distribution network prices $\lambda_{i,t}$ aiming to maximize the cluster's revenue. This model considers the operational constraints of internal distributed resources like photovoltaics, energy storage, and gas turbines, as well as load demand constraints, to construct the operating model for distributed resource clusters. For detailed modeling, refer to Appendix B.

$$\begin{aligned} &\max \sum_{t \in \mathcal{T}} U\left(p_{i,t}^{L}\right) \\ &\text{s.t.} p_{i,t}^{L}, p_{i,t}^{DER} \in DER \end{aligned} \tag{2}$$

3.3 *Carbon Emissions Analysis of the Distribution Network*

The total carbon emissions of the distribution network include carbon emissions from external main grid electricity inputs and those generated internally by distributed power sources.

(1) Carbon emissions from external main grid inputs to the distribution network are determined by the carbon emission intensity e^S of transmission grid nodes and the quantity of electricity purchased from the main grid, as shown in Eq. (3):

$$E_t^S = e^S p_t^S \tag{3}$$

(2) Carbon emissions from internal distributed resource clusters are generated by internal devices such as gas turbines, determined by the units' carbon emission factors and their electricity generation, as shown in Eq. (4), influenced by the incentive price λ.

$$E_t^j = \sum_{j \in \mathcal{DER}_i} e_j p_{j,t} \tag{4}$$

4 Constraint Learning Algorithms for Distributed Resource Clusters

Based on the problem modeling mentioned above, this chapter uses a constraint learning algorithm targeted at distributed resource clusters. Initially, the response characteristics of distributed resources are analyzed by constructing a dataset of distributed resources through data sampling. A data-driven response model based on neural networks is then constructed and learned. Ultimately, a neural network response constraint model is constructed using an equivalent transformation strategy based on the big-M method. The constraint learning algorithm includes two processes: training of the neural network model and the equivalent transformation of the network model into a constraint model. The relationship between the models is illustrated in Fig. 2.

4.1 Price Response Characteristics of Distributed Resources

Based on the distributed resource cluster model (2) and carbon emissions analysis (4), the electricity and carbon emissions of cluster i are determined by the upper-level distribution network's incentive price decision λ_i, thus constructing a cluster price response model.

(1) Price-Electricity Response Model: Based on (2), the mapping relationship between λ_i and the response electricity quantity $p^{der}{}_i$ is constructed using $f^p{}_i$, as shown in Eq. (5).

$$p_i^{der} = f_i^p(\lambda_i) \tag{5}$$

(2) Price-Carbon Emission Response Model: Based on (2) and (4), the mapping relationship between λ_i and the response carbon emission quantity $E^{der}{}_i$ is constructed using $f^e{}_i$, as shown in Eq. (6).

$$E_i^{der} = f_i^e(\lambda_i) \tag{6}$$

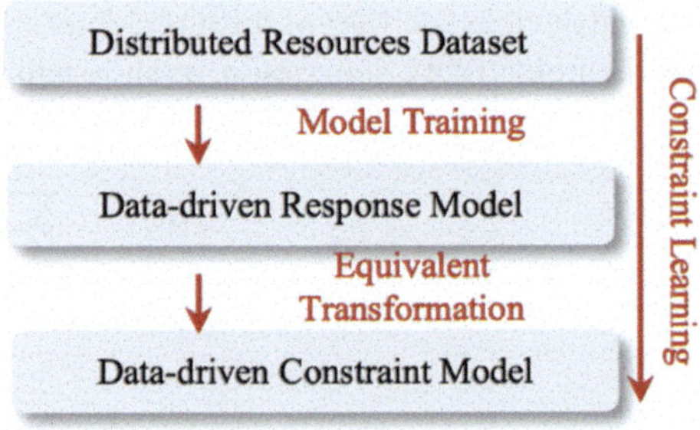

Fig. 2 Basic framework of constraint learning

The distributed resource response quantity $y^{der}{}_i := (p^{der}{}_i, E^{der}{}_i)$ is further defined, and combining the above two response models yields a unified response model (7).

$$y_i^{der} = (p_i^{der}, E_i^{der}) = f_i^{der}(\lambda_i) \tag{7}$$

(3) Data Generation: Using (7), by solving the offline model, $y^{der}{}_i$ under different incentive prices λ_i is obtained, constructing the response dataset D^{der}.

$$\mathcal{D}^{der} = \{\lambda_{i,n}, y_{i,n}^{der}\}_{n=1}^{N} \tag{8}$$

The mathematical models (2) and (4) for scheduling the distribution network expose operational parameters of the clusters, compromising user privacy; in contrast, sampled datasets reflect the response characteristics of users without revealing detailed cluster parameters, serving as the data basis for the data-driven response model.

4.2 Data-Driven Response Model Based on Neural Networks

Utilizing the response dataset D_{der} generated from the model and employing machine learning methods to achieve the aforementioned mapping relationships, considering the data-driven characteristics of distributed resource responses, a distributed resource cluster response model between the incentive price λ_i and the cluster response $y^{der}{}_i$ including electricity and carbon emissions is constructed, as shown in Eq. (9).

$$y_i^{der} = (p_i^{der}, e_i^{der}) = f_i^{ML}(\lambda_i; \theta) \tag{9}$$

where θ represents the parameters of the machine learning model.

To accurately model the complex response relationships of clusters from data, the chapter employs neural network models with strong fitting capabilities from the machine learning domain to learn the response relationships. In operational optimization, response relationships are essentially constraints; thus, this learning is referred to as constraint learning, leading to the construction of a data-driven response model.

(1) Basic Structure of Neural Networks: Considering the output $y^{der}{}_i$ as a continuous variable, learning the response relationship is essentially a regression problem. Therefore, a neural network based on the rectified linear activation function (ReLU) is constructed to learn the response relationship, as shown in Fig. 3.

The neural network model primarily includes an input layer (incentive price), hidden layers (neurons based on ReLU), and an output layer (response of distributed resource clusters), with the detailed mathematical model of (9) shown in Eq. (10):

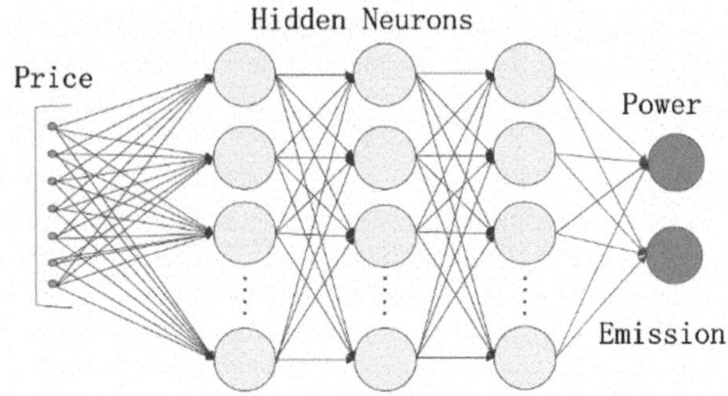

Fig. 3 Distributed energy resources constraint learning model

$$\begin{aligned} v_i^l &= \mathrm{ReLU}(\theta_{i0}^l + \sum_{j \in \mathcal{N}^{l-1}} \theta_{ij}^l v_j^{l-1}), l \in [L] \\ \hat{y}^{der} &= v^L = \theta_0^L + \sum_{j \in \mathcal{N}^{L-1}} \theta_j^L v_j^{L-1}, \lambda = v_0 \end{aligned} \tag{10}$$

where ReLU(x): = max(0, x), $v^l{}_i$ is the *i*-th neuron in the *l*-th layer of the network, θ represents the neural network parameters, λ is the incentive price, and ŷ_der is the response quantity output by the neural network.

(2) Loss Function: Defined based on the neural network structure, the loss function quantifies the distance between the neural network output and the true values. By optimizing the neural network parameters θ to make the predictions close to the real values, considering its essence as a regression problem, the mean square error is used to construct the loss function, as shown in Eq. (11).

$$L^{mse} = \frac{1}{N} \sum (\hat{y}^{der} - y^{der})^2 \tag{11}$$

(3) Model Training: Based on the neural network structure and loss function, with the goal of minimizing the loss function L_mse, the neural network parameters θ are optimized using the stochastic gradient descent (SGD) algorithm to enhance the accuracy of the inference values. The detailed steps of model training are referred to in Appendix C.

Through model training, a data-driven response model based on neural networks is constructed, achieving inference of the response characteristics of distributed resources. The trained network model does not depend on the operational parameters of distributed resources, thus protecting user privacy.

4.3 Data-Driven Response Constraint Model Based on Neural Networks

Due to the strong non-linearity of neural networks, it is challenging for the data-driven response model to directly participate in the operational scheduling of the distribution network. To address this issue, the chapter first analyzes the internal structure of neural networks and the connection characteristics of their hidden layers. Then, based on the characteristics of the data-driven response model, a data-driven response constraint model based on neural networks is constructed. The constraint model consists of mixed-integer constraints, suitable for joint scheduling of distributed resources and the distribution network.

As shown in Eq. (10), the nonlinear part of the neural network model is the ReLU activation function of hidden layer neurons. The essence of the ReLU function is to output zero for negative input values while maintaining positive values unchanged. Therefore, the big-M method is used to equivalently transform ReLU into mixed-integer linear constraints, as shown in Eq. (12).

$$\begin{aligned} & v \geq x, v \geq 0, \\ & v \leq x - M^L(1-t), \\ & v \leq M^U t, \quad t \in \{0,1\} \end{aligned} \tag{12}$$

where M_L is the minimum value, M_U is the maximum value, and t is an auxiliary 0–1 variable.

Equivalence Analysis: When the input x < 0, t takes 0, and the output v is 0; when the input $x > 0$, t takes 1, and the output v is x, verifying the equivalence of the constraint.

Applying the optimization relaxation strategy of the big-M method to all ReLU activation functions within the neural network, the data-driven response model (10) trained is then equivalently transformed into a set of mixed-integer linear constraints. Consequently, a neural network response constraint model for distributed resource clusters is constructed, as shown in Eq. (13).

$$\begin{aligned} & v_i^1 = \theta_{i0}^1 + \sum_{j \in \mathcal{X}} \theta_{ij}^l x_j, \; i \in [N_1] \\ & \hat{v}_i^l = \theta_{i0}^l + \sum_{j \in \mathcal{N}^{l-1}} \theta_{ij}^l v_j^{l-1}, \; l \in [L], i \in [N_l] \\ & v_i^l \geq \hat{v}_i^l, v_i^l \geq 0, \; l \in [L], i \in [N_l] \\ & v_i^l \leq \hat{v}_i^l - M_{l,i}^L(1-t_{l,i}), \; l \in [L], i \in [N_l] \\ & v_i^l \leq M_{l,i}^U t_{l,i}, \quad t_{l,i} \in \{0,1\}, l \in [L], i \in [N_l] \\ & \hat{y}_i^{der} = \theta_{i0}^L + \sum_{j \in \mathcal{N}^{L-1}} \theta_{ij}^L v_j^{L-1}, \; i \in [N_L] \end{aligned} \tag{13}$$

The core of constraint learning involves learning response characteristic models and equivalently transforming them into response constraint models. Through the analysis of distributed resource response characteristics and the data-driven response model, ultimately constructing the response constraint model lays the foundation for realizing the low-carbon operation of the distribution network integrating machine learning and optimization.

5 Fusing Learning and Optimization for Distribution Network Operation

Based on the constraint learning method for distributed resource clusters, a data-driven response constraint model based on neural networks is constructed. Subsequently, an integrated machine learning and optimization framework for distribution network operation (Fusing Learning and Optimization for Distribution Network Operation, FLO4DNO) is focused. Further considering carbon emission constraints to manage carbon emissions of the distribution network, the McCormick method is used to handle nonlinear terms. Combining with the distribution network operation model, the FLO4DNO low-carbon operation model is introduced, achieving distributed resource management and carbon emission management of the distribution network.

5.1 Overall Framework

The FLO4DNO operation framework, as shown in Fig. 4, includes optimization decision-making and machine learning parts. The optimization decision model provides a data basis for the machine learning model, and the machine learning model provides a data-driven constraint model for the optimization model. The two parts are integrated through data generation and the constraint learning process. The innovation of the FLO4DNO framework lies in integrating machine learning and optimization decision models to make distribution network decisions. The core is to equivalently transform the trained neural network-driven distributed resource response characteristic model into a neural network constraint model to achieve model integration, ultimately realizing low-carbon operation of the distribution network considering the complex characteristics of distributed resources.

The machine learning part mainly includes four sections: data sampling, model training, model equivalence, and learning evaluation.

(1) Data Sampling: Utilizing the distributed cluster response model under different incentive prices to form an offline training dataset;
(2) Model Training: Based on the sampled dataset, a neural network model is

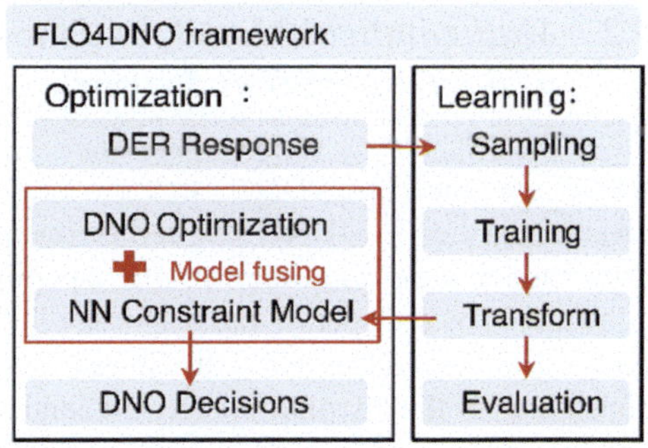

Fig. 4 Fusing learning and optimization for distribution network operation framework

constructed, initializing related parameters. Aiming to minimize inference errors, model parameters are optimized to build a data-driven response model;

(3) Model Equivalence: Based on the big-M equivalence transformation strategy, the trained data-driven response model is equivalently transformed into a set of constraints to construct a data-driven response constraint model;
(4) Learning Evaluation: Establishing relevant indicators to analyze the accuracy and effectiveness of the data model.

The optimization decision part mainly includes distributed resource cluster response models, distribution network decision models, neural network constraint models, and distribution network decision-making.

(1) Distributed Resource Cluster Response: According to Eq. (2), the distributed resource response characteristic model is constructed, calculating the response quantity under different incentive prices, generating a dataset for training (8);
(2) Distribution Network Decision Model: Based on Eq. (1), constructing the distribution network optimization scheduling model aiming for economic optimality;
(3) Neural Network Constraint Model: The data-driven response model based on neural networks is equivalently transformed into a response constraint model (13), used for joint optimization scheduling;
(4) Distribution Network Decision: Solving the joint model of the distribution network and constraint models to obtain the optimal scheduling solution for the distribution network.

Regarding the operation problem of distributed resources in the distribution network, i.e., the constraint learning method for distributed resource clusters. Considering the carbon emission constraints of the distribution network, integrating the response constraint model and the distribution network decision model ultimately optimizes the system's optimal decision.

5.2 Mathematical Model for Low-Carbon Operation of the Distribution Network

Based on the FLO4DNO framework, further considering carbon emission constraints, the FLO4DNO low-carbon operation mathematical model is constructed. This model aims to minimize the operating costs of the distribution network, constrained by the operational response relationships of distributed resource clusters, power flow constraints of the distribution network operation, overall carbon emission constraints of the distribution network, and safety constraints of nodes and branches.

(1) Carbon Emission Constraints of the Distribution Network: The overall carbon emissions of the distribution network are limited within a specified quota C, as shown in Eq. (14).

$$E_t^S + \sum_{i \in \mathcal{DER}} E_{i,t}^{DER} \leq C \tag{14}$$

(2) Handling Bilinear Terms Based on McCormick: In the objective function of Eq. (16), both $\lambda_{i,t}$ and ${P^{der}}_{i,t}$ are optimization variables, and their product is nonlinear and non-convex, making it challenging to optimize directly. Therefore, by introducing auxiliary variables $c_{i,t}$ and using the McCormick relaxation to relax the product terms, effective model solving is achieved, as shown in Eq. (15).

$$\begin{aligned}
c_i &\geq \lambda^L P + P^L \lambda_{i,t} - \lambda^L P^L \\
c_i &\geq \lambda^U P + P^U \lambda_{i,t} - \lambda^U P^U \\
c_i &\leq \lambda^U P + P^L \lambda_{i,t} - \lambda^U P^L \\
c_i &\leq \lambda^L P + P^U \lambda_{i,t} - \lambda^L P^U
\end{aligned} \tag{15}$$

where λ_L and λ_U are the lower and upper bounds of the incentive price, respectively, P_L and P_U are the lower and upper bounds of the response quantity, respectively.

Based on the operation model of the distribution network (16), replacing the bilinear terms with $c_{i,t}$, considering the relaxation constraints (15) and carbon emission constraints (16), integrating the distributed resource cluster response constraint model (16), the FLO4DNO low-carbon operation model is constructed. This model is a mixed-integer linear programming model, which can be effectively solved by the existing solvers' branch-and-bound method.

6 Case Study

The study, based on the IEEE-33 node standard distribution network system containing various types of distributed resource clusters, demonstrates the effectiveness of the algorithm and model. The model is built on Python 3.10, where the optimization model part is constructed using the CVXPY framework, and the constraint learning part based on neural networks is built using the Scikit-learn framework. CVXPY and Scikit-learn are programming libraries built on Python, applied in different fields, the former for optimization modeling and the latter for data analysis and machine learning. The FLO4DNO framework is built on the basis of the CVXPY and Scikit-learn frameworks, including optimization decision-making and machine learning parts, where the optimization decision part is based on the CVXPY framework, and the machine learning part is based on the Scikit-learn framework. The framework utilizes the features of both frameworks, integrating them to achieve low-carbon operation of the distribution network through fusing learning and optimization. Key parameter settings of the case study system are referred to in [21]. This section mainly includes two parts: constraint learning performance analysis for distributed resource clusters and analysis of the distribution network scheduling results integrating learning and optimization.

6.1 *Constraint Learning Analysis for Distributed Resource Clusters*

First, sampling the response characteristics of distributed resource clusters under different prices constructs the distributed resource response dataset D_{der}, building a neural network model and training the neural network-based response model aiming to minimize inference error. Specifically, Fig. 5 displays the training process of the distributed resource response characteristics.

As the training rounds increase, the inference error of the neural network for different distributed resource clusters gradually decreases and converges to a lower level after 8–9 rounds of training, demonstrating the effectiveness of the algorithm.

The neural network used has the lowest root mean squared error, and its mean absolute percentage error and mean error are also better than the linear regression model. Although the mean absolute percentage error is higher than the random forest

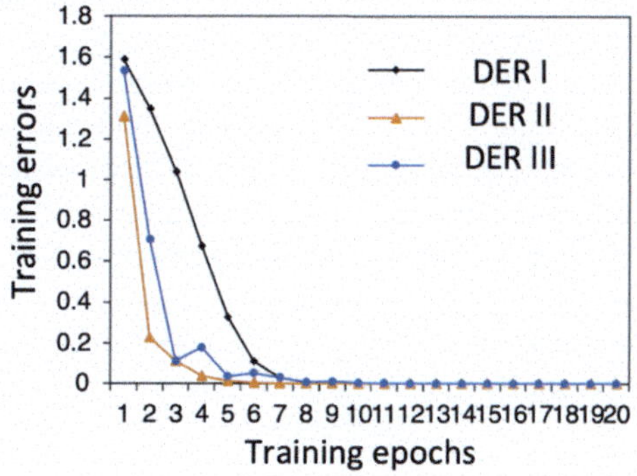

Fig. 5 The training process of the constraint learning approach

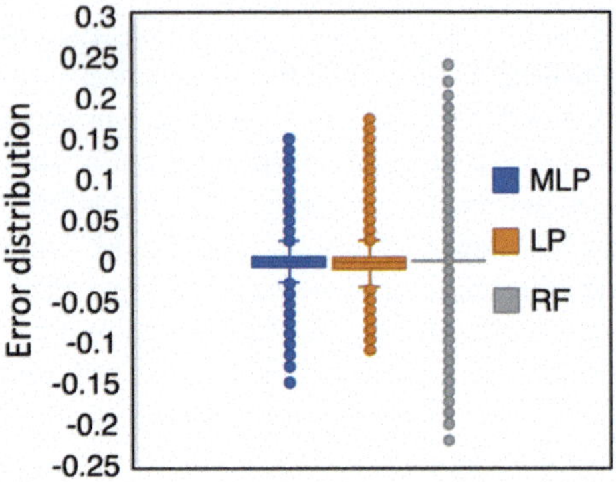

Fig. 6 Model error distribution analysis

model, the mean error is much lower than the random forest, indicating its stability in fitting.

Figure 6 analyzes the error distribution of different models, showing that the random forest model has more abnormal inferences, while the overall performance of the neural network model is superior to linear regression.

6.2 Analysis of Distribution Network Scheduling Integrating Learning and Optimization

The response constraint model based on the neural network's response characteristics is constructed to realize low-carbon operation scheduling of the distribution network operation integrating learning and optimization.

First, Fig. 7 analyzes the different incentive prices and response electricity quantities of distributed resources at different times. The integrated scheduling model gives different incentive prices to distributed resources at different times, effectively motivating distributed resource clusters to participate in the operation of the distribution network while considering user privacy.

To demonstrate the effectiveness of carbon emission constraints, the study compares the system operation under scenarios with carbon emission constraints for low-carbon scheduling and without carbon emission constraints for conventional scheduling. Figure 8 shows the carbon emissions under different models, where low-carbon scheduling effectively reduces system carbon emissions, demonstrating its ability to manage carbon emissions of the distribution network effectively.

Table 1 further quantifies the optimization scheduling costs and system carbon emissions of low-carbon scheduling, showing that low-carbon scheduling increases the operation cost by 4.58% and reduces the carbon emissions of the distribution network by 8.50%, demonstrating the effectiveness of the algorithm.

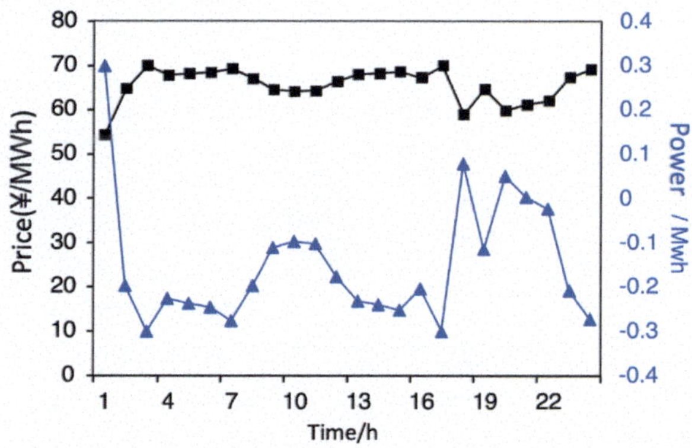

Fig. 7 Distribution network scheduling results analysis

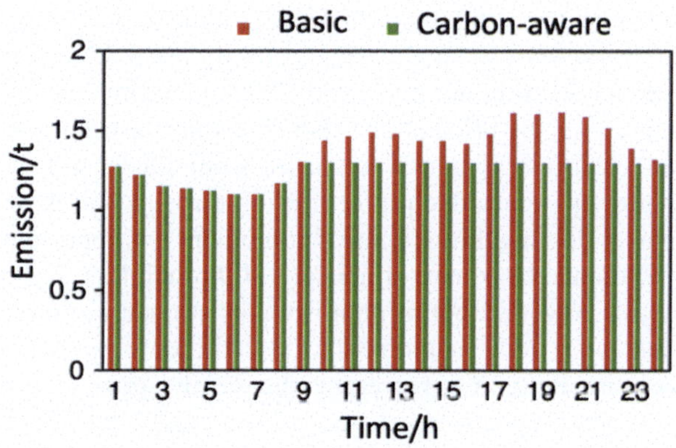

Fig. 8 Carbon emission analysis under various scenarios

Table 1 Carbon-aware scheduling analysis

Models	Cost/¥	Emission/ton
Proposed model	3717.29	30.10
Proposed model without carbon	3554.28	32.89

Acknowlegement This work was supported by the National Key Research and Development Program of China, Grant No. 2024YFB4206500).

Appendix A

Appendix A establishes a detailed operation model for the distribution network, as shown in Eq. (16).

$$
\begin{aligned}
&\min \sum_{t \in \mathcal{T}} \{\sigma_{S,t}^{P} P_t^{S} + \sum_{i \in \mathcal{DER}} \sigma_{S,t}^{P} P_{i,t}^{DER}\} \\
&\text{s.t.} \ \sum_{jk \in \mathcal{E}} P_{jk,t} - \sum_{ij \in \mathcal{E}} (P_{ij,t} - r_{ij} l_{ij,t}) = \hat{p}_{j,t}^{DER} - \hat{p}_{j,t}^{D} \\
&\sum_{jk \in \mathcal{E}} Q_{jk,t} - \sum_{ij \in \mathcal{E}} (Q_{ij,t} - x_{ij} l_{ij,t}) = \hat{q}_{j,t}^{reg} - \hat{q}_{j,t}^{D} \\
&v_{j,t} = v_{i,t} + (r_{ij}^2 + x_{ij}^2) l_{ij,t} - 2(r_{ij} P_{ij,t} + x_{ij} Q_{ij,t}) \forall ij \in \mathcal{E} \\
&||2P_{ij,t} \quad 2Q_{ij,t} \quad l_{ij,t} - v_{i,t}|| \le l_{ij,t} + v_{i,t} \\
&(V^{\min_{i,t}})^2 \le v_{i,t} \le (V^{\max_{i,t}})^2, \forall i \in \mathcal{N}_B \\
&(I^{\min_{ij,t}})^2 \le l_{ij,t} \le (I^{\max_{ij,t}})^2, \forall ij \in \mathcal{E}
\end{aligned}
\tag{16}
$$

where t represents the scheduling moment, T represents the planning time length, $\lambda^{S,t}$ represents the main grid electricity price, $P^{S}{}_{T}$ represents the electricity quantity purchased from the main grid, DER represents the set of distributed resources, $\lambda_{a,t}$ represents the price of distributed resource cluster a at moment t, $P^{DER}{}_{a,t}$ represents the electricity quantity purchased from cluster a, E represents the set of lines, P_{jkt} represents the active power flowing through line jk at moment t, P_{ijt} represents the power flowing through line ij at moment t, r_{ij} represents the resistance of line ij, I_{ijt}^2 represents the square of the current through line ij at moment t, P_{jt}^{PV} represents the active power output of photovoltaics at node j at moment t, D_{jt} represents the load demand at node j at moment t, Q_{jkt} represents the reactive power flowing through line jk at moment t, Q_{ijt} represents the reactive power flowing through line ij at moment t, x_{ij} represents the reactance of line ij, Q_{jt}^{GV} represents the reactive power output at node j at moment t, V_{jt}^2 represents the square of the voltage at node j at moment t, V_{it}^2 represents the square of the voltage at node i at moment t, V_{min}, V_{max} respectively represent the minimum and maximum values of voltage, NB represents the set of system nodes, $I^{ijt,min}$, $I^{ijt,max}$ respectively represent the minimum and maximum current through line ij at moment t.

Appendix B

Appendix B establishes a detailed operation model for distributed resource clusters, as shown in Eq. (17).

$$\max \sum_{t \in T} U(p_{i,t}^{L}) - \lambda_{i,t}^{S} p_{i,t}^{S} - C^{DG}(p_t^{DG})$$
$$p_t^{DG} + p_{i,t}^{S} + p_t^{res} = p_{i,t}^{L}$$
$$q_t^{DG} + q_{i,t}^{S} + q_t^{res} = q_{i,t}^{L} \quad (17)$$
$$p_t^{DG} \geq p_{t,min}^{DG}, p_t^{DG} \leq p_{t,max}^{DG}$$
$$-\Delta_{max}^{DG} \leq p_t^{DG} - p_{t-1}^{DG} \leq \Delta_{max}^{DG}$$
$$0 \leq p_t^{res} \leq p_{max}^{res}$$

where U represents the utility function of the load, CDG represents the cost function of distributed power sources in the cluster, referred to in [19]; D_{it} represents the load demand of user i, λ_t represents the price of the distribution network at moment t, CDG represents the cost function of distributed power sources in the cluster, P_t^{DG}, Q_t^{DG} respectively represent the active and reactive power output of distributed power sources at moment t; P_t^{in}, Q_t^{in} respectively represent the active and reactive power input to the distribution network, P_t^{RE}, Q_t^{RE} respectively represent the active and reactive power output of renewable energy sources, D_{it}^{P}, D_{it}^{Q} respectively represent the active and reactive demand of node i; P_t^{min}, P_t^{max} respectively represent the minimum and maximum power output of distributed power sources at moment t; ΔP_{max} represents the maximum ramp-up power of distributed power sources, $P^{RE}{}_{max}$ represents the maximum active power output of renewable energy sources.

Appendix C

The stochastic gradient descent algorithm for constraint learning mainly includes initialization, model training, and model output, as shown in Algorithm 1.

Algorithm 1: Stochastic gradient descent algorithm for constraint learning

1	Initialization: Initialize parameters θ, neural network $f_{i,\theta}^{der}$, learning rate α, dataset D, maximum iteration number T, t = 0;
2	Model Training:
3	While t < T:
4	Randomly select a batch of samples from the dataset D;
5	Calculate the gradient: ∇θ;
6	Update parameters: θ = θ - α∇θ;
7	t = t + 1
8	Output model: $f_{i,\theta}^{der}$

References

1. W.U Wenchuan, Z. Boming, S. Hongbin, et al., Energy management and distributed energy resources cluster control for active distribution networks. Autom. Electr. Power Syst. **44**(09), 111–118 (2020). (In Chinese)
2. P. Meiqi, H. Xing, A. Qian, T. Yuezhong, Research status and prospect of distributed energy resource dispatching in new distribution system. Power Syst. Technol. 1–16 (2023). (In Chinese)
3. P. Mingjie, X. Da, W. Xitian, Response strategy for virtual power plant considering difference in regulation characteristics of distributed energy resources. Autom. Electr. Power Syst. **46**(18), 108–117 (2022). (In Chinese)
4. W. Zhinong, Z. Side1, S. Guoqiang, Carbon trading based low-carbon economic operation for integrated electricity and natural gas energy system. Autom. Electr. Power Syst. **40**(15), 9–16 (2016). (In Chinese)
5. X. Xilin, S. Yiqun, Y. Liangzhong et al., Source-grid-load coordination method for active distribution network based on multi-level electricity price response mechanism. Autom. Electr. Power Syst. **42**(05), 9–41 (2018). (In Chinese)
6. Z. Renjun, S. Hong, T. Xiafei et al., Low-carbon economic dispatch based on virtual power plant made up o carbon capture unit and wind power under double carbon constraint, in *Proceedings of the CSEE* (2018), **38**(6), 9. (In Chinese)
7. X. Yunpeng, Z. Yuerong, W. Xiuli et al., Distributed resource expansion planning based on multi-prosumer equilibrium with peer-to-peer trading of electric power and green certificate, in *Proceedings of the CSEE,* (2023) 1–14. (in Chinese)
8. W. Jiye, Application and prospect of source-grid-load-storage coordination enabled by artificial intelligence, in *Proceedings of the CSEE*, (2022) **42**(21), 7667–7682. (In Chinese)
9. H. Fujia, W. Xiaohui, Q. Ji, S. Mengjie, P. Tianjiao, Review on artificial intelligence based load forecasting research for the new-type power system, in *Proceedings of the CSEE,* (2023) 1–24. (In Chinese)
10. W. Can, C. Wenkang, S. Yonghua, Probabilistic forecasting for power systems with renewable energy sources: basic concepts and mathematical principles, in *Proceedings of the CSEE*, (2021) **41**(19), 6493–6509. (In Chinese)
11. C. Xinhe, P. Wei, D. Wei, Z. Jian, W. Xinying, PU Tianjiao4. Data-driven virtual power plant dispatching characteristic packing method, in *Proceedings of the CSEE*, (2021) **41**(14), 13. (In Chinese)
12. L. Zhe, H. Zechun, S. Yonghua, Convex relaxation for optimal power flow problem: a recent review, in *Proceedings of the CSEE*, (2019) **39**(13), 3717–3728. (In Chinese)
13. L. Xie, X. Zheng, Y. Sun, T. Huang, T. Bruton, Massively digitized power grid: opportunities and challenges of use-inspired AI. Proc. IEEE **111**(7), 762–787 (2023)
14. L. Feng, W. Qi, H. Jianxiong, T. Yi, Combined data-driven and knowledge-driven methodology research advances and its applied prospect in power systems, in *Proceedings of the CSEE*, (2021) **41**(13)
15. L. Peng, H. Wenqi, W. Xin et al., Review on application of combined data-knowledge-driven artificial intelligence methods in power dispatching. Autom. Electr. Power Syst.1–16 (2023). (In Chinese)
16. G. Ruan, H. Zhong, G. Zhang et al., Review of learning-assisted power system optimization. CSEE J. Power Energy Syst. **7**(2), 221–231 (2021)
17. P. Tianjiao, Q. Ji, Z. Zixuan, et al., Research on interpretable methods of machine learning applied in intelligent analysis of power system (part I): basic concept and framework, in *Proceedings of the CSEE*. (In Chinese)
18. Y.I. Zhongkai, X.U. Yinliang, W.U. Wenchuan, Market clearing strategy for distribution system considering multiple power commodities offered by virtual power plant. Autom. Electr. Power Syst. **44**(22), 143–151 (2020). ((in Chinese))
19. Z. Yi, Y. Xu, W. Gu, W. Wu, A multi-time-scale economic scheduling strategy for virtual power plant based on deferrable loads aggregation and disaggregation. IEEE Trans. Sustain. Energy **11**(3), 1332–1346 (2020)

20. Z. Yi, Y. Xu, H. Wang, L. Sang, Coordinated operation strategy for a virtual power plant with multiple DER aggregators. IEEE Trans. Sustain. Energy **12**(4), 2445–2458 (2021)
21. L. Sang, Y. Xu, H. Sun, Encoding carbon emission flow in energy management: a compact constraint learning approach. IEEE Trans. Sustain. Energy. https://doi.org/10.1109/TSTE.2023.3274735
22. https://cloud.tsinghua.edu.cn/d/c7e99e58acec4bc2909d/

Distribution Network Reconfiguration

Chao Lei and Siqi Bu

Abstract Network reconfiguration is a classical optimal operation problem of electrical distribution networks. This problem aims at a specific topology optimization and chooses the optimal switch status of sectionalizing switches, tie-switches, and soft open points. The objective is maintaining real-time load balancing and loss reduction for distribution networks with the growing penetration of distributed energy resources and coordinating real-time transactive dispatch tasks between supply and demand at the market level of distribution networks. This chapter revisits the fundamentals of *DistFlow* equations and its convex relaxation formulations, and then introduces the disjunctive convex hull relaxation (DCHR) approach to tackle the disjunctive nature for distribution network reconfiguration problems. With continuous parent–child relationship variables as disjunctive variables, this DCHR approach is theoretically tighter than the McCormick linearization method and the Big-M method, and it is especially suitable for distribution networks with directional power flows. Through case studies, the chapter demonstrates the computing performance in terms of running time and iterations using a DCHR approach yields superior numerical performance than prior relaxation methods.

Keywords Distribution network reconfiguration · Disjunctive convex hull approach

1 Introduction

Distribution network reconfiguration (DNR) is a classical optimal operation problem of distribution networks over decades. It is a specific topology optimization about choosing the optimal switch status of sectionalizing switches (normally closed), tie-switches (normally open), and/or controllable power flows by soft open points [1, 2]. In contrast to passive distribution networks, two objectives are achievable

---

C. Lei (✉) · S. Bu
Department of Electrical and Electronic Engineering, The Hong Kong Polytechnic University, Hong Kong, China
e-mail: 21118924r@connect.polyu.hk

F. Shahnia and F. Wen (eds.), *Power and Energy Resources*, Power Systems,
https://doi.org/10.1007/978-981-96-2612-0_9

from performing DNR actions: (*i*) maintaining real-time load balancing and loss reduction for distribution networks with the growing penetration of distributed energy resources (DERs) and (*ii*) coordinating real-time transactive dispatch tasks between supply and demand at the market level of distribution networks. At present, these two objectives can be realizable due to the upgraded applications of high-speed switching devices. Therefore, this DNR research is the fundamental for the transition to active distribution networks involved with various requirements for the quick solvability, cyber-physical system security, privacy-preserving ability, and dispatch flexibility enhancement.

To address the quick solvability of DNR problems, heuristic methods and convex relaxations are two primary effective solving approaches for active distribution networks. In terms of heuristic methods, there is a rich set of optimization approaches in the power systems literature that tries to circumvent these problems. In fact, the non-linearity of AC power flow complicates solving a DNR problem. To handle this problem, researchers used DC power flow, a *DistFlow* model [3] that is a second-order mathematical programming model, and full AC power flow [4]. However, the DC power flow model is inaccurate while the *DistFlow* and the full AC power flow models are accurate but very time-consuming for a large system. Black-box heuristic methods, which push power flow calculations outside the optimization solver, have become very popular, owing to their broad applicability [5]. In summary, heuristic methods perform well in small systems but might converge slowly, especially in large-scale systems [6]. And the results obtained by heuristic methods for different runs might not be the same, which prevents them from being widely used in power system applications.

Regarding convex relaxations, efficacious convex relaxation approaches in existing studies for DNR are effective for fast solving. It mainly adopted the Big-M method [7], or McCormick linearization method [8, 9] to slack power flow constraints by switching status indicators, or constructing relaxation constraints with variables automatically satisfied for a disconnected line [10]. However, these convex relaxation methods suffer from loosened relaxation bounds, which render a large amount of computational time. Moreover, a DNR problem is cast as a combinatorial explosion in the number of total possible solutions as the number of branches increases, if the open/closed status of each branch is regarded as a binary variable. For instance, the number of total possible solutions is 2136 for a distribution network that has 136 circuit breakers. This induces that a DNR problem becomes high-dimensional, thus spending significant amounts of computational time [11]. To reduce the computational time caused by high-dimension binary variables, it is essential to exploit the DNR model characterization that can be used to accelerate this entire computation. On top of that, with a high penetration of DERs into distribution networks, the fluctuated power injections intricately trigger a complicated DNR decision-making process due to bi-directional power flows [12, 13]. This suggests that today's active distribution networks increase the difficulty level of quick solvability. In the light of loosened relaxation bounds, high-dimensional space of binary variables, and integrations of enrichable DERs, there is a research gap to explore the efficient and tight relaxation approach for DNR problems in large-scale active distribution networks.

Based on the convex hull of *DistFlow* equations [14], a disjunctive convex hull relaxation (DCHR) [15] can be further constructed to deal with these on/off constraints caused by switch status indicators. This chapter studies the DCHR approach for the classic DNR problem, which can be designed to tighten the voltage drop equation and quadratic equality of *DistFlow* equations with continuous parent–child relationship variables in spanning tree constraints.

In this chapter, we lay the theoretical foundation of *DistFlow* equations and its convex relaxation formulations. Initially, *DistFlow* equations can be derived from the branch flow model, and then linearized *DistFlow* equations can be obtained if the non-convex terms are negligible. Subsequently, the SOCO programming and semi-definite programming (SDP) form of *DistFlow* equations are formulated according to SOCO programming and SDP convex relaxation techniques. Additionally, the polyhedral approximation formulation is included for linearizing the SOC constraints. Subsequently, based on the theoretical foundation of *DistFlow* equations, we introduce the DCHR approach, which can be further used to deal with these on/off constraints caused by switch status indicators. This DCHR method can be designed to tighten the voltage drop equation and quadratic equality of *DistFlow* equations with continuous parent–child relationship variables in spanning tree constraints.

2 Branch Flow Model

The steady-state network power flow status can be described by branch flow model (BFM) of [3]. A arbitrary branch $l: = (m, n)$ in single-phase radial distribution networks (i.e., networks with a tree topology) is illustrated in Fig. 1. For this branch l, P_n^d and Q_n^d are fixed active and reactive power demands with $\tilde{S}_n^d = P_n^d + jQ_n^d$; z_{mn}^l refer to the impedance with $(z_{mn}^l)^2 = (r_{mn}^l)^2 + (x_{mn}^l)^2$; P_n^g and Q_n^g are active and reactive power generation at node n with $\tilde{S}_n^g = P_n^g + jQ_n^g$, respectively; P_{mn}^l and Q_{mn}^l are active and reactive power flows with $\tilde{S}_{mn}^l = P_{mn}^l + jQ_{mn}^l$, respectively.

In practice, we have one assumption that the shunt elements in distribution networks are assumed zero, namely $b_{mn}/2 = 0$. This assumption is reasonable for realistic distribution networks due to short distances of branches. Under this assumption, we can express the apparent power flow for this branch l as $\tilde{S}_{mn}^l = \dot{U}_m (I_{mn}^l)^*$, where $\dot{U}_m$ refers to the voltage phasor at node m and $(I_{mn}^l)^*$ refers to the conjugate of

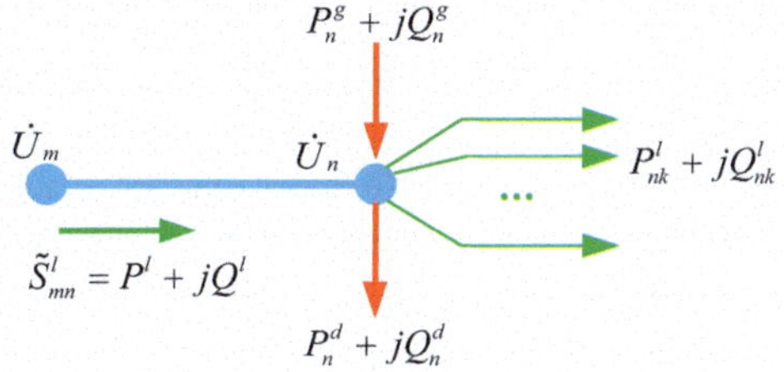

Fig. 1 Typical connection of a branch in distribution networks

current phasor between nodes m and n. However, if this branch is modeled by a series admittance y_{mn} with shunt elements, i.e., $b_{mn}/2 \neq 0$, then a shunt admittance $b_{mn}/2$ is included on each end of line (m, n) in the π model. Accordingly, the apparent power flow should be rearranged as $\tilde{S}^l_{mn} - j\frac{b_{mn}U_m^2}{2} = \dot{U}_m(I^l_{mn})^*$. Moreover, the voltage drop equation can be written as $\dot{U}_n = \dot{U}_m - z^l_{mn} \cdot I^l_{mn}$, where z^l_{mn} denotes the impedance of branch l. Therefore, when formulating distribution power flow equations in phasor form, we can achieve the BFM-based power flow equality, yielding

$$\begin{cases} \dot{U}_n = \dot{U}_m - z^l_{mn} \cdot I^l_{mn} \\ \tilde{S}^l_{mn} = \dot{U}_m(I^l_{mn})^* \\ \tilde{S}^l_{mn} + \tilde{S}^g_n - z^l_{mn}|I^l_{mn}|^2 = \sum\limits_{k \in \pi(n)} \tilde{S}^l_{nk} + \tilde{S}^d_n \end{cases}, \quad \forall l \in \mathrm{E}, \forall m, n \in \mathcal{N} \tag{1}$$

where $\pi(n)$ refers to the set of branches that connect the node n.

As observed in (1), distribution power flow equations are non-convex due to non-convex terms $\dot{U}_m(I^l_{mn})^*$. And $\tilde{S}^l_{mn} + S^g_n - z^l_{mn}|I^l_{mn}|^2 = \sum\limits_{k \in \pi(n)} \tilde{S}^l_{nk} + S^d_n$ is a nonlinear equation caused by $|I^l_{mn}|^2$. In other words, BFM-based power flow equality is not suitable for tractable computation, and thus we need to reformulate this BFM-based power flow equality.

3 *DistFlow* Equations

Due to this voltage drop equation in phasor form, we solve this equation with squares on both sides as $|\underset{n}{U}|^2 = |\underset{m}{U} - z^l_{mn} \cdot I^l_{mn}|^2$. After multiplying both sides by complex conjugate, the right-hand side of this voltage drop equation can be rearranged as

$$\begin{aligned} |\dot{U}_m - z^l_{mn} \cdot I^l_{mn}|^2 &= (\dot{U}_m - z^l_{mn} \cdot I^l_{mn})(\dot{U}_m - z^l_{mn} \cdot I^l_{mn})^* \\ &= |U_m|^2 + |z^l_{mn} \cdot I^l_{mn}|^2 - 2\mathrm{Re}(\dot{U}_m(z^l_{mn} \cdot I^l_{mn})^*) \\ &= |U_m|^2 + |z^l_{mn}|^2 \cdot |I^l_{mn}|^2 - 2\mathrm{Re}(\dot{U}_m \cdot z^{l*}_{mn} \cdot I^{l*}_{mn}) \\ &= |U_m|^2 + |z^l_{mn}|^2 \cdot |I^l_{mn}|^2 - 2\mathrm{Re}(z^{l*}_{mn} \cdot \tilde{S}^l_{mn}) \\ &= |U_m|^2 + |z^l_{mn}|^2 \cdot |I^l_{mn}|^2 - 2r^l_{mn} \cdot P^l_{mn} - 2x^l_{mn} \cdot Q^l_{mn} \end{aligned} \tag{2}$$

where r^l_{mn} and x^l_{mn} indicate the resistance and reactance of branch l. $|U_m|$ and $|I^l_{mn}|$ denote the voltage magnitude of voltage phasor $\dot{U}_m$ and current modulus of current phasor I^l_{mn}, respectively. Note that for any vector x, Re(x) refers to the real part of x.

It should be noted that (1) is actually a phase angle free equation, since all variables $|U_m|$, $|I^l_{mn}|$, P^l_{mn} and Q^l_{mn} are real-valued numbers. In mathematics, we call this equation as the angle relaxation. Indeed, we can consider the real-valued ($|U_m|$, $|I^l_{mn}|$, P^l_{mn}, Q^l_{mn})-space as a projection of complex-valued ($\dot{U}_m$, I^l_{mn}, $\tilde{S}^l_{mn}$)-space. This

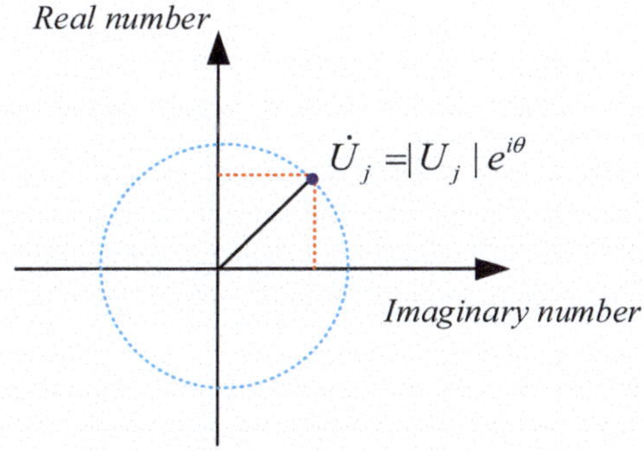

Fig. 2 Illustration of angle relaxation for (2)

implies that each variable $|I^l_{mn}|$ or $|U_m|$ is relaxed from a point in the complex plane to a circle with a radius equal to the distance of the point from the origin. The intuitive explanation for this angle relaxation can be taken $\dot{U}_m = |U_m|e^{i\theta}$ as shown in Fig. 2. Suppose that $\dot{U}_m$ on the complex coordinate space is a point. Then, a circle with a radius equal to $|U_m|$ is the set of solutions for any real-valued variable $|U_m|$. By $|\dot{U}_m|^2 = |U_m|^2$, a point is clearly relaxed to a circle.

Similarly, we take the modulus with squares on both sides of $\tilde{S}^l_{mn} = \dot{U}_m(I^l_{mn})^*$, yielding

$$|\dot{U}_m(I^l_{mn})^*|^2 = |\tilde{S}^l_{mn}|^2 \quad \Leftrightarrow \quad |U_m|^2 \cdot |I^l_{mn}|^2 = (P^l_{mn})^2 + (Q^l_{mn})^2 \tag{3}$$

As mentioned above, if a shunt admittance $b_{mn}/2$ is included, then $|\tilde{S}^l_{mn} - j\frac{b_{mn}U^2_m}{2}|^2 = |\dot{U}_m(I^l_{mn})^*|^2$ is a complicated quadratic equation that cannot be converted to this SOC form. In summary, it can be seen that (1) and (2) are two angle free equations after applying angle relaxation. According to the reference direction of power flow shown in Fig. 1, the left-hand (right-hand) side gathers total active and reactive power injected (withdrawn) in (from) node n. We can express the corresponding ***DistFlow*** equations with respect to unknown variables ($|U_m|$, $|I^l_{mn}|$, P^l_{mn}, Q^l_{mn}) based on nodal power flow balance equations (1, 2). In this chapter, the radial distribution networks are considered as a connected undirected tree $\mathcal{G}\ominus = (\mathcal{N}, \mathcal{E})$, where $\mathcal{N}\ominus = [1, 2, ..., |\mathcal{N}\ominus|]$ is the set of nodes and $\mathcal{E}$ is the set of branches in the dimension $|\mathcal{E}| = (|\mathcal{N}| - 1) \times 1$. The branch $l := (m, n)$, $l \in \mathcal{E}$ is between nodes (m, n) where $m, n \in \mathcal{N}$ and we assume that the point of common connection (PCC) node is 0. Therefore, *DistFlow* equations [3] are expressed as

$$P^l_{mn} + P^g_n - r^l_{mn}\ell^l_{mn} = \sum_{k\in\pi(n)} P^l_{nk} + P^d_n \quad \forall l \in \mathrm{E}, \forall m, n \in \mathcal{N} \tag{4}$$

$$Q^l_{mn} + Q^g_n - x^l_{mn}\ell^l_{mn} = \sum_{k\in\pi(n)} Q^l_{nk} + Q^d_n \quad \forall l \in \mathcal{E}, \forall m, n \in \mathcal{N} \tag{5}$$

$$v_n = v_m - 2(r^l_{mn}P^l_{mn} + x^l_{mn}Q^l_{mn}) + |z^l_{mn}|^2\ell^l_{mn} \quad \forall l \in \mathcal{E}, \forall m, n \in \mathcal{N} \tag{6}$$

$$\ell_{mn}^{l} \cdot v_m = (P_{mn}^{l})^2 + (Q_{mn}^{l})^2 \quad \forall l \in \mathcal{E}, \forall m, n \in \mathcal{N} \tag{7}$$

where $\ell_{mn}^{l} = |I_{mn}^{l}|^2$, $v_m = |U_m|^2$ and $v_n = |U_n|^2$.

In this set of ***DistFlow*** equations, (4, 5) are nodal active power and reactive power balancing conditions. (6) describes forward voltage drop on each branch obtained by (2) and (7) is derived from (3) which defines apparent power flow injection at the head node m of each branch. The squared voltage magnitude $v_0 = |U_0|^2$ at the PCC is a constant. In this regard, we can define a vector of variables as $\boldsymbol{x}$: $= (\mathbf{v}, \ell^{l}, \mathbf{P}^{\mathbf{l}}, \mathbf{Q}^{\mathbf{l}})$ for $\forall l \in \mathcal{E}$, the key feature of which is that $\boldsymbol{x}$ does not involve angles of voltage and current phasors. Most importantly, (4–7) in BFM are linear, and non-convexity only appears in branch flow equality (7). This BFM-based power flow equation is much different from bus injection model in which power balancing conditions render non-convex quadratic equalities.

It should be also noted that

- **Uniqueness of Solutions for *DistFlow* Equations**

There are $2|\mathcal{E}| + 2|\mathcal{E}| = 4|\mathcal{N}| - 4$ equations in $|\mathcal{N}| - 1 + 3|\mathcal{E}| = 4|\mathcal{N}| - 4$ real variables. It is clear that the number of variables is equal to the number of equations, which means that this ***DistFlow*** equations would have a unique solution or multiple solution. It is shown in [16, 17] that this set of equations have generally multiple solutions, but for radial distribution networks with $|U| \approx 1$ p.u. and small r_{mn}^{l} and x_{mn}^{l}, the solution of (4–7) is unique and same to the one produced by the traditional bus injection model in rectangle coordinate. It should be noted that for a connected mesh network ($|\mathcal{E}| = |\mathcal{N}|$), this set of (4–7) equations have $2|\mathcal{E}| + 2|\mathcal{E}| = 4|\mathcal{N}|$ equations in which the number of variables is $|\mathcal{N}| - 1 + 3|\mathcal{E}| = 4|\mathcal{N}| - 1$. Therefore, the solution is generally nonunique. Some of these solutions may be spurious, i.e., they do not correspond to a solution of the original branch flow equations.

- **Application to General Networks (Loop Networks)**

The ***DistFlow*** equations can be extended to be used for general networks that may contain cycles by introducing a cycle condition. We define the angle difference across for branch l as

$$\beta_{mn}(x) := \angle(\dot{U}_m - z_{mn}^{l} \cdot I_{mn}^{l}) \tag{8}$$

For convenience, we let $\beta(x) := (\beta_{mn}(x),\ l := (m, n),\ \forall l \in \mathcal{E})$. Thus, the ***DistFlow*** equations are extended to general networks as

$$(2.4) - (2.7),\ \exists \theta \in \mathbb{R},\ s.t.\ \beta(x) = A \cdot \theta \tag{9}$$

where A is a $|\mathcal{E}|$ by $|\mathcal{N}|$ branch-node incidence matrix in tree graph $\mathcal{G}\ominus$ with $A_{ml} =$ 1 if $l = m \rightarrow n$ for some n, $A_{ml} =$ -1 if $l = n \rightarrow m$ for some m, and 0 otherwise. This branch-node incidence matrix A will be discussed later. We refer to the condition

$\beta(x) = A \cdot \theta$ on x in (9) as the cycle condition, which can be enforced by introducing θ as additional variables. For general networks (4–7) can thus be interpreted as a relaxation of (9) where the cycle condition is ignored. When a network is radial, the cycle condition is vacuous and (9) reduces to (4–7).

- **Compact Matrix–Vector Form**

For brevity, l refers to a vector of branches if sending end nodes m, n are omitted; otherwise, l denotes an arbitrary branch $l: = (m, n)$, $l \in \mathcal{E}$ is between nodes (m, n) where $m, n \in \mathcal{N}$. This chapter assumes that the root node is 0. v_m and v_n are the squared voltage profiles at node m and n. It should be noted that $\boldsymbol{A}^l \cdot v = v_m - v_n$ holds where $\boldsymbol{A}$ is a $|\mathcal{E}|$ by $|\mathcal{N}|$ branch-node incidence matrix in tree graph $\mathcal{G}$ and $\boldsymbol{A}^l$ refers to the l-th row. Expressing with $A \cdot v$ is generally formulated in the matrix–vector form. To avoid heavy notations, the following steady-state network power flow equality (10–13) can be derived by the real-valued ***DistFlow*** equations for $\forall l \in \mathcal{E}$ in the compact matrix vector form.

$$-\boldsymbol{P}^g + \boldsymbol{P}^d = \boldsymbol{A}^T \boldsymbol{P}^l - \boldsymbol{D}_r \ell^l \tag{10}$$

$$-\boldsymbol{Q}^g + \boldsymbol{Q}^d = \boldsymbol{A}^T \boldsymbol{Q}^l - \boldsymbol{D}_x \ell^l \tag{11}$$

$$\boldsymbol{A}v - 2\boldsymbol{D}_r \boldsymbol{P}^l - 2\boldsymbol{D}_x \boldsymbol{Q}^l + \boldsymbol{D}_z \ell^l = 0 \tag{12}$$

$$\ell^l \cdot \boldsymbol{D}_v = |\boldsymbol{P}^l|^2 + |\boldsymbol{Q}^l|^2 \tag{13}$$

where P^l and Q^l refer to the vectors of sending-end active and reactive power flows with the moduli equal to $|P^l|$ and $|Q^l|$. P^g, Q^g and P^d, Q^d indicate the vectors of given nodal active and reactive power injections and active and reactive loads at nodes. Q^{cr} is the vector of nodal reactive power compensation. ℓ^l is the vector of squared current on branches. $\boldsymbol{D}_r$ and $\boldsymbol{D}_x$ indicate the diagonal matrices whose diagonal elements are the resistance vector and the reactance vector, respectively. $\boldsymbol{D}_z$ is the diagonal matrix whose diagonal elements are $|z^l| = |\boldsymbol{r}^l|^2 + |\boldsymbol{x}^l|^2$. $\boldsymbol{D}_v$ is the diagonal matrix whose diagonal elements are squared voltage profiles of the sending nodes for all branches.

Let us illustrate more about branch-node incidence matrix $\boldsymbol{A}$. For instance, the topology of a radial distribution network is displayed in Fig. 3, where $|\mathcal{N}| = 10$, $|\mathcal{E}| = 9$ and $|\mathcal{N}| = |\mathcal{E}|+1$. The line feeding node n is indexed as line n-th, e.g., the second branch is brn#2 where the ending node is 2.

For (12) and (13), we present the branch-node incidence matrix A as

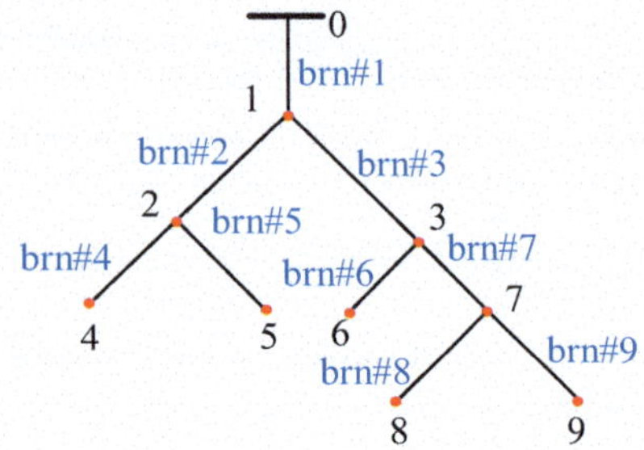

Fig. 3 Topology of a radial distribution network

$$
\begin{array}{r} \\ \\ \\ \\ \mathbf{A} = \end{array}
\begin{array}{r} node: \\ brn\#1: \\ brn\#2: \\ brn\#3: \\ brn\#4: \\ brn\#5: \\ brn\#6: \\ brn\#7: \\ brn\#8: \\ brn\#9: \end{array}
\begin{array}{cccccccccc} 0 & 1 & 2 & 3 & 4 & 5 & 6 & 7 & 8 & 9 \\
1 & -1 & & & & & & & & \\
 & 1 & -1 & & & & & & & \\
 & 1 & & -1 & & & & & & \\
 & & 1 & & -1 & & & & & \\
 & & 1 & & & -1 & & & & \\
 & & & 1 & & & -1 & & & \\
 & & & 1 & & & & -1 & & \\
 & & & & & & & 1 & -1 & \\
 & & & & & & & 1 & & -1 \end{array} \tag{14}
$$

As for (12), we substitute the given voltage magnitude of PCC at node 0 for the first brn #1. We can further simplify A as below

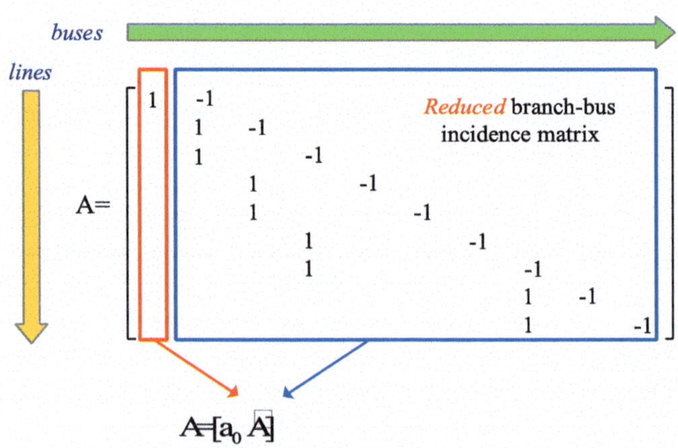

where $\tilde{A}$ refers to the reduced branch-node incidence matrix. Thus, (13) can be further rearranged as

$$-\tilde{A}v - 2D_rP^l - 2D_xQ^l + D_z\ell^l = a_0v_0 \tag{15}$$

According to Ohm's law, when each node is injected by the current in one unit, we can obtain

$$\boldsymbol{A} \cdot 1_{|\mathcal{E}|+1} = 0 \Rightarrow a_0 + \tilde{\boldsymbol{A}} \cdot 1_{|\mathcal{E}|} = 0 \Rightarrow 1_{|\mathcal{E}|} = -a_0 \cdot \tilde{\boldsymbol{A}}^{-1}$$

where $-a_0 \cdot \tilde{\mathbf{A}}^{-1}$ can be defined as the branch-branch incidence matrix and a_0 is a scalar. Given this branch-branch incidence matrix, we can obtain the current for any branches by the measured current from the PCC. For example, the branch-branch incidence matrix $-a_0 \cdot \tilde{\mathbf{A}}^{-1}$ can be equal to

$$\begin{array}{r} brn: \quad 1 \quad 2 \quad 3 \quad 4 \quad 5 \quad 6 \quad 7 \quad 8 \quad 9 \\ \begin{array}{r} brn\#1: \\ brn\#2: \\ brn\#3: \\ brn\#4: \\ brn\#5: \\ brn\#6: \\ brn\#7: \\ brn\#8: \\ brn\#9: \end{array} \begin{bmatrix} 1 & & & & & & & & \\ 1 & 1 & & & & & & & \\ 1 & & 1 & & & & & & \\ 1 & 1 & & 1 & & & & & \\ 1 & 1 & & & 1 & & & & \\ 1 & & 1 & & & 1 & & & \\ 1 & & 1 & & & & 1 & & \\ 1 & & 1 & & & & 1 & 1 & \\ 1 & & 1 & & & & 1 & & 1 \end{bmatrix} \end{array} \tag{16}$$

where $a_0 = 1$.

4 Linearized *DistFlow* Equations

The compact matrix–vector form of ***DistFlow*** equations can be simplified if dropping terms related to losses and quadratic equations are negligible. We can express the real-valued load distribution formula (LDF) equations for $\forall l \in \mathcal{E}$ in the compact matrix–vector notation using the LDF method

$$-\boldsymbol{P}^g + \boldsymbol{P}^d = \boldsymbol{A}^T \tilde{\boldsymbol{P}}^l \tag{17}$$

$$-\boldsymbol{Q}^g + \boldsymbol{Q}^d = \boldsymbol{A}^T \tilde{\boldsymbol{Q}}^l \tag{18}$$

$$-\boldsymbol{A}\tilde{\boldsymbol{v}} - 2\boldsymbol{D}_r \tilde{\boldsymbol{P}}^l - 2\boldsymbol{D}_x \tilde{\boldsymbol{Q}}^l = 0 \tag{19}$$

where $\tilde{P}^l$ and $\tilde{Q}^l$ refer to the vector of sending-end active and reactive power flows; $\tilde{v}$ denotes the vector of squared voltage profiles.

Equivalently, we rearrange this set of LDF equations as

$$\tilde{\boldsymbol{P}}^{l} = \boldsymbol{A}(-\boldsymbol{P}^{g} + \boldsymbol{P}^{d}) \tag{20}$$

$$\tilde{\boldsymbol{Q}}^{l} = \boldsymbol{A}(-\boldsymbol{Q}^{g} + \boldsymbol{Q}^{d}) \tag{21}$$

$$-\boldsymbol{A}\tilde{\boldsymbol{v}} - 2\boldsymbol{D}_{r}\boldsymbol{A}(-\boldsymbol{P}^{g} + \boldsymbol{P}^{d}) - 2\boldsymbol{D}_{x}\boldsymbol{A}(-\boldsymbol{Q}^{g} + \boldsymbol{Q}^{d}) = 0 \tag{22}$$

It should also be noted that there are several characteristics for this set of LDF equations

(i) All equations are linear and are only with respect to a vector of variables $\tilde{\boldsymbol{x}} := (\tilde{\boldsymbol{v}}, \tilde{\boldsymbol{P}}^{l}, \tilde{\boldsymbol{Q}}^{l})$, which is smaller than the size of $\boldsymbol{x}$ in ***DistFlow*** equations.
(ii) Voltage drop and line power flows are approximately linearly related to power injections.
(iii) LDF gives an over-estimator for squared voltage magnitudes.

Proof According to (22), we can formulate

$$-\tilde{\boldsymbol{A}}\boldsymbol{v} - 2\boldsymbol{D}_{r}\boldsymbol{P}^{l} - 2\boldsymbol{D}_{x}\boldsymbol{Q}^{l} + \boldsymbol{D}_{z}\ell^{l} = a_{0}v_{0} \tag{23}$$

Rearranging (23) as

$$\boldsymbol{v} = -(\tilde{\boldsymbol{A}})^{-1} \cdot [a_{0}v_{0} + 2\boldsymbol{D}_{r}\boldsymbol{P}^{l} + 2\boldsymbol{D}_{x}\boldsymbol{Q}^{l}] + (\tilde{\boldsymbol{A}})^{-1}\boldsymbol{D}_{z}\ell^{l} \tag{24}$$

Then, due to $(\tilde{\boldsymbol{A}})^{-1}\boldsymbol{D}_{z}\ell^{l} < 0$ and $-(\tilde{\boldsymbol{A}})^{-1} \cdot [a_{0}v_{0} + 2\boldsymbol{D}_{r}\boldsymbol{P}^{l} + 2\boldsymbol{D}_{x}\boldsymbol{Q}^{l}] > 0$, we can obtain this inequality as

$$\boldsymbol{v} < -(\tilde{\boldsymbol{A}})^{-1} \cdot [a_{0}v_{0} + 2\boldsymbol{D}_{r}\boldsymbol{P}^{l} + 2\boldsymbol{D}_{x}\boldsymbol{Q}^{l}] \tag{25}$$

The right-hand side of (25) is equal to $\tilde{\boldsymbol{v}} = -(\tilde{\boldsymbol{A}})^{-1} \cdot [a_{0}v_{0} + 2\boldsymbol{D}_{r}\boldsymbol{P}^{l} + 2\boldsymbol{D}_{x}\boldsymbol{Q}^{l}]$, which can be achieved by LDF. Thus, we prove that $\tilde{\boldsymbol{v}} > \boldsymbol{v}$.

(iv) LDF gives an under-estimator for line flows.

Proof According to (22), we can directly derive the following inequality

$$\boldsymbol{P}^{l} = \boldsymbol{A}(-\boldsymbol{P}^{g} + \boldsymbol{P}^{d} + \boldsymbol{D}_{r}\ell^{l}) > \boldsymbol{A}(-\boldsymbol{P}^{g} + \boldsymbol{P}^{d}) = \tilde{\boldsymbol{P}}^{l} \tag{26}$$

$$\boldsymbol{Q}^{l} = \boldsymbol{A}(-\boldsymbol{Q}^{g} + \boldsymbol{Q}^{d} + \boldsymbol{D}_{x}\ell^{l}) > \boldsymbol{A}(-\boldsymbol{Q}^{g} + \boldsymbol{Q}^{d}) = \tilde{\boldsymbol{Q}}^{l} \tag{27}$$

Thus, we prove that $\boldsymbol{P}^{l} > \tilde{\boldsymbol{P}}^{l}$ and $\boldsymbol{Q}^{l} > \tilde{\boldsymbol{Q}}^{l}$

(v) Approximation accuracy by LDF depends on loading conditions. The approximation error of this LDF can be acceptable if

- Voltage magnitudes close to unity, namely $|v| = 1 + \varepsilon$ with $|\varepsilon| \simeq 0$;
- Voltage angle differences across lines close to zero, i.e., $\theta_{nm} = \theta_n - \theta_m \simeq 0$;
- Ignoring line resistances and shunt elements.

In industrial applications, distribution networks generally have relatively flat voltage profiles maintained by automatic voltage control system. Due to flat voltage profiles, the small approximation error of LDF can only be about $\pm$ 1% [18]. Note that the squared voltage profile $|v_i|$ can be approximated from first-order Taylor series expansion around the normal voltage profile $|U_0| = 1$.

5 Convex Relaxation Formulation of *DistFlow* Equations

5.1 *Second-Order Conic Programming Formulation*

For the sake of realizing minimum power losses, we discuss the reactive power optimization problem for the fixed typology of distribution networks. The two convex relaxations of ***DistFlow*** equations are discussed in the section. Before incorporating convex relaxations, we at first discuss the conventional reactive power optimization model for distribution networks:

$$
\begin{aligned}
&\min_{\tilde{S}^l, I^l, \dot{U}, Q^{cr} \in \mathbb{C}} \quad \mathrm{Re}(\tilde{S}_0^l) \\
&s.t. \begin{cases}
\dot{U}_n = \dot{U}_m - z_{mn}^l \cdot I_{mn}^l, \quad \forall l \in \mathcal{E},\ \forall m, n \in \mathcal{N} \\
\tilde{S}_{mn}^l = \dot{U}_m (I_{mn}^l)^*, \quad \forall l \in \mathcal{E},\ \forall m, n \in \mathcal{N} \\
\tilde{S}_{mn}^l + \tilde{S}_n^g - z_{mn}^l |I_{mn}^l|^2 = \sum\limits_{k \in \pi(n)} \tilde{S}_{nk}^l + \tilde{S}_n^d, \quad \forall l \in \mathcal{E},\ \forall m, n \in \mathcal{N} \\
\underline{Q}_{inj}^l \le \mathrm{Im}(\tilde{S}_{inj}^l) \le \overline{Q}_{inj}^l \\
\underline{Q}^{cr} \le Q^{cr} \le \overline{Q}^{cr} \\
\underline{U} \le |\dot{U}| \le \overline{U}
\end{cases}
\end{aligned} \tag{28}
$$

where $\tilde{S}_0^l$ refers to the complex power on the first branch between the PCC and a PQ bus; $\mathrm{Re}(\tilde{S}_0^l)$ and $\mathrm{Im}(\tilde{S}_0^l)$ refer to the real part and imaginary part of $\tilde{S}_0^l$, respectively. Other symbols are illustrated in Sect. 1.

Recall that the conventional reactive power optimization model (28) is non-convex, we can further modify this conventional model based on *DistFlow* equations for the preparation of convex relaxations:

$$\min_{\mathbf{x^l} \in \mathbb{R}} \quad P_0^l$$

$$s.t. \begin{cases} P_{mn}^l + P_n^g - r_{mn}^l \ell_{mn}^l = \sum_{k \in \pi(n)} P_{nk}^l + P_n^d, \ \forall l \in \mathcal{E}, \forall m, n \in \mathcal{N} \\ Q_{mn}^l + Q_n^g - x_{mn}^l \ell_{mn}^l = \sum_{k \in \pi(n)} Q_{nk}^l + Q_n^d, \ \forall l \in \mathcal{E}, \forall m, n \in \mathcal{N} \\ v_n = v_m - 2(r_{mn}^l P_{mn}^l + x_{mn}^l Q_{mn}^l) + |z_{mn}^l|^2 \ell_{mn}^l, \ \forall l \in \mathcal{E}, \forall m, n \in \mathcal{N} \\ \ell_{mn}^l \cdot v_m = (P_{mn}^l)^2 + (Q_{mn}^l)^2, \forall l \in \mathcal{E}, \ \forall m, n \in \mathcal{N} \\ \underline{Q}_{inj}^l \le Q_{inj}^l \le \overline{Q}_{inj}^l \\ \underline{Q}^{cr} \le Q^{cr} \le \overline{Q}^{cr} \\ \underline{v} \le v \le \overline{v} \end{cases} \tag{29}$$

where P_0^l refers to the active power injection of the first branch between the PCC and a PQ bus.

For (29), all constraints are linear only except for $\ell_{mn}^l \cdot v_m = (P_{mn}^l)^2 + (Q_{mn}^l)^2$. Here, we equivalently slack this quadratic equality to a pair of two inequalities below

$$\ell_{mn}^l \cdot v_m \ge (P_{mn}^l)^2 + (Q_{mn}^l)^2 \quad \forall l \in \mathcal{E}, \forall m, n \in \mathcal{N} \tag{30}$$

$$\ell_{mn}^l \cdot v_m \le (P_{mn}^l)^2 + (Q_{mn}^l)^2 \quad \forall l \in \mathcal{E}, \forall m, n \in \mathcal{N} \tag{31}$$

After dropping (31), we can relax the non-convex quadratic equalities as an inequality below

$$\ell_{mn}^l \cdot v_m \ge (P_{mn}^l)^2 + (Q_{mn}^l)^2 \quad \forall l \in \mathcal{E}, \forall m, n \in \mathcal{N} \tag{32}$$

We can arrange this inequality by multiplying 4 on the both sides of (32)

$$(2P_{mn}^l)^2 + (2Q_{mn}^l)^2 - 2\ell_{mn}^l \cdot v_m \le 2\ell_{mn}^l \cdot v_m \quad \forall l \in \mathcal{E}, \forall m, n \in \mathcal{N} \tag{33}$$

After adding $(\ell_{mn}^l)^2 + (v_m)^2$ on both sides of (33), (33) can be rearranged as

$$(2P_{mn}^l)^2 + (2Q_{mn}^l)^2 + (\ell_{mn}^l - v_m)^2 \le (\ell_{mn}^l + v_m)^2 \quad \forall l \in \mathcal{E}, \forall m, n \in \mathcal{N} \tag{34}$$

Therefore, this inequality (34) is converted to a standard SOC form by

$$\left\| \begin{matrix} 2P_{mn}^l \\ 2Q_{mn}^l \\ \ell_{mn}^l - v_m \end{matrix} \right\|_2 \le (\ell_{mn}^l + v_m) \quad \forall l \in \mathcal{E}, \forall m, n \in \mathcal{N} \tag{35}$$

Note that (32) also refers to a rotated SOC inequality. The standard SOCO programming-based formulation of *DistFlow* equations is

$$P^l_{mn} + P^g_n - r^l_{mn}\ell^l_{mn} = \sum_{k\in\pi(n)} P^l_{nk} + P^d_n \quad \forall l \in \mathcal{E}, \forall m, n \in \mathcal{N} \tag{36}$$

$$Q^l_{mn} + Q^g_n - x^l_{mn}\ell^l_{mn} = \sum_{k\in\pi(n)} Q^l_{nk} + Q^d_n \quad \forall l \in \mathcal{E}, \forall m, n \in \mathcal{N} \tag{37}$$

$$v_n = v_m - 2(r^l_{mn}P^l_{mn} + x^l_{mn}Q^l_{mn}) + |z^l_{mn}|^2\ell^l_{mn} \quad \forall l \in \mathcal{E}, \forall m, n \in \mathcal{N} \tag{38}$$

$$\left\| \begin{matrix} 2P^l_{mn} \\ 2Q^l_{mn} \\ \ell^l_{mn} - v_m \end{matrix} \right\|_2 \le (\ell^l_{mn} + v_m) \quad \forall l \in \mathcal{E}, \forall m, n \in \mathcal{N} \tag{39}$$

According to SOCO programming-based *DistFlow* equations, we define the vector of optimization variables $\boldsymbol{x^l}$ for the minimization of real power loss. This is also called reactive power optimization problem for the fixed typology of distribution networks. The set of optimization variables involves a set of state variables $\left[P^l,\ Q^l,\ \ell^l,\ v_{PQ}\right]^T$ and a vector of controllable variables $\left[v_{PCC},\ Q^{cr}\right]^T$. Here, v_{PQ} refers to squared voltage profiles at PQ nodes and v_{PCC} is the adjustable squared voltage profile at PCC node, and Q^{cr} is a vector of the reactive power compensation sources. In addition, as observed in (39), two auxiliary variables can be incorporated to represent:

$$\begin{cases} w^l = \ell^l_{mn} - v_m \\ m^l = \ell^l_{mn} + v_m \end{cases} \tag{40}$$

Consequently, the vector of optimization variables $\boldsymbol{x^l}$ can be defined as

$$\mathbf{x^l} := [P^l,\ Q^l,\ \ell^l,\ v,\ Q^{cr},\ w^l,\ m^l]^T \tag{41}$$

The constraints are composed of *Distflow* equations and reactive power injection constraint of the first branch between the PCC and a PQ bus, reactive power capacity compensation constraints for all reactive power compensation sources as well as voltage security constraints for all nodes.

$$\underline{Q}^{cr} \le Q^{cr} \le \overline{Q}^{cr} \tag{42}$$

$$\underline{v} \le v \le \overline{v} \tag{43}$$

where $\underline{Q}^{cr}$ and $\overline{Q}^{cr}$ indicate the lower and upper limits of reactive power compensation sources. $\underline{v}$ and $\overline{v}$ denote the lower and upper squared voltage profile limits of PQ nodes.

The objective function of this minimization of real power loss can be converted to the minimal real power injection at PCC node. Thus, the reactive power optimization model is formulated as

$$\begin{aligned} &\min_{\mathbf{x}^l \in \mathbb{R}} \quad P_0^l \\ &s.t. \begin{cases} P_{mn}^l + P_n^g - r_{mn}^l \ell_{mn}^l = \sum\limits_{k \in \pi(n)} P_{nk}^l + P_n^d, \ \forall l \in \mathcal{E}, \forall m, n \in \mathcal{N} \\ Q_{mn}^l + Q_n^g - x_{mn}^l \ell_{mn}^l = \sum\limits_{k \in \pi(n)} Q_{nk}^l + Q_n^d, \ \forall l \in \mathcal{E}, \forall m, n \in \mathcal{N} \\ v_n = v_m - 2(r_{mn}^l P_{mn}^l + x_{mn}^l Q_{mn}^l) + |z_{mn}^l|^2 \ell_{mn}^l, \ \forall l \in \mathcal{E}, \forall m, n \in \mathcal{N} \\ \left\| \begin{matrix} 2P_{mn}^l \\ 2Q_{mn}^l \\ w^l \end{matrix} \right\|_2 \leq m^l, \forall l \in \mathcal{E}, \ \forall m, n \in \mathcal{N} \\ w^l = \ell_{mn}^l - v_m, \ \ m^l = \ell_{mn}^l + v_m, \quad \forall l \in \mathcal{E}, \ \forall m, n \in \mathcal{N} \\ \underline{Q}^{cr} \leq Q^{cr} \leq \overline{Q}^{cr}, \ \underline{v} \leq v \leq \overline{v} \end{cases} \end{aligned} \tag{44}$$

where P_0^l is the active power injection at the root node 0.

Exactness of SOCO Programming Convex Relaxation

However, is the optimal solution of (44) the same as the one solved by (1)? If and only if the convex relaxation techniques are exact, the obtained solution is the same as the optimal solution of original non-convex nonlinear optimization. In this section, we discuss whether SOCO programming relaxations for reactive power optimization problems are exact or not. Looking from case studies in previous subsections, it seems that all optimal solutions are satisfied with SOC equality. This suggests that the SOC relaxation is exact. Now, let us give the exactness of SOCO programming convex relaxation and its associated proof by contradiction [16, 17].

Theorem The SOCO programming-based reactive power optimization formulation (44) is convex. Moreover, it is exact, i.e., an optimal solution of (44) is also optimal for the original reactive power optimization problem (1).

Proof For this SOCO programming-based model (44), we rewrite the objective function min P_{inj}^l as min $\sum\limits_{l \in \mathcal{E}} r_{mn}^l \ell_{mn}^l$. To prove that the relaxation is exact, it suffices to show that any optimal solution of (1) has equality in $\ell_{mn}^l \cdot v_m = (P_{mn}^l)^2 + (Q_{mn}^l)^2$. Assume for the sake of contradiction that $x^{l\star} := (P_{ij}^{l\star}, Q_{ij}^{l n\star}, \ell^{l\star}, v^\star, Q^{cr\star})$ is optimal but has strict inequality, i.e.,

$$\ell_{mn}^{l\star} \cdot v_m^{\star} > (P_{mn}^{l\star})^2 + (Q_{mn}^{l\star})^2, \ \forall l \in \mathcal{E} \tag{45}$$

Now for some $\varepsilon > 0$, consider another point $\tilde{x}^l := (\tilde{P}_{ij}^l, \tilde{Q}_{ij}^l, \tilde{\ell}^l, \tilde{v}, \tilde{Q}^{cr})$, which is defined by

$$\begin{cases} \tilde{v} = v^{\star}, \ \tilde{Q}^{cr} = Q^{cr\star} \\ \tilde{P}_{mn}^l = P_{mn}^{l\star} - r_{mn}^l \varepsilon/2, \ \ \tilde{P}_{-mn}^l = P_{-mn}^{l\star} \\ \tilde{Q}_{mn}^l = Q_{mn}^{l\star} - x_{mn}^l \varepsilon/2, \ \ \tilde{Q}_{-mn}^l = Q_{-mn}^{l\star} \\ \tilde{\ell}_{mn}^l = \ell_{mn}^{l\star} - \varepsilon, \ \ \tilde{\ell}_{-mn}^l = \ell_{-mn}^{l\star} \\ \tilde{P}_n^d = P_n^d + r_{mn}^l \varepsilon/2, \ \ \tilde{P}_m^d = P_m^d + r_{mn}^l \varepsilon/2, \\ \tilde{Q}_n^d = Q_n^d + x_{mn}^l \varepsilon/2, \ \ \tilde{Q}_m^d = Q_m^d + x_{mn}^l \varepsilon/2, \\ \tilde{P}_{-n}^d = P_{-n}^d, \ \tilde{P}_{-m}^d = P_{-m}^d, \ \tilde{Q}_{-n}^d = Q_{-n}^d, \ \tilde{Q}_{-m}^d = Q_{-m}^d \end{cases} \tag{46}$$

where the negative indices mean excluding elements from a vector.

When $\tilde{\ell}_{mn}^l \cdot \tilde{v}_m = (\tilde{P}_{mn}^l)^2 + (\tilde{Q}_{mn}^l)^2$ holds for certain $\varepsilon > 0$, it can be verified that $\tilde{x}^l$ can satisfy

$$\begin{aligned} \tilde{v}_n &= \tilde{v}_m - 2(r_{mn}^l \tilde{P}_{mn}^l + x_{mn}^l \tilde{Q}_{mn}^l) + |z_{mn}^l|^2 \tilde{\ell}_{mn}^l \\ &= \tilde{v}_m - 2(r_{mn}^l P_{mn}^{l\star} + x_{mn}^l Q_{mn}^{l\star}) + |z_{mn}^l|^2 \ell_{mn}^{l\star} + (r_{mn}^l)^2 \varepsilon + (x_{mn}^l)^2 \varepsilon - |z_{mn}^l|^2 \varepsilon \\ &= \tilde{v}_m - 2(r_{mn}^l P_{mn}^{l\star} + x_{mn}^l Q_{mn}^{l\star}) + |z_{mn}^l|^2 \ell_{mn}^{l\star} \end{aligned} \tag{47}$$

For $\tilde{P}_{mn}^l + P_n^g - r_{mn}^l \tilde{\ell}_{mn}^l = \sum_{k \in \pi(n)} \tilde{P}_{nk}^l + \tilde{P}_n^d$, we can find $P_{mn}^{l\star} - r_{mn}^l \varepsilon/2 + P_n^g - r_{mn}^l \ell_{mn}^{l\star} + r_{mn}^l \varepsilon = \sum_{k \in \pi(n)} P_{nk}^{l\star} + P_n^d + r_{mn}^l \varepsilon/2$ holds and for $Q_{mn}^l + Q_n^g - x_{mn}^l \ell_{mn}^l = \sum_{k \in \pi(n)} Q_{nk}^l + Q_n^d$, this point $\tilde{x}^l$ still holds under this over-satisfaction of load. Please note that this over-satisfaction of load is needed because we have increased the loads $\tilde{P}_n^d$ and $\tilde{Q}_n^d$ on buses and to obtain the alternative feasible solution $\tilde{x}^l$.

This means that $\tilde{x}^l$ satisfies all constraints (1) and hence is a feasible point. Since $\tilde{\ell}_{mn}^l = \ell_{mn}^{l\star} - \varepsilon$, the $\tilde{x}^l$ has a strictly smaller objective value $\sum_{l \in \mathcal{E}} r_{mn}^l \ell_{mn}^l$ than $x^{l\star}$. This contradicts the optimality of $x^{l\star}$. This theorem indicates that the optimal solution satisfies SOC with equality and yield lower loss cost, if the objective function is strictly increasing in the power injections. Once there is an inverse power flow in distribution networks, i.e., DGs to generate power to grids, this relaxation may be inexact. So this exact SOCO programming relaxation condition holds under a load oversatisfaction assumption (i.e., for a node n, $\tilde{P}_n^d - P_n^d \geq 0$) and radial networks, while the objective is to minimize losses or minimize the total cost with respect to power injections. ■

5.2 Semi-Definite Programming Formulation

We firstly recall the general mathematic conversion between inequality constraints and semi-definite formulation. For example, we can find the following two formulations are equivalent:

$$\left\{(x,y)\in\mathbb{R}^2 : \boldsymbol{X}=\begin{bmatrix} x & 0 & y \\ 0 & 1 & -x \\ y & -x & 1 \end{bmatrix}\succcurlyeq 0,\ \text{rank}(\boldsymbol{X})=1\right\}$$
$$\Leftrightarrow \left\{(x,y)\in\mathbb{R}^2 : \begin{array}{c} 0\le x\le 1 \\ x\ge y^2 \\ x-x^3-y^2\ge 0 \end{array}\right\} \tag{48}$$

Similarly, we can apply this equivalent conversion for the quadratic constraints in *DistFlow* equations. The alternative formulation of non-convex quadratic equalities (7) can be expressed in SDP form. We establish $\boldsymbol{X}^l = \begin{bmatrix} \dot{U}_m \\ I^l_{mn} \end{bmatrix}\begin{bmatrix} \dot{U}^*_m & (I^l_{mn})^* \end{bmatrix} = \begin{bmatrix} v_m & \tilde{S}^l_{mn} \\ \tilde{S}^{l*}_{mn} & \ell^l_{mn} \end{bmatrix}$ that satisfies $\det(\boldsymbol{X}^l) = \ell^l_{mn}\cdot v_m - \tilde{S}^l_{mn}\tilde{S}^{l*}_{mn}$. If $\boldsymbol{X}^l$ is semi-definite as $\boldsymbol{X}^l \succcurlyeq 0$ and $\text{rank}(\boldsymbol{X}^l)=1$, then $\ell^l_{mn}\cdot v_m - \tilde{S}^l_{mn}\tilde{S}^{l*}_{mn}\ge 0$.

The non-convex quadratic equalities (7) are converted to

$$X^l = \begin{bmatrix} v_m & \tilde{S}^l_{mn} \\ \tilde{S}^{l*}_{mn} & \ell^l_{mn} \end{bmatrix}\succcurlyeq 0 \tag{49}$$

$$\text{rank}(\mathbf{X}^{\mathbf{l}})=1 \tag{50}$$

However, $\text{rank}(\mathbf{X}^{\mathbf{l}})=1$ is a non-convex constraint, and thus the standard SDP-based formulation omits this constraint. Therefore, the standard SDP-based formulation of *DistFlow* equations [3] is

$$P^l_{mn} + P^g_n - r^l_{mn}\ell^l_{mn} = \sum_{k\in\pi(n)} P^l_{nk} + P^d_n \quad \forall l\in\mathcal{E}, \forall m,n\in\mathcal{N} \tag{51}$$

$$Q^l_{mn} + Q^g_n - x^l_{mn}\ell^l_{mn} = \sum_{k\in\pi(n)} Q^l_{nk} + Q^d_n \quad \forall l\in\mathcal{E}, \forall m,n\in\mathcal{N} \tag{52}$$

$$v_n = v_m - 2\left(r^l_{mn}\frac{\tilde{S}^l_{mn}+\tilde{S}^{l*}_{mn}}{2} + x^l_{mn}\frac{\tilde{S}^l_{mn}-\tilde{S}^{l*}_{mn}}{2}\right) + |z^l_{mn}|^2\ell^l_{mn} \quad \forall l\in\mathcal{E}, \forall m,n\in\mathcal{N} \tag{53}$$

$$\mathbf{X}^{\mathbf{l}} = \begin{bmatrix} v_m & \tilde{S}^l_{mn} \\ \tilde{S}^{l*}_{mn} & \ell^l_{mn} \end{bmatrix}\succcurlyeq 0 \tag{54}$$

For the 2×2 complex-valued matrix $\boldsymbol{X}^l$, we can decompose it as a linear matrix inequality constraint

$$
\begin{aligned}
\mathbf{X}^{\mathbf{l}} &= \begin{bmatrix} v_m & \tilde{S}^l_{mn} \\ \tilde{S}^{l*}_{mn} & \ell^l_{mn} \end{bmatrix} \\
&= v_m \begin{bmatrix} 1 & 0 \\ 0 & 0 \end{bmatrix} + (P^l_{mn} + jQ^l_{mn}) \begin{bmatrix} 0 & 1 \\ 0 & 0 \end{bmatrix} + (P^l_{mn} - jQ^l_{mn}) \begin{bmatrix} 0 & 0 \\ 1 & 0 \end{bmatrix} + \ell^l_{mn} \begin{bmatrix} 0 & 0 \\ 0 & 1 \end{bmatrix} \\
&= v_m \begin{bmatrix} 1 & 0 \\ 0 & 0 \end{bmatrix} + P^l_{mn} \begin{bmatrix} 0 & 1 \\ 1 & 0 \end{bmatrix} + Q^l_{mn} \begin{bmatrix} 0 & j \\ -j & 0 \end{bmatrix} + \ell^l_{mn} \begin{bmatrix} 0 & 0 \\ 0 & 1 \end{bmatrix} \succcurlyeq 0
\end{aligned} \tag{55}
$$

According to SDP-based *DistFlow* equations, we define the vector of optimization variables $\boldsymbol{x}^l$ for the minimization of real power loss. By applying the linear matrix inequality constraint, the complex-valued matrix $\boldsymbol{X}^l$ is rearranged to a constraint with respect to variables P^l_{mn}, Q^l_{mn}, ℓ^l_{mn}, and v_m. Therefore, the vector of optimization variables $\boldsymbol{x}^l$ can be defined as

$$
\boldsymbol{x}^l := [\boldsymbol{P}^l, \boldsymbol{Q}^l, \ell^l, \boldsymbol{v}, \boldsymbol{Q}^{cr}]^T \tag{56}
$$

Consequently, the complete reactive power optimization model is formulated as

$$
\begin{aligned}
&\min_{\mathbf{x}^{\mathbf{l}} \in \mathbb{R}} \; P^l_0 \\
&s.t. \begin{cases}
P^l_{mn} + P^g_n - r^l_{mn}\ell^l_{mn} = \sum\limits_{k \in \pi(n)} P^l_{nk} + P^d_n, \; \forall l \in \mathcal{E}, \forall m, n \in \mathcal{N} \\
Q^l_{mn} + Q^g_n - x^l_{mn}\ell^l_{mn} = \sum\limits_{k \in \pi(n)} Q^l_{nk} + Q^d_n, \; \forall l \in \mathcal{E}, \forall m, n \in \mathcal{N} \\
v_n = v_m - 2(r^l_{mn}P^l_{mn} + x^l_{mn}Q^l_{mn}) + |z^l_{mn}|^2 \ell^l_{mn}, \; \forall l \in \mathcal{E}, \forall m, n \in \mathcal{N} \\
v_m \begin{bmatrix} 1 & 0 \\ 0 & 0 \end{bmatrix} + P^l_{mn} \begin{bmatrix} 0 & 1 \\ 1 & 0 \end{bmatrix} + Q^l_{mn} \begin{bmatrix} 0 & j \\ -j & 0 \end{bmatrix} + \ell^l_{mn} \begin{bmatrix} 0 & 0 \\ 0 & 1 \end{bmatrix} \succcurlyeq 0, \; \forall l \in \mathcal{E}, \forall m, n \in \mathcal{N} \\
\underline{Q}^{cr} \leq Q^{cr} \leq \overline{Q}^{cr} \; \underline{v} \leq v \leq \overline{v}
\end{cases}
\end{aligned} \tag{57}
$$

Note that there is another formulation of this model with respect to complex-valued vector $\mathbf{x}^{\mathbf{l}} := [\tilde{S}^l, \ell^l, v, Q^{cr}]^T$. In this vein, we formulate this complex-valued reactive power optimization model below

$$\min_{\mathbf{x}^l \in \mathbb{C}} \frac{\text{Tr}(S_0^l + S_0^{l*})}{2}$$

$$\text{s.t.} \begin{cases} P_{mn}^l + P_n^g - r_{mn}^l \ell_{mn}^l = \sum_{k \in \pi(n)} P_{nk}^l + P_n^d, \ \forall l \in \mathcal{E}, \forall m, n \in \mathcal{N} \\ Q_{mn}^l + Q_n^g - x_{mn}^l \ell_{mn}^l = \sum_{k \in \pi(n)} Q_{nk}^l + Q_n^d, \ \forall l \in \mathcal{E}, \forall m, n \in \mathcal{N} \\ \qquad v_n = v_m - 2(r_{mn}^l \frac{\tilde{S}_{mn}^l + (\tilde{S}_{mn}^l)^*}{2} + x_{mn}^l \frac{\tilde{S}_{mn}^l - (\tilde{S}_{mn}^l)^*}{2}) \\ + |z_{mn}^l|^2 \ell_{mn}^l, \ \forall l \in \mathcal{E}, \forall m, n \in \mathcal{N} \\ v_m \begin{bmatrix} 1 & 0 \\ 0 & 0 \end{bmatrix} + S_{mn}^l \begin{bmatrix} 0 & 1 \\ 0 & 0 \end{bmatrix} + S_{mn}^{l*} \begin{bmatrix} 0 & 0 \\ 1 & 0 \end{bmatrix} + \ell_{mn}^l \begin{bmatrix} 0 & 0 \\ 0 & 1 \end{bmatrix} \succcurlyeq 0, \ \forall l \in \mathcal{E}, \forall m, n \in \mathcal{N} \\ \underline{Q}^{cr} \leq Q^{cr} \leq \overline{Q}^{cr} \ \underline{v} \leq v \leq \overline{v} \end{cases} \tag{58}$$

where $\text{Tr}(S_0^l + S_0^{l*})$ refers to the trace of $S_0^l + S_0^{l*}$.

6 Polyhedral Approximation Formulation

It is evident that the quadratic equality of *DistFlow* equations can be relaxed in a SOC formulation according to Sect. 4.1. In this subsection, we investigate how can a SOC cone can be approximated into a set of linearizations. Mathematically, the second-order cone is also called a Lorentz cone or an ice-cream cone. The general mathematical definition is cast as

$$L^n := \{(x, t) \in \mathbb{R}^n \times \mathbb{R} \ |||x||_2 \leq t\} \tag{59}$$

For instance, if $n = 2$, then the geometry of a Lorentz cone L^2 is displayed in Fig. 4.

The feasible region of a second-order cone can be well-approximated by a polyhedral cone, as presented in Fig. 5. This polyhedral approximation makes a SOC constraint that becomes a series of linear constraints. The approximation accuracy depends on the number of outer polyhedral linearizations.

Definition Let $\mathcal{F}$ be a feasible region of a SOCO programming, i.e., $\mathcal{F} := \{x : Ax \geq b, ||Ax - b||_2 \leq cx - d\}$ and let $\mathcal{F}_\varepsilon := \{x : Ax \geq b, ||Ax - b||_2 \leq (cx - d)(1 + \varepsilon)\}$. A polyhedral cone $\boldsymbol{P}$ is an ε-approximation of $\mathcal{F}$ if $\mathcal{F} \subseteq \mathrm{P} \subseteq \mathcal{F}_\varepsilon$.

Next, we seek to find an ε-approximation of $\mathcal{F}$. Recall that a three-dimensional Lorentz cone $L^2 = \{(x, t) \in \mathbb{R}^2 \times \mathbb{R} : ||x|| \leq t\}$. According to $\tilde{x} = (\tilde{x}_1, \tilde{x}_2, \tilde{t}) \in \mathbb{R}^3$, how can determine this any point lies in L^2. Assume w.l.o.g. $\tilde{t} = 1, \tilde{x} \geq 0$. Ideally, rotating the point $(\tilde{x}_1, \tilde{x}_2, 1)$ to $(\tilde{x}_1\prime, 0, 1)$ is shown in Fig. 6. It is clear that $\tilde{x} \in L^2$

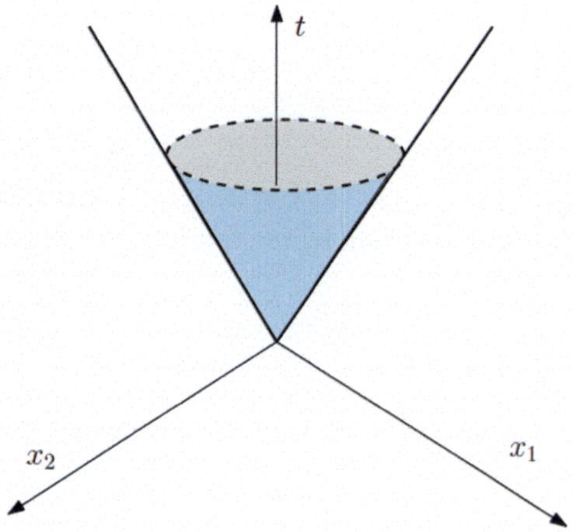

Fig. 4 Geometry of a Lorentz cone L^2

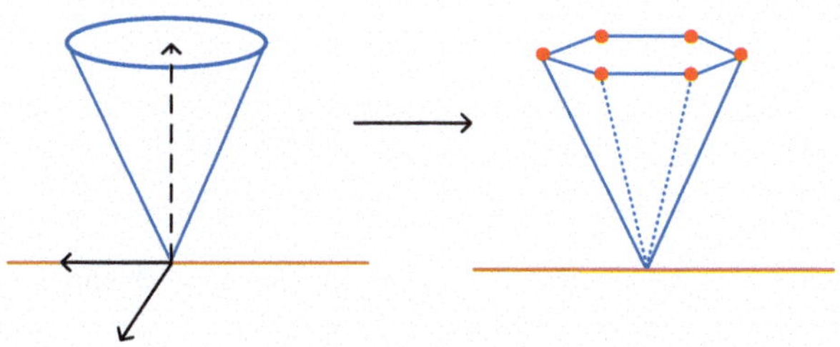

Fig. 5 Approximation of a Lorentz cone as a polyhedral cone

iff $\tilde{x}_1\prime \leq 1$. If this angle is arbitrarily small, then we can check whether the rotated point $\tilde{x}_1\prime \in L^2_\varepsilon$ for an appropriate small ε, since the component $\tilde{x}_2\prime$ is very close to 0.

This polyhedral approximation method is proposed in [19]. For $\sqrt{x_1^2 + x_2^2} \leq t$ can be approximated by a system of linear homogeneous equalities and inequalities in terms of x_1, x_2, t, and $2(v+1)$ variables ς^j, η^j for $j = 0, 1, 2, \ldots v$ where v is a parameter of the polyhedral $\varepsilon(v)$ relaxed approximation such that

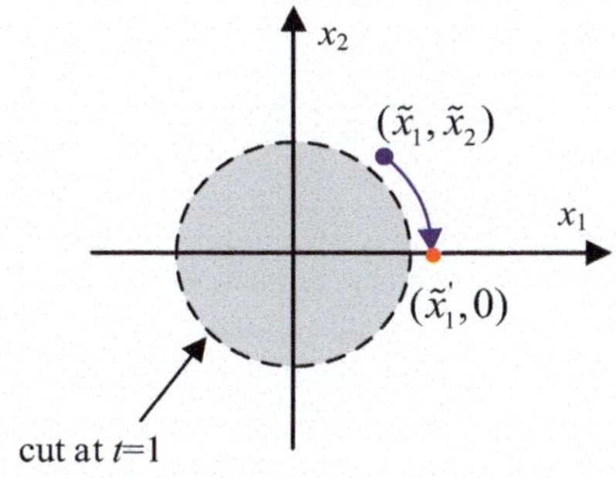

Fig. 6 ε-approximation of a Lorentz cone L^2

$$\varepsilon(v) = \frac{1}{\cos(\frac{\pi}{2^{v+1}})} - 1 \tag{60}$$

This gives $\varepsilon(v) \approx 3 \times 10^{-7}$ when $v = 11$; the relaxed approximation in $(1+\varepsilon)t \geq \sqrt{x_1^2 + x_2^2}$ will have $(1 + \varepsilon(v))^2 - 1 \approx 6 \times 10^{-7}$. The system of linear homogeneous equations is given by the following polyhedron T.

$$\begin{aligned} T := \{(x_1, x_2, \varsigma^0, \eta^0, ..., \varsigma^v, \eta^v, t) : \ & \varsigma^0 \geq |x_1| \\ & \eta^0 \geq |x_2| \\ & \varsigma^j = \cos(\frac{\pi}{2^{j+1}})\varsigma^{j-1} + \sin(\frac{\pi}{2^{j+1}})\eta^{j-1}, \quad j = 1, ..., v \\ & \eta^j \geq -\sin(\frac{\pi}{2^{j+1}})\varsigma^{j-1} + \cos(\frac{\pi}{2^{j+1}})\eta^{j-1}, \quad j = 1, ..., v \\ & \eta^j \geq \sin(\frac{\pi}{2^{j+1}})\varsigma^{j-1} - \cos(\frac{\pi}{2^{j+1}})\eta^{j-1}, \quad j = 1, ..., v \\ & \varsigma^v \leq t \\ & \eta^v \leq \tan(\frac{\pi}{2^{v+1}})\varsigma^v \end{aligned} \tag{61}$$

where (ξ_0, η_0) rotates (x_1, x_2) so that $x_1, x_2 \geq 0$, and we have two additional variables ξ_j, η_j for each iteration $j = 1,\ldots,v$.

The polyhedral approximation given by (61) can be reduced by using the linear equality constraints in $\varsigma^j = \cos(\frac{\pi}{2^{j+1}})\varsigma^{j-1} + \sin(\frac{\pi}{2^{j+1}})\eta^{j-1}, j = 1,\ldots,v$ to solve for $\varsigma^j, j = 1, ..., v$ in terms of $\varsigma^0, \eta^{j-1}, j = 1,\ldots,v$ and then substitute ς^j out of the system (61) by this linear equality constraint. The resulting system will only have linear inequality constraints in terms of the variables $x_1, x_2, x_3, \varsigma^0$ and the $(v+1)$ variables for η^j for $j = 0,\ldots,v$.

The system of linear homogeneous equalities and inequalities (61) is known as the polyhedral approximation for a Lorentz cone L^2. Inspired by this idea, we can extend this polyhedral approximation for an arbitrary Lorentz cone L^n. Assume w.l.o.g $n = 2^k$ for some $k \in \mathbb{Z}$. We show that we can give an extended formulation for

$$L^n := \left\{(x_{0,1}, \cdots, x_{0,n}, t) \in \mathbb{R}^n \times \mathbb{R} \middle| ||(x_{0,1}, \cdots, x_{0,n})||_2 \leq t\right\} \tag{62}$$

We split all the original variables into pairs $(x_{0,2j-1}, x_{0,2j})$ for $j \in [n/2]$ and associate each pair to a new "1st level" variable $x_{1,j}$ such that $||(x_{0,2j-1}, x_{0,2j})||_2 \leq x_{1,j}$, where $x_{1,j}$ is a new introduced variable for a pair $(x_{0,2j-1}, x_{0,2j})$. Then, introducing "2nd" level variables $x_{2,j}$ is subject to $||(x_{1,2j-1}, x_{1,2j})||_2 \leq x_{2,j}$ in a similar way. We therefore construct a binary tree, where each i-th level variable has as children two (i-1)-th level variables. The same structure is preserved up to the level $\log_2(n)$, which has a single node $x_{\log_2(n)} = t$. At each level of the tree, we have an ε-approximation

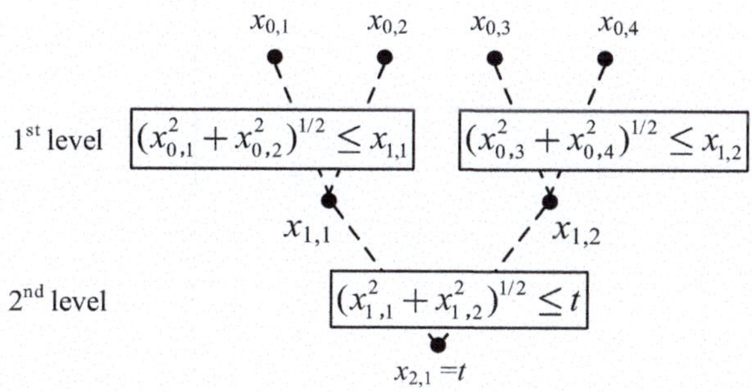

Fig. 7 A ε-approximation flowchart of a Lorentz cone L^2

of the L^2 cones with the construction from the previous section. This gives in total a $((1+\varepsilon)^{\log_2(n)} - 1)$-approximation for L^n.

For explanation, let us exemplify this idea by

$$L^4 := \left\{(x_{0,1}, x_{0,2}, x_{0,3}, x_{0,4}, t) \in \mathbb{R}^4 \times \mathbb{R} \middle| ||(x_{0,1}, x_{0,2}, x_{0,3}, x_{0,4})||_2 \le t\right\} \quad (1.63)$$

The approximation process can be summarized in Fig. 7.

Consider the following SOCO programming problem by the polyhedral approximation method.

$$\begin{aligned} \min \quad & x_1 + x_2 + 3x_3 \\ s.t. \quad & x_1 + x_2 + 2x_3 = 3 \\ & x_3 \ge \sqrt{x_1^2 + x_2^2} \\ & 0 \le x_1, x_2, x_3 \le 1 \end{aligned} \quad (64)$$

This SOCO programming-based model can obtain the exact optimum. The minimum objective value is 3.8787 and the optimal vector $x = (0.6213, 0.6213, 0.8787)^{\mathrm{T}}$ is achieved. The SOC constraint satisfies with equality as observed by the gap error 3.1666×10^{-7}.

Subsequently, we linearize this SOC constraint with respect to x_1, x_2, x_3 and 2(3+1)=8 variables ς^0, ς^1, ς^2, ς^3, η^0, η^1, η^2, η^3 The system of polyhedral approximation constraints is expressed below

$$
\begin{cases}
-\varsigma^0 \le x_1 \le \varsigma^0 \\
-\eta^0 \le x_2 \le \eta^0 \\
\varsigma^1 - \cos(\frac{\pi}{2^2})\varsigma^0 - \sin(\frac{\pi}{2^2})\eta^0 = 0 \\
\varsigma^2 - \cos(\frac{\pi}{2^3})\varsigma^1 - \sin(\frac{\pi}{2^3})\eta^1 = 0 \\
\varsigma^3 - \cos(\frac{\pi}{2^4})\varsigma^2 - \sin(\frac{\pi}{2^4})\eta^2 = 0
\end{cases}
\begin{cases}
-\sin(\frac{\pi}{2^2})\varsigma^0 + \cos(\frac{\pi}{2^2})\eta^0 - \eta^1 \le 0 \\
\sin(\frac{\pi}{2^2})\varsigma^0 - \cos(\frac{\pi}{2^2})\eta^0 - \eta^1 \le 0 \\
-\sin(\frac{\pi}{2^3})\varsigma^1 + \cos(\frac{\pi}{2^3})\eta^1 - \eta^2 \le 0 \\
\sin(\frac{\pi}{2^3})\varsigma^1 - \cos(\frac{\pi}{2^3})\eta^1 - \eta^2 \le 0 \\
-\sin(\frac{\pi}{2^4})\varsigma^2 + \cos(\frac{\pi}{2^4})\eta^2 - \eta^3 \le 0 \\
\sin(\frac{\pi}{2^4})\varsigma^2 - \cos(\frac{\pi}{2^4})\eta^2 - \eta^3 \le 0
\end{cases}
$$

$$
\begin{cases}
\varsigma^3 \le x_3 \\
\eta^3 \le \tan(\frac{\pi}{2^4})\varsigma^3
\end{cases}
\tag{65}
$$

The vector of optimization variables is $\boldsymbol{x} = [x_1, x_2, x_3, \varsigma^0, \varsigma^1, \varsigma^2, \varsigma^3, \eta^0, \eta^1, \eta^2, \eta^3]^{\mathrm{T}}$, the total of which is $3 + 2(3 + 1) = 11$. By solving this linear programming model, the minimum objective value is 3.8787 and the optimal vector $x = (0.6213, 0.6213, 0.8787)^{\mathrm{T}}$ is achieved. The SOC constraint satisfies inequality by the gap error -0.0301. Improving the approximation accuracy is just to set larger linearization segments, e.g., $n = 10$. Thus, a more accurate solution with a gap error equal to -1.8167×10^{-6} can be provided.

7 Radiality Constraints

In this thesis, we suppose that each branch of distribution networks has a sectionalizing switch or tie switch. Each bus in distribution networks is connected by a branch with a switch, and one tie switch can form a loop with other sectionalizing switches. The DNR problem can be deemed as a combinatorial issue about choosing optimal switch status of sectionalizing switches (normally closed) and tie-switches (normally open). The DNR provides an optimal network structure to realize minimum power losses and achieve better load balancing, or to be used for post-outage restoration and planned maintenance, etc. In this section, we discuss the relaxation techniques for DNR problems based on *DistFlow* equations.

7.1 Virtual Commodity Flow Constraints

In the traditional distribution networks, DNR is not a frequent operation. However, with the application of high-speed switching devices in distribution networks, the

DNR is developing toward real-time reconfiguration for maintaining the optimal operation condition of distribution networks. The ability to switch between different topologies enables a class of radial grids subject to single commodity flow constraints. Let u^l be the binary state vector of circuit breakers, which is zero if the switch is open and one if closed, namely $u^l \in \{0, 1\}, u^l \in \mathbb{Z}$.

One of the most popular radiality models ensures connectivity of loads to DERs via the power flow equations, and connects DERs to the substation via flows of a virtual commodity [7]. The tightness of a linear programming relaxation for this model has also been recent. Upon the commodity flow approach, this section introduces a more succinct model with fewer variables and constraints. We consider a virtual single commodity flowing on the network graph $\mathcal{G}\ominus$ and set a demand of one unit on all non-substation buses. This is to say, all load nodes to have nonzero power injection (no transfer nodes in distribution networks). Consequently, all demands supplied by substation and all nodes have a path to substation ensuring connectivity. Virtual commodity flow constraints can be expressed with respect to the branch-bus incidence matrix **A**, by forcing the virtual flows in f^l to be zero for open lines:

$$\boldsymbol{A}^T f = 1 \tag{66}$$

$$-N \cdot u^l \leq f^l \leq N \cdot u^l, \quad \forall l \in \mathcal{E} \tag{67}$$

$$1^T u^l = |\mathcal{N}| - 1 \tag{68}$$

where f^l refers to the vector of virtual flows on each branch.

In this set of virtual commodity flow constraints, the virtual flow variable f^l does not relate to the actual line flows and is introduced only to enforce connectivity. Given that flows are allowed only on active lines, enforcing flow balance on a node with only one unit flow injection, which results in $\boldsymbol{A}^T f = 1$. Moreover, $-N \cdot u^l \leq f^l \leq N \cdot u^l$ can be used to maintain radiality of distribution networks by linking f^l and u^l. Lastly, $\mathbf{1}^T u^l = |\mathcal{N}| - 1$ guarantees that the network reconfiguration solution will be an acyclic network. It should be worthwhile to note that when there are no transfer nodes in distribution networks, $\mathbf{1}^T u^l = |\mathcal{N}| - 1$ and power flow equations can be effective to maintain radiality of distribution networks; otherwise, virtual commodity flow constraints should be considered.

7.2 *Spanning Tree Constraints*

The spanning tree constraints are used to formulate a linear program of the minimum spanning tree problem [20]. For a DNR problem, we can encode every branch as a directed arc with respect to an acyclic network rooted at node, which spanning tree constraints can be applied for network reconfiguration. To enforce a radial typology, a pair of variables β^l_{mn} and β^l_{nm} for branch l is ranged in [0, 1], indicating the parent–child relationship between nodes m and n for the radiality of networks. For instance, $\beta^l_{mn} = 1$ and $\beta^l_{nm} = 0$ means that n is the parent node of m, otherwise $\beta^l_{mn} = 0$ and $\beta^l_{nm} = 1$ implies that m is the parent node of n.

$$\beta^l_{mn} + \beta^l_{nm} = u^l \tag{89}$$

$$\beta^l_{mn} = 0, \quad if\ m = S \tag{70}$$

$$\sum_{n:(m,n)\in E} \beta^l_{mn} = 1, \quad \forall\, m \in N \backslash S \tag{71}$$

$$\beta^l_{mn} \in [0, 1] \tag{72}$$

where S is the set of source nodes.

Flowing on the branch with nodes (m,n) can only be in one direction, which explicitly is set to a combination of variables β^l_{mn} and β^l_{nm}. The parent–child relationship variables β^l are continuous, i.e.,$\beta^l \in \mathbb{R}^n$. After applying this set of spanning tree constraints,β^l can be proved to get converged as 0 or 1. Reference [7] proves that any feasible β^l_{mn} must be zeros and ones, and for each source node, it describes the edges of an unweighted, directed tree graph with a source node.

Please note that if only spanning tree constraints are considered in the DNR model, then spanning tree constraints are inadequate to represent the radiality. Take the following Fig. 8 as an example from reference [21] where we suppose that node 1 is a point of common coupling (PCC) node.

By spanning tree constraints, we give the following β^l values for this figure in Table 1.

Fig. 8 An example of inadequate spanning tree constraints

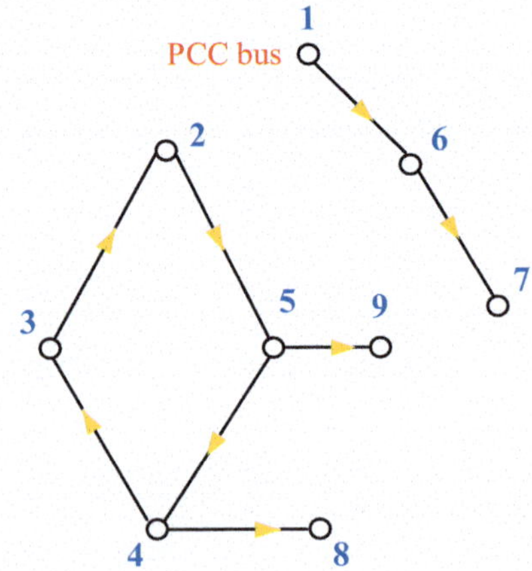

Table 1 Values of parent–child relationship variable β^l

Connected branches		β_{mn}^l	β_{nm}^l	Parent–child relationship
Node m	Node n			
6	1	1	0	1-parent, 6-child
7	6	1	0	6-parent, 7-child
3	**2**	0	1	3-parent, 2-child
4	**3**	0	1	4-parent, 3-child
5	**4**	0	1	5-parent, 4-child
5	**2**	1	0	2-parent, 5-child
9	5	1	0	5-parent, 9-child
8	4	1	0	4-parent, 8-child

It can be inferred from Table 1 that all β^l values satisfy spanning tree constraints (25–28). Unfortunately, there is a cycle topology that is not a radial network. For a DNR problem, power flow equations are hard constraints in this DNR model. Clearly, this disconnected network in Fig. 8 leads to an infeasible power flow solution. In other words, the above-mentioned example cannot satisfy the nodal power flow balance at node 7 under assumed power flow directions in Fig. 8.

But from our perspective, the power flow equations are necessary but insufficient conditions for the connectivity constraint in DNR model. As displayed in Fig. 9, we can show this counter-example based on Fig. 8.

In this case, we can find that spanning tree and power flow equations constraints cannot guarantee a radial topology when there is a DG source at node 2 in this distribution system. It may result in an unconnected graph containing loops with nodes 2, 5, 4, and 3. This illustrates that power flow equations cannot entirely guarantee the connective topology. Thus, it is suggested in [20] that single-commodity flow constraints and spanning tree constraints can guarantee the radial topology.

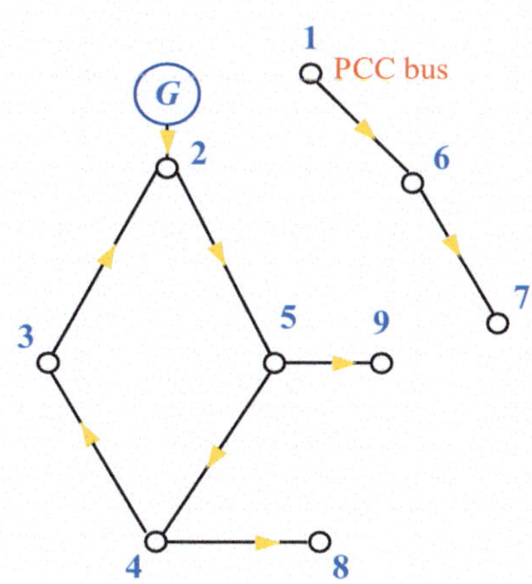

Fig. 9 An example of inadequate spanning tree and power flow constraints

8 Conventional DNR Models

By the angle relaxation, *DistFlow* equations can be cast as a second-order conic (SOCO) programming or a SDP formulation for *DistFlow* equations. By these convex relaxations, the DNR model can be modeled as a mixed-integer SOCO programming (MISOCP) or a mixed integer semi-definite programming problem in the following. These convex models exploit the optimal distribution topology with reduced computational complexity. In this subsection, we only discuss SOCO programming-based model of DNR problems with Big-M relaxation method.

8.1 MISOCP-Based DNR Model with Big-M Relaxation Method

For the sake of the minimal power loss in reconfigurable distribution networks, the set of optimization variables involves a set of operational variables $\left[P^l, Q^l, \ell^l, v_{PQ}\right]^T$ and a vector of controllable variables $\left[v_0, Q^{cr}, f^l, \beta^l_{mn}, \beta^l_{nm}, u^l\right]^T$. Here, v_0 refers to the squared voltage profile at the root node 0. Accordingly, once controllable variables are provided, operational variables can be automatically determined just by power flow equations. For this DNR optimization problem, we combine operational and controllable variables together as $x^l := [P^l, Q^l, \ell^l, v, Q^{cr}, w^l, m^l, f^l, \beta^l_{mn}, \beta^l_{nm}, u^l]^T$ for all branches of distribution networks, where $(P^l, Q^l, \ell^l, v, Q^{cr}, w^l, m^l, f^l, \beta^l_{mn}, \beta^l_{nm}) \in \mathbb{R}$ and $u^l \in \mathbb{Z}$.

With *DistFlow* equations, the big-M method is utilized to relax equality $v_n - v_m + 2(r^l_{mn}P^l_{mn} + x^l_{mn}Q^l_{mn}) - |z^l_{mn}|^2\ell^l_{mn} = 0$ as two inequalities with a large-enough positive scalar $\boldsymbol{M}$. The optimization variables are expressed as $x^l := \left[P^l, Q^l, \ell^l, v, Q^{cr}, w^l, m^l, f^l, \beta^l_{mn}, \beta^l_{nm}, u^l\right]^T, \forall l \in \mathcal{E}$. Thus, the entire network reconfiguration model with the loss minimization is cast as

$$
\min_{\mathbf{x^l}} \quad P_0^l
$$

$$
s.t. \begin{cases}
P_{mn}^l + P_n^g - r_{mn}^l \ell_{mn}^l = \sum\limits_{k \in \pi(n)} P_{nk}^l + P_n^d, \ \forall l \in \mathcal{E}, \forall m, n \in \mathcal{N} \\
Q_{mn}^l + Q_n^g - x_{mn}^l \ell_{mn}^l = \sum\limits_{k \in \pi(n)} Q_{nk}^l + Q_n^d, \ \forall l \in \mathcal{E}, \forall m, n \in \mathcal{N} \\
v_n - v_m + 2(r_{mn}^l P_{mn}^l + x_{mn}^l Q_{mn}^l) - |z_{mn}^l|^2 \ell_{mn}^l - (1 - u^l) M \le 0, \ \forall l \in \mathcal{E}, \forall m, n \in \mathcal{N} \\
v_n - v_m + 2(r_{mn}^l P_{mn}^l + x_{mn}^l Q_{mn}^l) - |z_{mn}^l|^2 \ell_{mn}^l + (1 - u^l) M \ge 0, \ \forall l \in \mathcal{E}, \forall m, n \in \mathcal{N} \\
\left\| \begin{matrix} 2P_{mn}^l \\ 2Q_{mn}^l \\ w^l \end{matrix} \right\|_2 \le m^l, \forall l \in \mathcal{E}, \ \forall m, n \in \mathcal{N} \\
w^l = \ell_{mn}^l - v_m, \ \ m^l = \ell_{mn}^l + v_m, \ \ \forall l \in \mathcal{E}, \ \forall m, n \in \mathcal{N} \\
-\overline{Q}_0^l \le Q_0^l \le \overline{Q}_0^l \\
-u^l \Gamma_{\max}^l P^l, Q^l u^l \Gamma_{\max}^l \\
\underline{Q}^{cr} \le Q^{cr} \le \overline{Q}^{cr} \\
\underline{v} \le v \le \overline{v} \\
\beta_{mn}^l + \beta_{nm}^l = u^l \\
\beta_{mn}^l = 0, \ \text{if } m = S \\
\sum\limits_{n:(m,n) \in \mathcal{E}} \beta_{mn}^l = 1, \forall m \in \mathcal{N} \ominus \backslash S \\
\beta_{mn}^l \ge 0 \\
\mathbf{A}^T f = 1 \\
-N \cdot u^l \le f^l \le N \cdot u^l, \ \ \forall l \in \mathcal{E} \\
1^T u^l = |\mathcal{N}| - 1 \\
u^l \in \{0, 1\}
\end{cases} \tag{73}
$$

where P_0^l is the active power injection at the root node S.

8.2 MISOCP-Based DNR Model Using McCormick Linearization Method

In the previous subsection, the SOCO programming-based model of DNR problems can be formulated with Big-M relaxation method. However, we can further strengthen this bound relaxation using McCormick Linearization Method [22]. Let $y^l = (v_n - v_m) \cdot u^l$. Then, the McCormick linearization method is utilized to relax equality $v_n - v_m + 2(r_{mn}^l P_{mn}^l + x_{mn}^l Q_{mn}^l) - |z_{mn}^l|^2 \ell_{mn}^l = 0$ as

$$
y^l + 2(r_{mn}^l P_{mn}^l + x_{mn}^l Q_{mn}^l) - |z_{mn}^l|^2 \ell_{mn}^l = 0 \tag{74}
$$

$$(\underline{v}-\overline{v})\cdot u^l y^l \ominus (\overline{v}-\underline{v})\cdot u^l \tag{75}$$

$$v_n - v_m - (\overline{v}-\underline{v})\cdot(u^l-1)y^l \ominus v_n - v_m - (\underline{v}-\overline{v})\cdot(u^l-1) \tag{76}$$

The optimization variables are expressed as $\boldsymbol{x}^l := \left[\boldsymbol{P}^l, \boldsymbol{Q}^l, \ell^l, \boldsymbol{v}, \boldsymbol{Q}^{cr}, \boldsymbol{w}^l, \boldsymbol{m}^l, \boldsymbol{y}^l, \boldsymbol{f}^l, \boldsymbol{\beta}^l_{mn}, \boldsymbol{\beta}^l_{nm}, \boldsymbol{u}^l\right]^T, \forall l \in \mathcal{E}$. Thus, the entire network reconfiguration model with the loss minimization is simply reformulated as

$$\min_{\mathbf{x}^l} \ P_0^l$$

$$s.t.\begin{cases} P^l_{mn} + P^g_n - r^l_{mn}\ell^l_{mn} = \sum\limits_{k\in\pi(n)} P^l_{nk} + P^d_n, \ \forall l \in \mathcal{E}, \forall m,n \in \mathcal{N} \\ Q^l_{mn} + Q^g_n - x^l_{mn}\ell^l_{mn} = \sum\limits_{k\in\pi(n)} Q^l_{nk} + Q^d_n, \ \forall l \in \mathcal{E}, \forall m,n \in \mathcal{N} \\ y^l + 2(r^l_{mn}P^l_{mn} + x^l_{mn}Q^l_{mn}) - |z^l_{mn}|^2\ell^l_{mn} = 0, \ \forall l \in \mathcal{E}, \forall m,n \in \mathcal{N} \\ (\underline{v}-\overline{v})\cdot u^l y^l \ominus (\overline{v}-\underline{v})\cdot u^l, \ \forall l \in \mathcal{E}, \forall m,n \in \mathcal{N} \\ v_n - v_m - (\overline{v}-\underline{v})\cdot(u^l-1)y^l \ominus v_n - v_m - (\underline{v}-\overline{v})\cdot(u^l-1) \\ \left\| \begin{matrix} 2P^l_{mn} \\ 2Q^l_{mn} \\ w^l \end{matrix} \right\|_2 \le m^l, \forall l \in \mathcal{E}, \ \forall m,n \in \mathcal{N} \\ w^l = \ell^l_{mn} - v_m, \ \ m^l = \ell^l_{mn} + v_m, \ \ \forall l \in \mathcal{E}, \ \forall m,n \in \mathcal{N} \\ -\overline{Q}^l_0 \le Q^l_0 \le \overline{Q}^l_0 \\ -u^l\Gamma^l_{\max}P^l, Q^l u^l \Gamma^l_{\max} \\ \underline{Q}^{cr} \le Q^{cr} \le \overline{Q}^{cr} \\ \underline{v} \le v \le \overline{v} \\ \beta^l_{mn} + \beta^l_{nm} = u^l \\ \beta^l_{mn} = 0, \ \text{if } m = 0 \\ \sum\limits_{n:(m,n)\in\mathcal{E}} \beta^l_{mn} = 1, \forall m \in \mathcal{N} \ominus \backslash 0 \\ \beta^l_{mn} \ge 0 \\ \mathbf{A}^T f = 1 \\ -N\cdot u^l \le f^l \le N\cdot u^l, \ \ \forall l \in \mathcal{E} \\ 1^T u^l = |\mathcal{N}| - 1, \ \ u^l \in \{0,1\} \end{cases} \tag{77}$$

8.3 Quadratic DNR Model with Big-M and McCormick Linearization Methods

With LDF-based equations, the Big-M method is utilized to relax equality $v_n - v_m + 2(r^l_{mn}P^l_{mn} + x^l_{mn}Q^l_{mn}) = 0$ as two inequalities with a large-enough positive scalar $\boldsymbol{M}$. Therefore, a linearized DNR model with Big-M relaxation method is developed in [7]. The optimization variables are expressed as $\boldsymbol{x^l} := \left[\boldsymbol{P^l}, \boldsymbol{Q^l}, \boldsymbol{v}, \boldsymbol{Q^{cr}}, \boldsymbol{f^l}, \boldsymbol{\beta^l_{mn}}, \boldsymbol{\beta^l_{nm}}, \boldsymbol{u^l}\right]^T$, $\forall l \in \mathcal{E}$. Since network loss is omitted in this LDF-based equations, we rewrite the objective function in the DNR model as $\sum_{l\in\mathcal{E}} r^l_{mn}((P^l_{mn})^2 + (Q^l_{mn})^2)$ where voltage profiles are all assumed as one unit. Due to this quadratic objective function, the entire linearized DNR model with the loss minimization is reformulated as a mixed integer quadratic programming problem.

$$
\begin{aligned}
&\min_{\mathbf{x^l}} \sum_{l\in\mathcal{E}} r^l_{mn}((P^l_{mn})^2 + (Q^l_{mn})^2)\\
&s.t. \begin{cases}
P^l_{mn} + P^g_n = \sum\limits_{k\in\pi(n)} P^l_{nk} + P^d_n, \ \forall l \in \mathcal{E}, \forall m, n \in \mathcal{N}\\
Q^l_{mn} + Q^g_n = \sum\limits_{k\in\pi(n)} Q^l_{nk} + Q^d_n, \ \forall l \in \mathcal{E}, \forall m, n \in \mathcal{N}\\
v_n - v_m + 2(r^l_{mn}P^l_{mn} + x^l_{mn}Q^l_{mn}) - (1-u^l)M \le 0, \ \forall l \in \mathcal{E}, \forall m, n \in \mathcal{N}\\
v_n - v_m + 2(r^l_{mn}P^l_{mn} + x^l_{mn}Q^l_{mn}) + (1-u^l)M \ge 0, \ \forall l \in \mathcal{E}, \forall m, n \in \mathcal{N}\\
-\overline{Q}^l_0 \le Q^l_0 \le \overline{Q}^l_0\\
-u^l\Gamma^l_{\max}P^l, Q^l u^l \Gamma^l_{\max}\\
\underline{Q}^{cr} \le Q^{cr} \le \overline{Q}^{cr}\\
\underline{v} \le v \le \overline{v}\\
\beta^l_{mn} + \beta^l_{nm} = u^l\\
\beta^l_{mn} = 0, \ \text{if } m = S\\
\sum\limits_{n:(m,n)\in\mathcal{E}} \beta^l_{mn} = 1, \forall m \in \mathcal{N}\ominus\backslash S\\
\beta^l_{mn} \ge 0\\
\mathbf{A}^T f = 1\\
-N \cdot u^l \le f^l \le N \cdot u^l, \quad \forall l \in \mathcal{E}\\
1^T u^l = |\mathcal{N}| - 1\\
u^l \in \{0, 1\}
\end{cases}
\end{aligned}
\tag{78}
$$

8.4 Polyhedral Approximation of DNR Model Using Big-M Relaxation Method

For SOC constraints in the DNR model, the polyhedral approximation can be used to simplify the SOC constraints to be a family of linear constraints. We can express $\| (2P^l_{mn} \; 2Q^l_{mn} \; w^l)^T \|_2 \leq m^l, \forall l \in \mathcal{E}, \; \forall m, n \in \mathcal{N}$ as a polyhedral cone, which is visualized in Fig. 10. The error of approximation can be regarded as the distance between the exact optimal point ★ and the approximate optimal point ●.

Moreover, the Big-M method is utilized to relax equality $v_n - v_m + 2(r^l_{mn} P^l_{mn} + x^l_{mn} Q^l_{mn}) - |z^l_{mn}|^2 \ell^l_{mn} = 0$ as two inequalities with a large-enough positive scalar M. Therefore, a polyhedral approximation DNR model with Big-M relaxation method is established [10]. The optimization variables are expressed as $\boldsymbol{x}^l :=$ $\left[\boldsymbol{P}^l, \boldsymbol{Q}^l, \boldsymbol{v}, \boldsymbol{Q}^{cr}, \boldsymbol{w}^l, \boldsymbol{m}^l, \boldsymbol{S}^l, \boldsymbol{\varsigma}, \boldsymbol{\eta}, \boldsymbol{\omega}, \boldsymbol{\kappa}, \boldsymbol{f}^l, \boldsymbol{\beta}^l_{mn}, \boldsymbol{\beta}^l_{nm}, \boldsymbol{u}^l\right]^T$, $\forall l \in \mathcal{E}$, where $\varsigma, \eta, \omega, \kappa$ are variables in the polyhedral approximation of SOC constraints. Thus, the entire DNR model with the polyhedral approximation of SOC constraints for the loss minimization is reformulated as

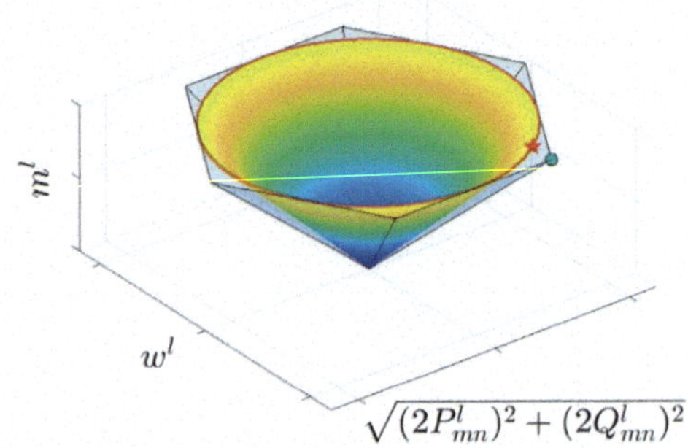

Fig. 10 Polyhedral cone for SOC constraint

$$\min_{\mathbf{x}^l} \quad P_0^l$$

$$\text{s.t.} \begin{cases} P_{mn}^l + P_n^g - r_{mn}^l \ell_{mn}^l = \sum\limits_{k \in \pi(n)} P_{nk}^l + P_n^d, \ \forall l \in \mathcal{E}, \forall m, n \in \mathcal{N} \\ Q_{mn}^l + Q_n^g - x_{mn}^l \ell_{mn}^l = \sum\limits_{k \in \pi(n)} Q_{nk}^l + Q_n^d, \ \forall l \in \mathcal{E}, \forall m, n \in \mathcal{N} \\ v_n - v_m + 2(r_{mn}^l P_{mn}^l + x_{mn}^l Q_{mn}^l) - (1 - u^l)M \le 0, \ \forall l \in \mathcal{E}, \forall m, n \in \mathcal{N} \\ v_n - v_m + 2(r_{mn}^l P_{mn}^l + x_{mn}^l Q_{mn}^l) + (1 - u^l)M \ge 0, \ \forall l \in \mathcal{E}, \forall m, n \in \mathcal{N} \\ w^l = \ell_{mn}^l - v_m, \ m^l = \ell_{mn}^l + v_m, \quad \forall l \in \mathcal{E}, \ \forall m, n \in \mathcal{N} \\ \varsigma^0 \ge |2P_{mn}^l|, \ \forall l \in \mathcal{E} \\ \eta^0 \ge |2Q_{mn}^l|, \ \forall l \in \mathcal{E} \\ \varsigma^j = \cos(\frac{\pi}{2^{j+1}})\varsigma^{j-1} + \sin(\frac{\pi}{2^{j+1}})\eta^{j-1}, \quad j = 1, \ldots, \upsilon \\ \eta^j \ge |-\sin(\frac{\pi}{2^{j+1}})\varsigma^{j-1} + \cos(\frac{\pi}{2^{j+1}})\eta^{j-1}|, \quad j = 1, \ldots, \upsilon \\ \varsigma^\upsilon \le S_{mn}^l, \ \forall l \in \mathcal{E} \\ \eta^\upsilon \le \tan(\frac{\pi}{2^{\upsilon+1}})\varsigma^\upsilon \\ \omega^0 \ge |S_{mn}^l|, \ \forall l \in \mathcal{E} \\ \kappa^0 \ge |w^l|, \ \forall l \in \mathcal{E} \\ \omega^j = \cos(\frac{\pi}{2^{j+1}})\omega^{j-1} + \sin(\frac{\pi}{2^{j+1}})\kappa^{j-1}, \quad j = 1, \ldots, \upsilon \\ \kappa^j \ge |-\sin(\frac{\pi}{2^{j+1}})\omega^{j-1} + \cos(\frac{\pi}{2^{j+1}})\kappa^{j-1}|, \quad j = 1, \ldots, \upsilon \\ \omega^\upsilon \le m^l, \ \forall l \in \mathcal{E} \\ \kappa^\upsilon \le \tan(\frac{\pi}{2^{\upsilon+1}})\omega^\upsilon \\ -\overline{Q}_0^l \le Q_0^l \le \overline{Q}_0^l \\ \underline{Q}^{cr} \le Q^{cr} \le \overline{Q}^{cr} \\ \underline{v} \le v \le \overline{v} \\ \beta_{mn}^l + \beta_{nm}^l = u^l \\ \beta_{mn}^l = 0, \ \text{if } m = 0 \\ \sum\limits_{n:(m,n) \in \mathcal{E}} \beta_{mn}^l = 1, \forall m \in \mathcal{N} \ominus \backslash 0 \\ \beta_{mn}^l \ge 0 \\ \mathbf{A}^T f = 1 \\ -N \cdot u^l \le f^l \le N \cdot u^l, \quad \forall l \in \mathcal{E} \\ 1^T u^l = |\mathcal{N}| - 1 \\ u^l \in \{0, 1\} \end{cases} \tag{79}$$

where υ is the parameter of the polyhedral $\varepsilon(\upsilon)$ relaxed approximation in (60).

9 Disjunctive Convex Hull Approach for DNR Formulation

For the sake of the minimal power loss in reconfigurable distribution networks, the set of optimization variables involves a set of operation variables $\boldsymbol{x}^l := [\boldsymbol{P}^l, \boldsymbol{Q}^l, \ell^l, \boldsymbol{v}, \boldsymbol{Q}^{cr}]^T, \boldsymbol{x}^l \in \mathbb{R}^n$, the continuous parent–child relationship variable $\boldsymbol{\beta}^l \in \mathbb{R}^{2\varepsilon}$ in spanning tree constraints, and binary state vector of circuit breakers $\boldsymbol{u}^l \in \{0, 1\}, \boldsymbol{u}^l \in \mathbb{Z}^{\varepsilon}$, where $\boldsymbol{u}^l$ is zero if the switch is open and one if closed. The DNR model is to seek the loss minimization over *DistFlow* equations. In reconfigurable distribution networks, if a branch $l := (m, n)$ is unconnected, $v_n - v_m + 2(r^l_{mn}P^l_{mn} + x^l_{mn}Q^l_{mn}) - |z^l_{mn}|^2\ell^l_{mn}=0$ in (6) can hold for $P^l_{mn}=Q^l_{mn}=0$, $\ell^l = 0$ and $\underline{v} \le v \le \overline{v}$, but $v_n - v_m + 2(r^l_{mn}P^l_{mn} + x^l_{mn}Q^l_{mn}) - |z^l_{mn}|^2\ell^l_{mn}=0$ may not stand with this solution due to $\boldsymbol{A}^l \cdot v \neq \boldsymbol{0}$. Suppose that no additional active/reactive power injections from reactive compensation or DERs are available to this branch l, and as long as $\boldsymbol{A}^l \cdot v = v_m - v_n > 0$ holds, then it is inferred that m should be the parent node of n.

According to [7], continuous parent–child relationship variables β^l are proved to get converged as binary solutions from real-valued continuous numbers. Inspired by this characteristic, we can ideally link disjunctive parent–child relationship variables β_l with “on/off” equality (6) and (7) via the DCHR approach. In other words, if an arbitrary branch $l := (m, n)$ with active/reactive power flow directions depending on parent–child relationship variables of distribution networks, then $v_m - v_n > 0$ is consistent with $\beta l\ nm = 1$ and $\beta l\ mn = 0$, or $v_m - v_n < 0$ is consistent with $\beta l\ mn = 1$ and $\beta l\ nm = 0$. Mathematically, we have

Theorem Let Ω^l be the feasible set of the DNR problem with respect to optimization variables (x^l, β^l, u^l). For an arbitrary branch $l := (m, n)$ for $\forall (m, n) \in \mathcal{N}$, if the signs of $Pl\ mn$ and $Ql\ mn$ simultaneously depend on the combination of $\beta l\ nm$ and $\beta l\ mn$, then Ω^l can be expressed as

$$\Omega^l = \left\{ (\boldsymbol{x}^l, \boldsymbol{\beta}^l, \boldsymbol{u}^l) \in \mathbb{R}^{n+2|\varepsilon|} \times \mathbb{Z}^{2|\varepsilon|} \left| \begin{array}{l} (2.4) - (2.7) \\ (3.4) - (3.7) \\ (v_m - v_n) \cdot \beta^l_{nm} > 0 \\ \vee (v_m - v_n) \cdot \beta^l_{nm} < 0 \\ \vee (v_m - v_n) \cdot u^l_{mn} = 0 \end{array} \right. \right\} \tag{80}$$

where $\vee$ denotes the logic operator for disjunction.

Proof Suppose that there exists a directional branch power flow from parent node n to child node m such that $v_m - v_n > 0$ for $\beta l\ mn = 1$ and $\beta l\ nm = 0$. Due to $\ell l\ mn > 0$, then we have $- |zl\ mn|^2 \ell l\ mn < 0$. And directional branch power flow from n to node m means that $Pl\ mn < 0$ and $Ql\ mn < 0$. Subsequently, it is derived that $v_m - v_n = 2R_{mn}Pl\ mn + 2X_{mn}Ql\ mn - |zl\ mn|^2\ell l\ mn < 0$ where R_{mn} and X_{mn} refer to the resistance and reactance of branch l. This induces $v_m - v_n < 0$, which contradicts $v_m - v_n > 0$ as the initial assumption. Similarly, this theorem can be also proven for which $v_m - v_n < 0$ is consistent with $\beta l\ mn = 1$ and $\beta l\ nm = 0$. If $u^l_{mn} = 0$, then $\beta^l_{nm} = \beta^l_{mn} = 0$ by (80) and $v_m - v_n$ can be free. ■

With this theorem's framework, it is found that no additional disjunctive variables are incorporated for tighter relaxations, but it can guarantee a more rapid convergence than using Big-M and McCormick linearization method. Indeed, if there exists a distribution network without any reactive compensation or DERs injection, the enforced disjunctive constraint naturally holds for each branch since the voltage profile at the root node is the only highest. For a branch $l := (m, n)$ with additional reactive compensation or DERs injection at the ending node n, this theorem may not hold due to $v_m - v_n < 0$ and $\beta l\ mn = 1$, $\beta l\ nm = 0$. It is known that this l is a connected branch with parent node m and child node n, but $(v_m - v_n) \cdot u^l_{mn} \neq 0$, $(v_m - v_n) \cdot \beta^l_{mn}$=0 and $(v_m - v_n) \cdot \beta^l_{nm} < 0$ disobeys this **Theorem**. Therefore, if a branch has a bi-directional power flow caused by reactive compensation or DERs injection, any enforced disjunctive constraints between parent–child relationship variables and voltage drops in (80) should be relaxed. Otherwise, additional disjunctive variables should be incorporated to replace parent–child relationship variables in (80). Excluding branches with the possibility of bi-directional power flows, a disjunctive CH based on parent–child relationship variables can be utilized to construct tighter convex relaxations of (6) and (7) and disjunctive constraints in (80) for the rest of the branches. Please also note that the simple way is to estimate which is the farthest load node from the center of each reactive compensation and DERs injection node via calculating their capacity accommodation. Thus, we can easily find these possible branches and relax their corresponding disjunctive constraints between parent–child relationship variables and voltage drops in (6).

For (6), we define $\Omega_{1,\ \mathrm{mn}}$ and $\Omega_{1,\ nm}$ as the disjunctive convex sets for (80) integrated with $\beta l\ mn = 1$ and $\beta l\ nm = 1$ for branch $l := (m, n)$, respectively. Subsequently, $\Omega_{1,\ mn}$ can be written as a union of two convex sets Ω0 1,*mn* and Ω1 1,nm corresponding to $\beta l\ nm = 1$ and $\beta l\ mn = 0$. This is similar to $\Omega_{1,\ nm}$ under $\beta l\ nm = 1$ and $\beta l\ nm = 0$.

$$\Omega_{1,mn} = \mathrm{Conv}(\Omega^0_{1,mn} \cup \Omega^1_{1,mn})\ \ \Omega_{1,mn} = \mathrm{Conv}(\Omega^0_{1,nm} \cup \Omega^1_{1,nm}) \tag{81}$$

$$\Omega^0_{1,mn}, \Omega^0_{1,nm} = \{\mathbf{x}^\mathbf{l} \in \mathbb{R}^n | \underline{\mathbf{v}} \leq \mathbf{v} \leq \overline{\mathbf{v}},\ \ell^l_{mn}, P^l_{mn}, Q^l_{mn} = 0\} \tag{82}$$

$$\Omega^1_{1,mn}, \Omega^1_{1,nm} = \{\mathbf{x}^\mathbf{l} \in \mathbb{R}^n | (1.6), \underline{\mathbf{v}} \leq \mathbf{v} \leq \overline{\mathbf{v}},\ 0 \leq \ell^l_{mn} \leq \overline{\ell}^l\} \tag{83}$$

In this vein, we sketch the geometry closure of $\Omega_{1,mn}$ and $\Omega_{1,nm}$ with a continuous variation of $\beta l\ nm, \beta l\ mn \in [0, 1]$ on ($v_m - v_n$, $2R_{mn}Pl\ mn + 2X_{mn}Ql\ mn - |zl\ mn|^2\ell l\ mn$, β^l)-space in Fig. 11a. The geometry closure of $\Omega_{1,\ mn}$ and $\Omega_{1,\ nm}$ are disjunctive polyhedral sets with vertexes $V_1 - V_4$ and V_1, $V_5 - V_7$ in blue and green CHs, respectively. Note that the former vertical coordinate axis is upward $\beta l\ nm$ while the latter vertical axis is downward $\beta l\ mn$. Clearly, the hyperplane ψ_1 with vertexes V_1, V_2, and V_4 is at the bottom of the blue polyhedral hull, whereas hyperplane ψ_2 with vertexes V_4, V_5, and V_7 is at the top of the green polyhedral hull. With this definition of bi-directional vertical axis, ψ_1 and ψ_2 are bottom hyperplanes of blue and green polyhedral CHs, respectively. As a result, with the hyperplanes ψ_1 and ψ_2, we explicitly express two constraints in (84) that are obtained from the coordinates of vertexes V_1, V_2, V_4, V_5, and V_7, which are $\{\overline{v}-\underline{v}, \overline{v}-\underline{v}, 1\}$, $\{0, 0, 1\}$, $\{0, 0, 0\}$, $\{\underline{v}-\overline{v}, \underline{v}-\overline{v}, 1\}$, and $\{0, 0, 1\}$. In terms of top hyperplane, both disjunctive CH $\Omega_{1,mn}$ and $\Omega_{1,nm}$ have the same upper bound. Subsequently, the third constraint in (84) using $u^l_{mn} = \beta^l_{mn} + \beta^l_{nm}$ can represent the top hyperplane with vertexes V_1, V_2, V_3 or V_5, V_6, V_7. These linear constraints from equations of the above-mentioned hyperplane:

$$\begin{cases} \dfrac{1}{\overline{v}-\underline{v}} \cdot \boldsymbol{A} \cdot \boldsymbol{v} - \dfrac{1}{\overline{v}-\underline{v}} \cdot (2\boldsymbol{D}_r\boldsymbol{P}^l + 2\boldsymbol{D}_x\boldsymbol{Q}^l - |z^l|^2\ell^l) - \beta^l_{nm} \geq -1 \\ \dfrac{1}{\overline{v}-\underline{v}} \cdot \boldsymbol{A} \cdot \boldsymbol{v} - \dfrac{1}{\overline{v}-\underline{v}} \cdot (2\boldsymbol{D}_r\boldsymbol{P}^l + 2\boldsymbol{D}_x\boldsymbol{Q}^l - |z^l|^2\ell^l) - \beta^l_{mn} \geq -1 \\ \dfrac{1}{\overline{v}-\underline{v}} \cdot \boldsymbol{A} \cdot \boldsymbol{v} - \dfrac{1}{\overline{v}-\underline{v}} \cdot (2\boldsymbol{D}_r\boldsymbol{P}^l + 2\boldsymbol{D}_x\boldsymbol{Q}^l - |z^l|^2\ell^l) + u^l_{mn} \leq 1 \end{cases} \quad (84)$$

As a DCHR formation via $u^l = \beta l\ mn + \beta l\ nm$, the disjunctive CH $\Omega_1 = \text{Conv}(\Omega_{1,\,nm} \cup \Omega_{1,\,mn})$ of (6) and the enforced disjunctive constraint in (80) yields

$$\Omega_1 = \{(\boldsymbol{x}^l, \boldsymbol{\beta}^l, \boldsymbol{u}^l) \in \mathbb{R}^{n+2|\varepsilon|} \times \mathbb{Z}^{|\varepsilon|} | (1.84) \text{ and } (1.86) - (1.88)\} \quad (85)$$

where corresponding perspective linear cuts are imposed to improve bounds of v and ℓ^l by

$$(1-\beta^l_{nm})(\underline{v}-\overline{v}) \leq v_m - v_n \leq \beta^l_{nm}(\overline{v}-\underline{v}) \quad (86)$$

$$\beta^l_{mn}(\underline{v}-\overline{v}) \leq v_m - v_n \leq (1-\beta^l_{mn})(\overline{v}-\underline{v}) \quad (87)$$

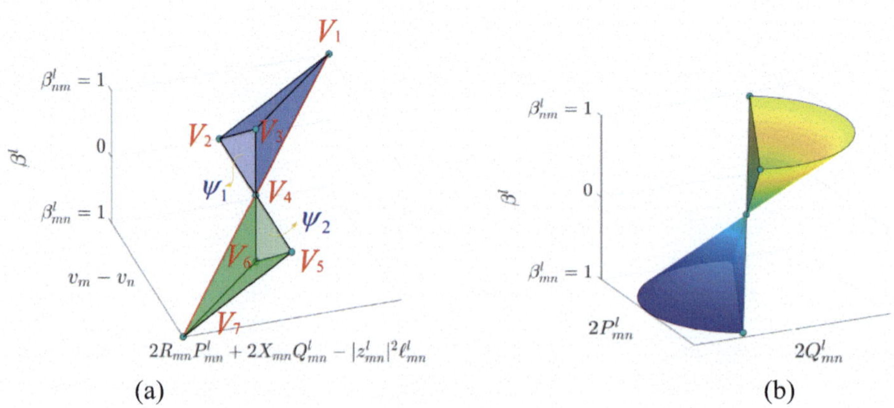

Fig. 11 **a** Disjunctive closure of Ω_1; **b** disjunctive closure of Ω_2

$$\ell^l = \ell^l_+ + \ell^l_-, \ \ 0 \le \ell^l_+ \le \beta^l_{mn}\overline{\ell}^l, \ 0 \le \ell^l_- \le \beta^l_{nm}\overline{\ell}^l \tag{88}$$

For (7) of the branch l: $= (m, n)$, the possible CH of both (7) and the enforced disjunctive constraint in (80) can be expressed as the union of two points $(x^l, u^l) =$ $(0, 0, 0, \underline{v}, 0, 0)$ and $(0, 0, 0, \overline{v}, 0, 0)$ and a continuous SOCO set, where this SOC set is drawn from a continuous SOC relaxation of the quadratic equality (7), yielding

$$||(2P^l_{mn}, \ 2Q^l_{mn}, \ v_m - \ell^l_{mn})^T||_2 \le v_m + \ell^l_{mn}, \ \forall m, n \in \mathcal{N} \tag{89}$$

With the above-mentioned theorem, we visualize the geometry closure of this SOC-representable set by (89) as a disjunctive CH on ($2Pl\ mn$, $2Ql\ mn$, β^l)-space as shown in Fig. 11b. In this figure, the intersection between the second-order cone and a valid cutting plane by either $v_m - v_n = 2Pl\ mn + 2Ql\ mn - |zl\ mn|^2 \ell l\ mn \geqslant 0$ or $v_m - v_n \leqslant 0$ contributes to forming a nearly half second-order cone. Evidently, this disjunctive closure is tighter than a full second-order cone. Let Ω_2 be a DCHR formation with respect to $\beta l\ mn$ and $\beta l\ nm$, which is expressed as

$$\Omega_2 = \{\ (\boldsymbol{x}^l, \boldsymbol{\beta}^l, \boldsymbol{u}^l) \in \mathbb{R}^{n+2|\varepsilon|} \times \mathbb{Z}^{|\varepsilon|} | (1.89) \text{ and } (1.91) - (1.92)\} \tag{90}$$

where corresponding perspective linear cuts are given as

$$0 \le 2\boldsymbol{D}_r\boldsymbol{P}^l + 2\boldsymbol{D}_x\boldsymbol{Q}^l - \boldsymbol{D}_z\ell^l \le (\overline{v} - \underline{v}) \cdot \boldsymbol{\beta}^l_{nm} \tag{91}$$

$$(\underline{v} - \overline{v}) \cdot \boldsymbol{\beta}^l_{mn} \le 2\boldsymbol{D}_r\boldsymbol{P}^l + 2\boldsymbol{D}_x\boldsymbol{Q}^l - \boldsymbol{D}_z\ell^l \le 0 \tag{92}$$

By the observation of Ω_1 and Ω_2, this DCHR approach is tighter than the McCormick linearization method and the Big-M method for any branch l under Theorem, which corresponding proof can be provided here.

Proof For McCormick linearization method, we suppose $\boldsymbol{y}^l = \boldsymbol{u}^l \cdot \boldsymbol{A} \cdot \boldsymbol{v}$ and then $\boldsymbol{y}^l - 2\boldsymbol{D}_r\boldsymbol{P}^l - 2\boldsymbol{D}_x\boldsymbol{Q}^l + \boldsymbol{D}_z\ell^l = 0$ automatically holds also for connected and/or unconnected branches subject to $\boldsymbol{A}\cdot\boldsymbol{v} + (\overline{v}-\underline{v})\cdot(\boldsymbol{u}^l - 1) \le \boldsymbol{y}^l \le \boldsymbol{A}\cdot\boldsymbol{v} - (\overline{v}-\underline{v})\cdot(\boldsymbol{u}^l - 1)$. It is clear that, due to the large-enough positive scalar number $M \ge (\overline{v}-\underline{v})$, McCormick linearization method has tighter relaxation bounds of (6) than Big-M method.

For an arbitrary branch l, the lower bound of (6) by McCormick linearization method is denoted as L_1^{lower}, yielding

$$L_1^{lower} = \boldsymbol{A} \cdot \boldsymbol{v} + (\overline{v} - \underline{v}) \cdot (\boldsymbol{u}^l - 1) - 2\boldsymbol{D}_r\boldsymbol{P}^l - 2\boldsymbol{D}_x\boldsymbol{Q}^l + \boldsymbol{D}_z\ell^l \tag{93}$$

However, in terms of DCHR, the lower bound of (6) for branch l denoted as L_2^{lower} is derived from the intersection set of first two inequality in (94), which is given by

$$L_2^{lower} = \boldsymbol{A} \cdot \boldsymbol{v} - 2\boldsymbol{D}_r\boldsymbol{P}^l - 2\boldsymbol{D}_x\boldsymbol{Q}^l + \boldsymbol{D}_z\ell^l \tag{94}$$

Hence, we can achieve the difference in algebra between L_1^{lower} and L_2^{lower} by $L_1^{lower} - L_2^{lower} = (\overline{v} - \underline{v}) \cdot (\boldsymbol{u}^{\mathbf{l}} - 1)$. Since $u^l \in \{0, 1\}$, this clearly renders $L_1^{lower} - L_2^{lower} \leq 0$. Similarly, we can also achieve the upper bound of (6) by two methods are equal.

Moreover, this McCormick linearization method for (7) adopts a Lorentz cone $\boldsymbol{L}$ defined from (89). Let Φ_1 be the feasible set of this Lorentz cone $\boldsymbol{L}$, whereas DCHR has the feasible region

$$\Phi_2 = \{\ \Phi_1 \cap (v_m - v_n > 0)\} \cup \{\ \Phi_1 \cap (v_m - v_n < 0)\} \cup \{\ \Phi_1 \cap (v_m - v_n = 0)\} \tag{95}$$

This implies that Φ_2 is smaller than Φ_1 since $v_m - v_n$ only holds for either $v_m - v_n > 0$ or $v_m - v_n < 0$ or $v_m - v_n = 0$. In light of tighter bounds in (6) and (7), we have proved that DCHR has tighter relaxation bounds than the McCormick linearization method and Big-M method. ■

With tighter bounds, this DCHR approach can also converge at the optimal point with achieving SOC equality for a load oversatisfaction assumption and radial networks as proved in Sect. 2.4. Making use of the above, the loss minimization of DNR is a non-convex mixed-integer quadratically constrained programming (MIQCP) problem. Now this model is enabled to a tractable MISOCP-based DNR reformulation with optimization variables set of $\boldsymbol{x}^{\boldsymbol{l}} := \left[\boldsymbol{P}^{\boldsymbol{l}}, \boldsymbol{Q}^{\boldsymbol{l}}, \ell^l, \boldsymbol{v}, \boldsymbol{Q}^{\boldsymbol{cr}}, \boldsymbol{w}^{\boldsymbol{l}}, \boldsymbol{m}^{\boldsymbol{l}}, \boldsymbol{f}^{\boldsymbol{l}}, \boldsymbol{\beta}^{\boldsymbol{l}}_{\boldsymbol{mn}}, \boldsymbol{\beta}^{\boldsymbol{l}}_{\boldsymbol{nm}}, \boldsymbol{u}^{\boldsymbol{l}}\right]^T$, $\forall l \in \mathcal{E}$ subject to system operational constraints below. Leveraging Ω_1 and Ω_2 to represent (6) and (7), we have

$$
l\min_{\mathbf{x}^{\mathbf{l}}} \quad P_0^l
$$

$$
s.t. \begin{cases}
P_{mn}^l + P_n^g - r_{mn}^l \ell_{mn}^l = \sum\limits_{k\in\pi(n)} P_{nk}^l + P_n^d, \ \forall l \in \mathcal{E}, \forall m, n \in \mathcal{N} \\
Q_{mn}^l + Q_n^g - x_{mn}^l \ell_{mn}^l = \sum\limits_{k\in\pi(n)} Q_{nk}^l + Q_n^d, \ \forall l \in \mathcal{E}, \forall m, n \in \mathcal{N} \\
v_n - v_m + 2(r_{mn}^l P_{mn}^l + x_{mn}^l Q_{mn}^l) - (1-u^l)M \le 0, \ \forall l \in \mathcal{E}, \forall m, n \in \mathcal{N} \\
v_n - v_m + 2(r_{mn}^l P_{mn}^l + x_{mn}^l Q_{mn}^l) + (1-u^l)M \ge 0, \ \forall l \in \mathcal{E}, \forall m, n \in \mathcal{N} \\
w^l = \ell_{mn}^l - v_m, \ m^l = \ell_{mn}^l + v_m, \quad \forall l \in \mathcal{E}, \ \forall m, n \in \mathcal{N} \\
\varsigma^0 \ge |2P_{mn}^l|, \ \forall l \in \mathcal{E} \\
\eta^0 \ge |2Q_{mn}^l|, \ \forall l \in \mathcal{E} \\
\varsigma^j = \cos(\frac{\pi}{2^{j+1}})\varsigma^{j-1} + \sin(\frac{\pi}{2^{j+1}})\eta^{j-1}, \quad j = 1, ..., \upsilon \\
\eta^j \ge |-\sin(\frac{\pi}{2^{j+1}})\varsigma^{j-1} + \cos(\frac{\pi}{2^{j+1}})\eta^{j-1}|, \quad j = 1, ..., \upsilon \\
\varsigma^\upsilon \le S_{mn}^l, \ \forall l \in \mathcal{E} \\
\eta^\upsilon \le \tan(\frac{\pi}{2^{\upsilon+1}})\varsigma^\upsilon \\
\omega^0 \ge |S_{mn}^l|, \ \forall l \in \mathcal{E} \\
\kappa^0 \ge |w^l|, \ \forall l \in \mathcal{E} \\
\omega^j = \cos(\frac{\pi}{2^{j+1}})\omega^{j-1} + \sin(\frac{\pi}{2^{j+1}})\kappa^{j-1}, \quad j = 1, ..., \upsilon \\
\kappa^j \ge |-\sin(\frac{\pi}{2^{j+1}})\omega^{j-1} + \cos(\frac{\pi}{2^{j+1}})\kappa^{j-1}|, \quad j = 1, ..., \upsilon \\
\omega^\upsilon \le m^l, \ \forall l \in \mathcal{E} \\
\kappa^\upsilon \le \tan(\frac{\pi}{2^{\upsilon+1}})\omega^\upsilon \\
-\overline{Q}_0^l \le Q_0^l \le \overline{Q}_0^l \\
\underline{Q}^{cr} \le Q^{cr} \le \overline{Q}^{cr} \\
\underline{v} \le v \le \overline{v} \\
\beta_{mn}^l + \beta_{nm}^l = u^l \\
\beta_{mn}^l = 0, \ \text{if } m = 0 \\
\sum\limits_{n(m,n)\in\mathcal{E}} \beta_{mn}^l = 1, \forall m \in \mathcal{N} \ominus \backslash 0 \\
\beta_{mn}^l \ge 0 \\
\mathbf{A}^T f = 1 \\
-N \cdot u^l \le f^l \le N \cdot u^l, \quad \forall l \in \mathcal{E} \\
1^T u^l = |\mathcal{N}| - 1 \\
u^l \in \{0, 1\}
\end{cases} \tag{96}
$$

where P_0^l is the active power injection at the root node 0.

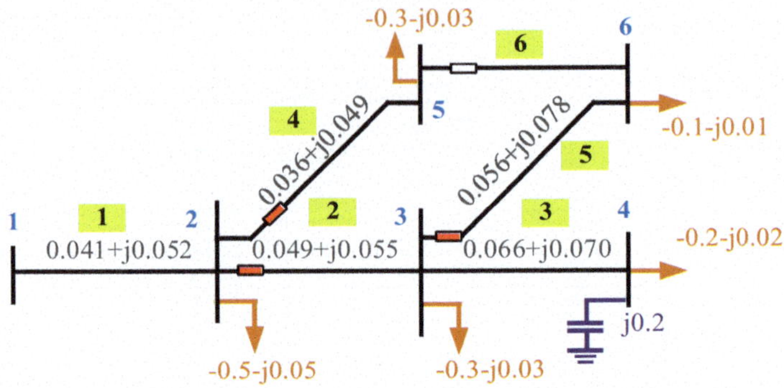

Fig. 12 Distribution topology of node-6 system

10 Case Study

10.1 Simple 6-node DN

The following simple 6-node distribution network is used to exemplify this polyhedral approximation of DNR model using Big-M relaxation method. The network is shown in Fig. 12. The node 1 is the PCC bus, while nodes 2–6 are PQ buses. In other words, $\mathcal{N} = \{1, 2, 3, 4, 5, 6\}$ and $\mathcal{E} = \{1, 2, 3, 4, 5\}$, where branch labels are displayed in yellow-filled square boxes. The impedance of each branch, the load demand of each node, and the reactive power compensation capacity of installed capacitors are labeled in this figure. The voltage allowance band of each node is set to 0.97–1.07 p.u. The switch-off circuit breakers are marked with "" and the rest are switch-on marked by "▪". For a fixed topology, we focus on reactive power optimization on the active branches for the minimization of real power loss.

According to this DCHR-based DNR model (96), the size of $[P^l, Q^l, \ell^l, v, Q^{cr}]$ is $6{\cdot}4 + 1 = 25$; $[w^l, m^l, \ell^{l+}, \ell^{l-}]$ has the size $6{\cdot}4 = 24$ and $[f^l, \beta^l_{mn}, \beta^l_{nm}]$ has the size $3{\cdot}6 = 18$.The remaining u^l is a vector of $u^l = [u_{12}, u_{23}, u_{34}, u_{25}, u_{36}, u_{56}]$. Thus, the total number of variables is $25 + 24 + 18 + 6 = 73$. The single-commodity flow constraints and spanning tree constraints are the same as the MISOCP-based DNR model with Big-M relaxation method. The tree-shaped voltage profiles before and after DNR operations are presented in Fig. 13. The output solutions are shown in Tables 2 and 3 after running the above codes.

10.2 Large-Scale Distribution Networks

To validate the effectiveness of DCHR, we show the voltage profiles before and after DNR operations by the DCHR approach. The tree-shaped voltage profiles are

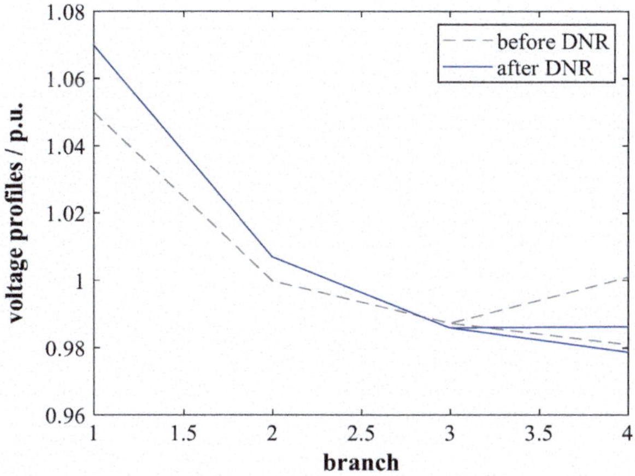

Fig. 13 Tree-shaped voltage profiles of simple 6-node distribution network

Table 2 Optimal DNR objective of the 6-node distribution network

Solvers	Minimal injected real power at PCC node (p.u.)	Real power loss (p.u.)	Algorithm iterations	Computational time (seconds)
MOSEK	1.504	0.1035	7	0.031
Baron	1.504	0.1035	6	0.030

Table 3 Optimal DNR solutions of the 6-node distribution network

Solvers	Circuit breakers	
	Switch-on status of branch no	Switch-off status of branch no
MOSEK	5	1, 2,3, 4, 6
Baron	5	1, 2,3, 4, 6

presented in Fig. 14a–d for different distribution networks: (a) 16-node; (b) 33-node; (c) 123-node; and (d) 1060-node. The voltage profiles before and after DNR operations are enhanced to reduce the power loss, where this phenomenon can be observed by the blue lines as compared to grey dashed lines in Fig. 14a–d. This demonstrates that this DCHR approach can successfully solve this DNR problem with voltage profile enhancements.

To compare the relaxation tightness of DCHR, we employ the DNR models using four methods: (*i*) the Big-M method for (73) and continuous SOC relaxation for (7) (M1) [7], (*ii*) the McCormick linearization method for (6) and continuous SOC relaxation for (78) (M2) [8], and (*iii*) the polyhedral approximation method with 32 segments of linearizations (M3) [10], and (*iv*) the DCHR approach (M4). These comparable M1-M4 are implemented for the low and the high penetrations of reactive power capacitors and DERs for cases I and II, respectively. The computational performance in terms of CPU time in seconds and algorithm iterations are given in

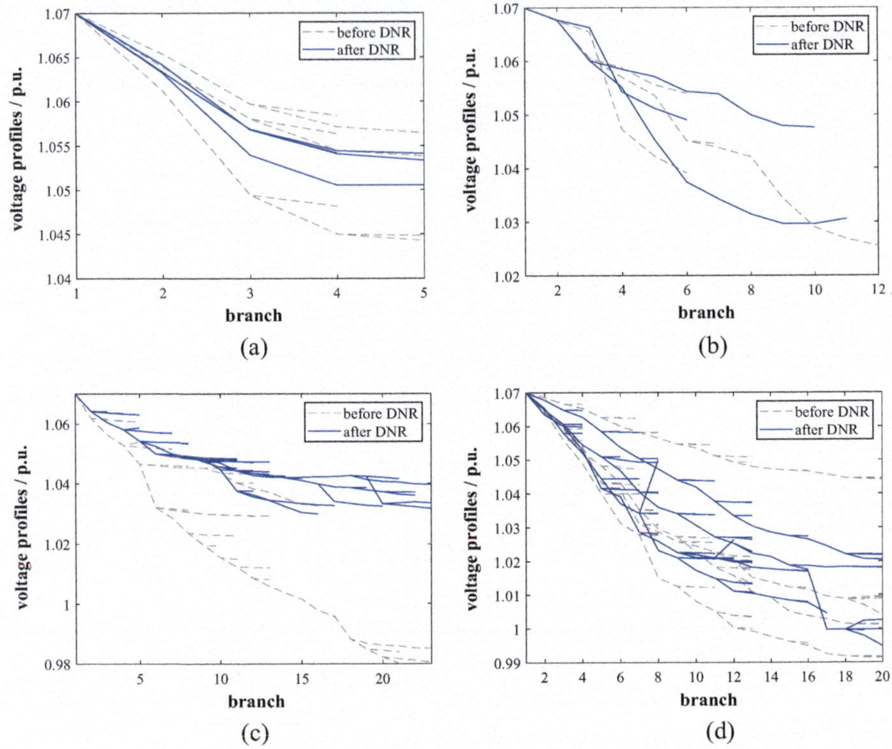

Fig. 14 Tree-shaped voltage profiles of different distribution networks: **a** 16-node; **b** 33-node; **c** 123-node; and **d** 1060-node

Table 4 for M1-M4. The convergence performance of duality gaps is presented in Fig. 15 by M1-M4.

Table 4 shows that M4 significantly outperforms M1 and M2 with the less CPU running time and M4's solutions are more accurate than M3 in theory for all systems of two cases. Regarding the number of iterations, it suggests that the relaxation bounds of DNR models by M4 is more effective than M1-M3 as justified with the less running time. Since a large-enough positive scalar in M1 leads to loosing bounds, it is tricky for M1 to find the optimal configuration scheme quickly. Between M2 and M4, M4 can induce fast convergence with achieving SOC equality than M2, since SOC constraints in M4 are coupled with parent–child relationship variables. Additionally, M4 is superior for case I than case II, due to a larger number of branches that may have bi-directional power flows in case II. These conclusions are also justified by the fast convergence in two test systems in Fig. 15a–d.

Table 4 Computational performance among M1-M4

Cases syst		CPU time (s)				Iterations			
		Convex	Relaxations		Approx	Convex	Relaxations		Approx
		M1	M2	M4	M3	M1	M2	M4	M3
I	16-node	0.281	0.250	0.234	0.297	202	172	113	99
	33-node	2.906	1.985	1.187	1.781	3452	2259	949	1290
	123-node	1.613	1.488	1.171	1.875	237	329	135	275
	1060-node	329.437	56.922	18.187	20.328	116,663	16,114	3764	4510
II	16-node	0.309	0.297	0.228	0.275	175	151	133	109
	33-node	3.703	2.141	1.432	1.890	2021	1432	635	1136
	123-node	1.703	1.641	1.218	1.918	237	329	179	152
	1060-node	686.813	97.969	39.094	46.938	127,315	13,929	5567	5936

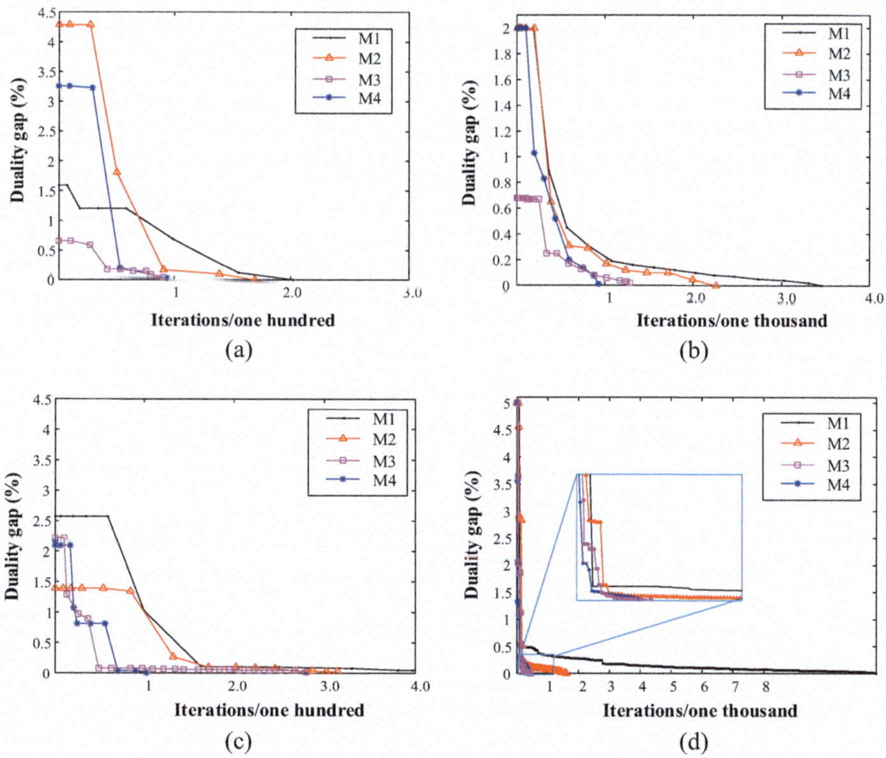

Fig. 15 Convergence performance: **a** 33-node; **b** 33-node; **c** 123-node; **d** 1060-node

References

1. X. Jiang, Y. Zhou, W. Ming, P. Yang, J. Wu, An overview of soft open points in electricity distribution networks. IEEE Trans. on Smart Grid **13**(3), 1899–1910 (2022)
2. C. Lei, S. Bu, J. Zhong, Q. Chen, Q. Wang, Distribution network reconfiguration: a disjunctive convex hull approach. IEEE Trans. on Power Syst. (2023)
3. M.E. Baran, F.F. Wu, Network reconfiguration in distribution systems for loss reduction and load balancing. IEEE Trans. Power Deli. **4**(2) (1989)
4. H.D. Chiang, R. Jean-Jumeau, Optimal network reconfigurations in distribution systems: a new formulation and a solution methodology. IEEE Trans. Power Deli. **5**(4) (1990)
5. S. Toune, H. Fudo, T. Genji, Y. Fukuyama, Y. Nakanishi, Comparative study of modern heuristic algorithms to service restoration in distribution systems. IEEE Trans. Power Syst. **17**(1) (2002)
6. S. Mishra, D. Das, S. Paul, A comprehensive review on power distribution network reconfiguration. Energy Syst. **8**(2), 227–284 (2017)
7. J.A. Taylor, F.S. Hover, Convex models of distribution system reconfiguration. IEEE Trans. on Power Syst. **27**(3), 1407–1413 (2012)
8. M.K. Singh, V. Kekatos, S. Taheri, K.P. Schneider, C.-C. Liu, Enforcing radiality constraints for der-aided power distribution grid reconfiguration (2019). arXiv:1910.03020
9. M. Lavorato, J.F. Franco, M.J. Rider, R. Romero, Imposing radiality constraints in distribution system optimization problems. IEEE Trans. Power Syst. **27**(1), 172–180 (2012)
10. R.A. Jabr, R. Singh, B.C. Pal, Minimum loss network reconfiguration using mixed-integer convex programming. IEEE Trans. on Power Syst. **27**(2), 1106–1115 (2012)
11. J.-P. Chiou, C.-F. Chang, C.-T. Su, Variable scaling hybrid differential evolution for solving network reconfiguration of distribution systems. IEEE Trans. Power Syst. **20**(2), 668–674 (2005)
12. A. Asrari, T. Wu, S. Lotfifard, The impacts of distributed energy sources on distribution network reconfiguration. IEEE Trans. Power Electron. **31**(2), 606–613 (2016)
13. J.F. Franco, M.J. Rider, M. Lavorato, R. Romero, A mixed integer LP model for the reconfiguration of radial electric distribution systems considering distributed generation. Elect. Power Syst. Res. **97**, 51–60 (2013)
14. Q. Li, V. Vittal, Convex hull of the quadratic branch ac power flow equations and its application in radial distribution networks. IEEE Trans. on Power Syst. **33**(1), 839–850 (2017)
15. S. Wang, J. Liu, R. Bo, Y. Chen, Approximating input-output curve of pumped storage hydro plant: a disjunctive convex hull method. IEEE Trans. on Power Syst. (2022)
16. M. Farivar, S. Low, Branch flow model: relaxations and convexification—part I. IEEE Trans. Power Syst. **28**(3) (2013)
17. S.H. Low, Convex relaxation of optimal power flow part I: formulations and equivalence. IEEE Trans. Control Netw. Syst. **1**(1), 15–27 (2014)
18. H. Zhu, H.J. Liu, Fast local voltage control under limited reactive power: optimality and stability analysis. IEEE Trans. Power Syst. **31**(5), 3794–3803 (2016)
19. Ben-Tal, A. Nemirovski, On polyhedral approximations of the second-order cone. Math. Oper. Res. **26**(2), 193–205 (2001)
20. Y. Wang, Y. Xu, J. Li, J. He, X. Wang, On the radiality constraints for distribution system restoration and reconfiguration problems. IEEE Trans. Power Syst. **35**(4), 3294–3296 (2020)
21. H. Ahmadi, J.R. Martí, Mathematical representation of radiality constraint in distribution system reconfiguration problem. Int. J. Electr. Power Energy Syst. **64**, 293–299 (2015)
22. G. McCormick, Computability of global solutions to factorable nonconvex programs: part I convex underestimating problems. Math. Program. **10**, 146–175 (1976)

The manufacturer's authorised representative in the EU is Springer Nature Customer Service Centre GmbH, Europaplatz 3, 69115 Heidelberg, Germany. If you have any concerns regarding our products, please contact ProductSafety@springernature.com

Printed and bound by CPI Group (UK) Ltd, Croydon, CR0 4YY
30/09/2026
02225302-0008